Lecture Notes in Control and Information Sciences 429

Editors

Series Advisory Board

For further volumes:
http://www.springer.com/series/642

Alexandra Grancharova and Tor Arne Johansen

Explicit Nonlinear Model Predictive Control

Theory and Applications

Springer

Authors
Assoc. Prof. Dr. Alexandra Grancharova
Institute of System Engineering and Robotics
Bulgarian Academy of Sciences
Sofia
Bulgaria

Prof. Tor Arne Johansen
Department of Engineering Cybernetics
Norwegian University of Science and Technology
Trondheim
Norway

ISSN 0170-8643 e-ISSN 1610-7411
ISBN 978-3-642-28779-4 e-ISBN 978-3-642-28780-0
DOI 10.1007/978-3-642-28780-0
Springer Heidelberg New York Dordrecht London

Library of Congress Control Number: 2012933325

Printed on acid-free paper

Springer is part of Springer Science+Business Media (www.springer.com)

To my parents Assoc. Prof. Dr. Siyka Popova and Prof. D.Sc. Ivan Grancharov for their love and support
(Alexandra Grancharova)

Preface

Model predictive control (MPC) has become the accepted methodology to solve complex control problems related to process industries. It allows the design of multi-input multi-output (MIMO) control systems that minimize a certain performance index in the presence of input and output constraints. The Nonlinear Model Predictive Control (NMPC) is an optimization-based method for control which involves the solution at each sampling instant of a finite horizon optimal control problem subject to the *nonlinear* system dynamics and input and output constraints imposed on the system. However, the solution of an on-line *nonlinear* optimization problem is often computationally complex and time consuming and the real-time NMPC implementation is usually limited to slow processes where the sampling time is sufficient to support the computational needs. The on-line computational complexity can be circumvented with an *explicit* approach to NMPC, where an *explicit* approximate representation of the solution is computed using multi-parametric Nonlinear Programming (mp-NLP).

Motivation

The main motivation behind *explicit* MPC is that an *explicit* state feedback law avoids the need for executing a numerical optimization algorithm in real time, and is therefore potentially useful for applications where MPC has not traditionally been used. It has been shown that the feedback solution to MPC problems for constrained *linear* systems has an *explicit* representation as a piecewise linear state feedback defined on a polyhedral partition of the state space. The benefits of an *explicit* solution, in addition to the efficient on-line computations, include also verifiability of the implementation (which is an essential issue in safety-critical applications) and the possibility to design embedded control systems with low software and hardware complexity. For *nonlinear* MPC the prospects of *explicit* solutions are even higher than for *linear* MPC, since the benefits of computational efficiency and verifiability are even more important.

The main reasons to develop methods for *explicit* NMPC can be summarized as follows:

- Dramatical reduction in online computations, since online *nonlinear* numerical optimization is avoided and replaced by piecewise function evaluation. This may lead to significant reduction in the requirements to real-time embedded computer hardware.
- NMPC optimization depends on appropriate initialization in order to avoid local minima, and appropriate formulation of constraints in order to avoid infeasibility. With *explicit* NMPC the validation of initialization procedures and infeasibility handling can be conducted based on a complete and *explicit* solution.
- Significant reduction in online software complexity since the code for piecewise function evaluation is much simpler than a *nonlinear* numerical optimization solver. This may lead to formal software verification being a feasible practical tool.
- Approximate *explicit* solutions with reduced complexity, and with guaranteed levels of sub-optimality, may be computed offline. Formal analysis of performance, sub-optimality and stability may be possible since an *explicit* representation of the controller is known.
- Formulations such as stochastic NMPC and robust NMPC may not lead to increased online computations in an *explicit* NMPC approach, compared to a nominal NMPC formulation, although they will require more offline computations.

Main contributions of the book
This book considers the mp-NLP approaches to *explicit* approximate NMPC of constrained *nonlinear* systems, developed by the authors, as well as their applications to various NMPC problem formulations and several case studies. The proposed mp-NLP methods are based on orthogonal partition of the state space and they are general in sense that they can be applied to solve both convex and non-convex optimization problems. The following types of *nonlinear* systems are considered, resulting in different NMPC problem formulations:

- *Nonlinear* systems described by first-principles models and *nonlinear* systems described by black-box models;
- *Nonlinear* systems with continuous control inputs and *nonlinear* systems with quantized control inputs;
- *Nonlinear* systems without uncertainty and *nonlinear* systems with uncertainties (polyhedral description of uncertainty and stochastic description of uncertainty);
- *Nonlinear* systems, consisting of interconnected *nonlinear* sub-systems.

The proposed mp-NLP approaches to *explicit* solution of various NMPC problems are illustrated with applications to several case studies, which present mathematical models, NMPC formulations, mp-NLP computational results, and closed loop simulations. They are taken from diverse areas such as automotive mechatronics, compressor control, combustion plant control, reactor control, pH maintaining system control, cart and spring system control, and diving computers.

Intended audience

The book is intended to support graduate courses and the study of Ph.D. and advanced M.Sc. students in *nonlinear* control and optimization. Readers should be familiar with the basics of linear model predictive control, numerical optimization methods, and linear and nonlinear control theory. The book could be also useful for academic researchers working in the field of NMPC, as well as researchers from industrial companies, including automotive and aerospace, whose responsibilities include the development of embedded optimal control systems.

Book organization

The book is structured as follows:

- In **Chapter 1**, basic theory and algorithms to find an *explicit* approximate solution of mp-NLP problems, based on orthogonal ($k-d$ tree) partition of the parameter space, are described by considering both the convex and the non-convex case. Procedures and heuristic rules for efficient splitting of a region in the parameter space and for handling the infeasible cases are formulated.
- In **Chapter 2**, the main aspects of formulation of the NMPC optimization problem are considered, which is an essential part of the control design and involves numerous decisions that are important for the control performance, feasibility, stability, and robustness as well as the computational complexity and the numerical challenges of computing the solution.
- In **Chapter 3**, an algorithm for *explicit* NMPC, which locally approximates the mp-NLP problem with a multi-parametric quadratic program is described. The approach is applied to a case study.
- **Chapter 4** considers the design of *explicit* NMPC controllers for several case studies by applying the approximate mp-NLP algorithms, described in Chapter 1. The case studies present mathematical models, NMPC formulations, mp-NLP computational results, and closed loop simulations. They are taken from diverse areas such as automotive mechatronics, compressor control, and diving computers. In this chapter, it is also shown that bounding the approximation error of the *explicit* approximate solution to convex regulation NMPC problems ensures the asymptotic stability of the suboptimal closed-loop system.
- **Chapter 5** presents an approximate multi-parametric Nonlinear Integer Programming (mp-NIP) approach to design *explicit* NMPC controllers for constrained nonlinear systems with quantized control inputs. The approach is applied to two case studies.
- In **Chapter 6**, two approaches to *explicit* min-max NMPC of constrained nonlinear systems in the presence of bounded disturbances and/or parameter uncertainties are considered. The first approach is based on an open-loop min-max NMPC problem statement, while the second approach adopts a closed-loop min-max NMPC formulation. With the latter approach, conditions for guaranteeing the l_2-stability of the closed-loop system are derived. Two case studies are considered.
- In **Chapter 7**, two approaches to *explicit* stochastic NMPC of constrained nonlinear systems in the presence of disturbances and/or parameter uncertainties with known probability distributions are presented. The first approach constructs

explicit approximate NMPC solution for systems, described by stochastic parametric models, while the second approach considers systems, described by Gaussian process models. The approaches are applied to two case studies.

- **Chapter 8** considers an approximate mp-NLP approach to *explicit* solution of output-feedback NMPC problems for constrained nonlinear systems described by neural network NARX models. A dual-mode control strategy is proposed in order to achieve an offset-free closed-loop response in the presence of bounded disturbances and/or model errors. One case study is considered.
- In **Chapter 9**, a suboptimal approach to distributed NMPC for systems consisting of nonlinear subsystems with linearly coupled dynamics, subject to both state and input constraints, is considered. The approach is based entirely on distributed on-line optimization and can be applied to large-scale nonlinear systems. Also, a *semi-explicit* NMPC approach to efficiently solve the distributed NMPC problem for small- and medium-scale systems is proposed. Both distributed NMPC approaches are applied to an example nonlinear system.

Alexandra Grancharova has been the main contributor to Chapters 1 and 4 – 9, and Tor Arne Johansen has been the main contributor to Chapters 2 and 3.

Sofia,
Trondheim,
January 2012

Alexandra Grancharova
Tor Arne Johansen

Contents

Chapter 1
Multi-parametric Programming

Abstract. This chapter presents an overview of the approaches to solve multi-parametric programming problems. It is organized as follows. In Section 1.1, a general multi-parametric nonlinear programming (mp-NLP) problem is formulated and the Karush-Kuhn-Tucker (KKT) optimality conditions are presented. Then, the three main groups of methods to find a local minimum of a NLP problem for a given parameter vector are reviewed (Newton-type methods, penalty function methods and direct search methods). The Basic Sensitivity Theorem, which addresses the local regularity conditions for the optimal solution as function of the parameters is reviewed. Then, algorithms to find an approximate explicit solution of mp-NLP problems are described, which are based on an orthogonal (k–d tree) partition of the parameter space. Both convex and non-convex mp-NLP problems are considered. Procedures and heuristic rules for efficient splitting of a region in the parameter space and for handling the infeasible cases are formulated. In Section 1.2, a multi-parametric quadratic programming (mp-QP) problem is formulated and two approaches to find its exact explicit solution are described.

1.1 Multi-parametric Nonlinear Programming

There are two ways to address the parameter variations in mathematical programs: *sensitivity analysis*, which characterizes the change of the solution with respect to small perturbations of the parameters, and *parametric programming*, where the characterization of the solution is found for a full range of parameter values. Mathematical programs which depend only on one scalar parameter are referred to as *parametric programs*, while problems depending on a vector of parameters are referred to as *multi-parametric programs*.

The basic results within multi-parametric nonlinear programming (mp-NLP) can be found in [26]. Main topics in [26] include local regularity conditions, local sensitivity results and calculation of the parameter derivatives of the optimal solution vector.

A. Grancharova & T.A. Johansen: Explicit Nonlinear Model Predictive Ctrl., LNCIS 429, pp. 1–37.
springerlink.com

1.1.1 Problem Formulation

Consider the nonlinear mathematical program dependent on a parameter x appearing in the objective function and in the constraints:

$$V^*(x) = \min_z f(z,x) \tag{1.1}$$

$$\text{s.t. } g(z,x) \leq 0 , \tag{1.2}$$

where $z \in \mathbb{R}^s$ is the vector of optimization variables, $x \in \mathbb{R}^n$ is the vector of parameters, $f : \mathbb{R}^s \times \mathbb{R}^n \mapsto \mathbb{R}$ is the objective function, and $g : \mathbb{R}^s \times \mathbb{R}^n \mapsto \mathbb{R}^q$ is the constraints function. In (1.1), it is supposed that the minimum exists. It should be noted that the problem (1.1)–(1.2) includes only inequality constraints, and we remark that equality constraints can be incorporated with a straightforward modification since they are always included in the optimal active set.

Let X be a closed polytopic set of parameters, defined by $X = \{x \in \mathbb{R}^n \mid Ax \leq b\}$. In multi-parametric programming, it is of interest to characterize the solution or solutions of the mp-NLP problem (1.1)–(1.2) for the set X [26]. As described in [2], the solution of an mp-NLP problem is a triple $(V^*(x), Z^*(x), X_f)$, where:

i. the *set of feasible parameters* X_f is the set of all $x \in X$ for which the problem (1.1)–(1.2) admits a solution, i.e.:

$$X_f = \{x \in X \mid g(z,x) \leq 0 \text{ for some } z \in \mathbb{R}^s\} ; \tag{1.3}$$

ii. the *optimal value function* $V^* : X_f \mapsto \mathbb{R}$ associates with every $x \in X_f$ the corresponding optimal value of (1.1)–(1.2);

iii. the *optimal set* $Z^*(x)$ associates to each parameter $x \in X_f$ the corresponding set of optimizers $Z^*(x) = \{z \in \mathbb{R}^s \mid f(z,x) = V^*(x)\}$ of problem (1.1)–(1.2). If $Z^*(x)$ is a singleton for all $x \in X_f$, then $z^*(x) \triangleq Z^*(x)$ is called the *optimizer function*.

In this book we will assume that X_f is closed and $V^*(x)$ is finite for every $x \in X_f$. We denote by $g_i(z,x)$ the i-th component of the vector valued function $g(z,x)$.

Let z be a feasible point of (1.1)–(1.2) for a given parameter x. The *active constraints* are the constraints that fulfill (1.2) at equality, while the remaining constraints are called *inactive constraints*. The *active set* $\mathscr{A}(z,x)$ is the set of indices of the active constraints, i.e.:

$$\mathscr{A}(z,x) \triangleq \{i \in \{1, 2, \dots, q\} \mid g_i(z,x) = 0\} . \tag{1.4}$$

The *optimal active set* $\mathscr{A}^*(x)$ is the set of indices of the constraints that are active for all $z \in Z^*(x)$, for a given $x \in X$, i.e.:

$$\mathscr{A}^*(x) \triangleq \{i \mid i \in \mathscr{A}(z,x), \ \forall z \in Z^*(x)\} . \tag{1.5}$$

Given an index set $\mathscr{A} \subseteq \{1, 2, \dots, q\}$, the *critical region* $CR_{\mathscr{A}}$ is the set of parameters for which the optimal active set is equal to $\mathscr{A}$, i.e.:

$$CR_{\mathscr{A}} \triangleq \{x \in X \mid \mathscr{A}^*(x) = \mathscr{A}\} \ . \tag{1.6}$$

As it will be shown in Section 1.2, for strictly convex quadratic function f and linear constraints g, the critical regions $CR_{\mathscr{A}}$ are polyhedrons and the optimizer z^* is unique, piecewise affine, and continuous. However, for general nonlinear functions f and g, the exact solution of the multi-parametric programming problem (1.1)–(1.2) can not be found, and suboptimal methods for approximating its optimizer function $z^*(x)$ (or selection in case the optimizer function is not unique) are described in Section 1.1.5.

1.1.2 Optimality Conditions

For a given $x_0 \in X$, a local minimum z_0 of problem (1.1)–(1.2) has to satisfy the well known Karush-Kuhn-Tucker (KKT) first-order conditions [56]:

$$\nabla_z L(z_0, x_0, \lambda_0) = 0 \tag{1.7}$$
$$\mathrm{diag}(\lambda_0) g(z_0, x_0) = 0 \tag{1.8}$$
$$\lambda_0 \geq 0 \tag{1.9}$$
$$g(z_0, x_0) \leq 0 \,, \tag{1.10}$$

with associated Lagrange multiplier λ_0 and the Lagrangian defined as:

$$L(z, x, \lambda) \triangleq f(z, x) + \lambda^T g(z, x) \ . \tag{1.11}$$

Here, sufficient regularity (smoothness) is assumed, and this will be discussed later in Section 1.1.4.

Consider the optimal active set $\mathscr{A}_0$ at x_0, i.e. a set of indices to active constraints in (1.10). The above conditions are sufficient provided the following second order condition holds [56]:

$$v^T \nabla^2_{zz} L(z_0, x_0, \lambda_0) v > 0, \ \forall v \in \mathscr{F} - \{0\} \tag{1.12}$$

with $\mathscr{F}$ being the set of all directions where it is not given from the first order conditions if the objective function will increase or decrease:

$$\begin{aligned} \mathscr{F} = \ & \{v \in \mathbb{R}^s \mid \nabla_z g_{\mathscr{A}_0}(z_0, x_0) v \geq 0, \\ & \nabla_z g_i(z_0, x_0) v = 0, \ \text{for all } i \text{ with } (\lambda_0)_i > 0\} \ . \end{aligned} \tag{1.13}$$

The notation $g_{\mathscr{A}_0}$ means the rows of g with indices in $\mathscr{A}_0$.

1.1.3 Nonlinear Programming Methods

There exist various methods to numerically compute a local minimum z_0 of the problem (1.1)–(1.2) for a given $x_0 \in X$. The most commonly used methods can be classified in the following three groups.

1.1.3.1 Newton-Type Methods

The Newton type methods [20] appear to be the most widely used optimization methods. They try to find a point satisfying the KKT conditions (1.7)–(1.10) by using successive linearizations of the problem functions. The motivation behind this is that the linearized KKT system can be solved by using standard numeric linear algebra tools. Depending on how the conditions (1.8)–(1.10) (related to the imposed constraints) are treated, the two main groups of Newton type methods are the Sequential Quadratic Programming (SQP) methods and the Interior Point (IP) methods.

- *Sequential Quadratic Programming (SQP) methods.*
  The SQP methods iteratively solve the KKT system (1.7)–(1.10) by linearizing the nonlinear functions included in it. The resulting linearized KKT system at the $k+1$-th iteration can be considered as the KKT conditions of the following quadratic program (QP):

$$V_{qp}^*(z^k, x_0) = \min_z f_{qp}(z, z^k, x_0) \tag{1.14}$$

$$\text{s.t.} \;\; g(z^k, x_0) + \nabla_z g(z^k, x_0)(z - z^k) \leq 0 \,, \tag{1.15}$$

  with the quadratic objective function given by:

$$f_{qp}(z, z^k, x_0) = \nabla_z f(z^k, x_0)^T z + \frac{1}{2}(z - z^k)^T \nabla_z^2 L(z^k, x_0, \lambda^k)(z - z^k) \,. \tag{1.16}$$

  Here, z^k and λ^k represent, respectively, the values of optimization variables and Lagrange multipliers, which solve the k-th sequential iteration of the KKT system (1.7)–(1.10). It is assumed that an initial guess z^0 is provided. In the case when the Hessian matrix $\nabla_z^2 L(z^k, x_0, \lambda^k)$ is positive semi-definite, the QP problem (1.14)–(1.16) is convex and its unique solution can be found.

  Typically, the QP sub-problem (1.14)–(1.16) is solved by using an Active Set (AS) method [56, 49], which identifies the active set of its solution z^*. The method begins with finding a feasible initial guess $\mathscr{A}_0(z, z^k, x_0) = \{i \in \{1, 2, ..., q\} \mid g_i(z^k, x_0) + \nabla_z g_i(z^k, x_0)(z - z^k) = 0\}$ of the active set by solving a linear programming problem [56]. In the next iteration, $\mathscr{A}_0(z, z^k, x_0)$ is refined by deleting a constraint from $\mathscr{A}_0(z, z^k, x_0)$ or by adding a constraint to $\mathscr{A}_0(z, z^k, x_0)$. In this way, the active set is refined iteratively until the optimal active set $\mathscr{A}^*(z^k, x_0)$ is found.

  There are several SQP methods which use approximations of the Hessian matrix $\nabla_z^2 L(z^k, x_0, \lambda^k)$ and the constraints Jacobian matrix $\nabla_z g(z^k, x_0)$, and they are referred to as quasi-Newton methods. They usually lead to slower convergence rates, but computationally less expensive iterations, in comparison to the exact SQP method. One of the quasi-Newton SQP methods is the method by Powell [61]. It uses exact constraints Jacobian matrix, but replaces the Hessian matrix $\nabla_z^2 L(z^k, x_0, \lambda^k)$ by an approximation H_k. Each new Hessian approximation H_{k+1} is obtained from the previous approximation H_k by an update formula that uses the difference of the Lagrange gradients, $\psi = \nabla_z L(z^{k+1}, x_0, \lambda^{k+1}) - \nabla_z L(z^k, x_0, \lambda^{k+1})$, and the step $\tau = z^{k+1} - z^k$ in order to obtain second order information in H_{k+1}. The most widely used update formula is the one by Broyden-Fletcher-Goldfarb-Shanno (BFGS) [56]:

$$H_{k+1} = H_k + \frac{\psi\psi^T}{\psi^T\tau} - \frac{H_k\tau\tau^T H_k}{\tau^T H_k \tau} \,. \tag{1.17}$$

Another successful quasi-Newton SQP method is the constrained Gauss-Newton method [21]. It uses approximations of the Hessian matrix, based on some Jacobian, and is applicable when the objective function is a sum of squares.

- *Interior Point (IP) methods.*
  The IP methods represent an alternative way to solve the KKT system (1.7)–(1.10), which consists in replacing the nonsmooth KKT condition (1.8) by a smooth nonlinear approximation [16, 76]:

$$\nabla_z L(z_0, x_0, \lambda_0) = 0 \tag{1.18}$$

$$\lambda_{0,i} g_i(z_0, x_0) = \rho \;,\; i = 1, 2, ..., q \tag{1.19}$$

$$\lambda_0 \geq 0 \tag{1.20}$$

$$g(z_0, x_0) \leq 0 \,, \tag{1.21}$$

where $\rho > 0$ is a slack variable and $g_i(z_0, x_0)$ is the i-th constraint function. This system is then solved with a Newton-type method. The obtained solution is not a solution to the original NLP problem (1.1)–(1.2), but to the following problem [16, 76, 8]:

$$Q^*(x_0, \rho) = \inf_z [\, f(z, x_0) + \rho B(z, x_0) \,] \,. \tag{1.22}$$

Here, $B(z, x_0)$ is the so called barrier function, which is nonnegative and continuous over the region $\{z \in \mathbb{R}^s \,|\, g(z, x_0) < 0\}$ and approaches ∞ as the boundary of the feasible region $\{z \in \mathbb{R}^s \,|\, g(z, x_0) \leq 0\}$ is approached from the interior. Thus, the function $B(z, x_0)$ sets a barrier against leaving the feasible region. The solution of the barrier problem (1.22) requires for the optimization to start from a point inside the region $\{z \in \mathbb{R}^s \,|\, g(z, x_0) < 0\}$. The IP methods are also referred to as barrier function methods. They generate a sequence of feasible points whose limit is an optimal solution to the original problem (1.1)–(1.2) [8]. If the optimal solution occurs at the boundary of the feasible region, the procedure moves from the interior to the boundary. Typically, the barrier function $B(z, x_0)$ has the form [8]:

$$B(z,x_0) = \sum_{i=1}^{q} \frac{-1}{g_i(z,x_0)} \quad \text{or} \quad B(z,x_0) = -\sum_{i=1}^{q} \ln[-g_i(z,x_0)] \,. \tag{1.23}$$

The solution of problem (1.22) is closer to the true solution the smaller ρ gets. An important feature of the IP methods is that once a solution for a given ρ is found, the parameter ρ can be reduced by a constant factor and an accurate solution of the original NLP problem (1.1)–(1.2) is obtained after a limited number of Newton iterations [16, 76]. The relation between the original problem (1.1)–(1.2) and the barrier problem (1.22) is given by [8]:

$$V^*(x_0) = \lim_{\rho \to 0^+} Q^*(x_0,\rho) = \inf_{\rho > 0} Q^*(x_0,\rho) \,. \tag{1.24}$$

1.1.3.2 Penalty Function Methods

Methods using penalty functions transform a constrained problem into a single unconstrained problem or into a sequence of unconstrained problems [8]. The constraints are placed into the objective function via a penalty parameter in a way that penalizes any violation of the constraints. The penalty function methods are also referred to as the exterior penalty function methods, since they generate a sequence of infeasible points whose limit is an optimal solution to the original problem [8]. Consider the problem (1.1)–(1.2) for a given $x_0 \in X$. A penalty is desired only if the point z is not feasible, i.e., if $g(z,x_0) > 0$. A suitable unconstrained problem is therefore given by [8]:

$$J^*(x_0,\eta) = \inf_z [\, f(z,x_0) + \eta p(z,x_0) \,] \ \text{ s.t. } z \in \mathbb{R}^s \,, \tag{1.25}$$

where $p(z,x_0) = \sum_{i=1}^{q} [\max\{0,\, g_i(z,x_0)\}]^l$ is the so called penalty function, $l \geq 2$ is an integer, and $\eta > 0$ is a penalty parameter. If $g_i(z,x_0) \leq 0$, $\forall i = 1,2,\ldots,q$ then $\max\{0,\, g_i(z,x_0)\} = 0$, $\forall i = 1,2,\ldots,q$ and no penalty is incurred, i.e., $p(z,x_0) = 0$. On the other hand, if $g_i(z,x_0) > 0$, for some i, then $\max\{0,\, g_i(z,x_0)\} > 0$ and the penalty term $\eta p(z,x_0)$ is realized [8]. The condition $l \geq 2$ ensures that the penalty function $p(z,x_0)$ will be differentiable.

An important issue in the penalty function methods is the selection of the penalty parameter η. Consider the penalty problem [8]:

$$W^* = \sup_{\eta > 0} W(\eta) \,, \tag{1.26}$$

where $W(\eta) = J^*(x_0,\eta)$. The relation between the primal problem (1.1)–(1.2) and the penalty problem (1.26) is given by [8]:

$$V^*(x_0) = \sup_{\eta > 0} W(\eta) = \lim_{\eta \to \infty} W(\eta) \tag{1.27}$$

From this result it is clear that we can get arbitrarily close to the optimal objective value of the primal problem (1.1)–(1.2) by computing $W(\eta)$ for a sufficiently large

η. However, as pointed out in [8], there are computational difficulties associated with large penalty parameters, due to ill-conditioning. Therefore, most algorithms using penalty functions solve a sequence of problems (1.25) for an increasing sequence of penalty parameters. With each new value of the penalty parameter, an optimization technique is employed, starting with the optimal solution of problem (1.25) obtained for the parameter value chosen previously. Such an approach is sometimes referred to as *a sequential unconstrained minimization technique* [8]. More details about the penalty function methods can be found in [8].

For a given η, the optimization problem (1.25) can be solved by applying the steepest descent method [18]. Let $h(z,x_0,\eta) = f(z,x_0) + \eta p(z,x_0)$. Then, the steepest descent direction from z is $-\nabla_z h(z,x_0,\eta)$. With the method of steepest descent [18], the values of optimization variables at the $k+1$-th iteration are obtained by the formula:

$$z^{k+1} = z^k - \alpha \nabla_z h(z^k, x_0, \eta) \,, \tag{1.28}$$

where $\alpha > 0$ is the step length. In order for the steepest descent method to be successful, it is important to choose the step length α. One way to do this is to let $\alpha = \beta^m$, where $\beta \in (0,1)$ and $m \geq 0$ is the smallest nonnegative integer such that there is a sufficient decrease in $h(z,x_0,\eta)$. This means that:

$$h(z^k - \alpha \nabla_z h(z^k, x_0, \eta), x_0, \eta) - h(z^k, x_0, \eta) < -\mu \nabla_z h(z^k, x_0, \eta) \,, \tag{1.29}$$

where $\mu \in (0,1)$. This strategy, introduced in [3], is an example of a line search in which one searches on a ray from z^k in a direction in which $h(z,x_0,\eta)$ is locally decreasing. More details about the method of steepest descent can be found in [44]. Unfortunately, the methods based on steepest descent have slow local convergence, even for very simple functions [44]. This is due to the fact that the steepest descent direction scales with $h(z,x_0,\eta)$ and therefore the speed of convergence depends on conditioning and scaling. A good alternative to the steepest descent method is the conjugate gradient method [28, 1], which has improved local convergence properties. Also, the Newton-type methods can be successfully applied to solve the optimization problem (1.25).

1.1.3.3 Direct Search Methods

The direct search methods do not use or approximate the objective function's gradient, i.e. they represent derivative-free methods for optimization. These methods use values of the objective function and constraints taken from a set of sample points and use that information to continue the sampling. More precisely, the direct search methods consist in a sequential examination of trial solutions involving comparison of each trial solution with the best obtained up to that time together with a strategy for determining (as a function of earlier results) what the next trial solution will be [35]. There is a number of direct search methods for unconstrained optimization (see for example [44, 51]). However, here, the most widely used direct search methods for constrained nonlinear optimization are outlined.

- *The method of Box (the Complex method).*
  The Complex method of Box [15] has been developed from the Simplex method [66, 54]. It requires for the NLP problem to be of the form:

$$V^*(x_0) = \min_z f(z, x_0) \tag{1.30}$$

subject to :

$$z_{l,i} \leq z_i \leq z_{u,i}\ ,\ i = 1, 2, ..., s \tag{1.31}$$

$$g_j(z, x_0) \leq 0\ ,\ j = 1, 2, ..., q \tag{1.32}$$

where z_i is the i-th optimization variable, and $z_{l,i}$ and $z_{u,i}$ are the lower and upper bound on this variable.

It is assumed that an initial point z^1, which satisfies both constraints (1.31) and (1.32) is available. In this method, a set of $m \geq s+1$ points is used (referred to as complex), of which one is the given point z^1 (recall that s is the dimension of the optimization vector z). The further $(m-1)$ points required to set up the initial configuration are obtained one at a time by the use of pseudo-random numbers and ranges for each of the independent variables, i.e., $z_i = z_{l,i} + r_i(z_{u,i} - z_{l,i})$, where r_i is a pseudo-random deviate rectangularly distributed over the interval $(0,1)$ [15]. A point so selected must satisfy the bound constraints (1.31), but need not satisfy all the functional constraints (1.32). If a functional constraint is violated, the trial point is moved halfway towards the centroid of those points already selected (where the given initial point is included) [15]. Ultimately, a satisfactory point will be found. It is assumed that the feasible region is convex. Proceeding in this way, $(m-1)$ points are found which satisfy all the constraints.

The function is evaluated at each vertex of the complex, and the vertex of the worst (maximal) function value is replaced by a point $\gamma \geq 1$ times as far from the centroid of the remaining points as the reflection of the worst point in the centroid (the new point is collinear with the rejected point and the centroid of the retained vertices) [15]. If this trial point is also the worst, it is moved halfway towards the centroid of the remaining points to give a new trial point. The above procedure is repeated until some constraint is violated. If a trial vertex does not satisfy the lower or the upper bound on some optimization variable z_i, that variable is reset to a value $z_{l,i} + \varepsilon$ or value $z_{u,i} - \varepsilon$ (depending on which bound has been violated), with ε being a small positive number. If a functional constraint $g_j(z, x_0)$ is violated, the trial point is moved halfway towards the centroid of the remaining points. Ultimately, a permissible point is found. Thus, as long as the complex has not collapsed into the centroid, progress will continue.

The idea of the Box's method is illustrated in Fig. 1.1 for the case when $s = 2$ and the number of points is $m = s + 1 = 3$, i.e., for a simplex of points. The point z^2 is considered to be the worst point and c is the center of mass of the other two points (z^1 and z^3).
- *DIRECT method.*
  The DIRECT algorithm (DIViding RECTangles) is a direct search method for *global* optimization, which was first introduced in [43, 42]. In [30, 27],

Fig. 1.1 The simplex with reflection of the point z^2 into point z^4 and two consecutive contractions (z^5, z^6) due to infeasibility.

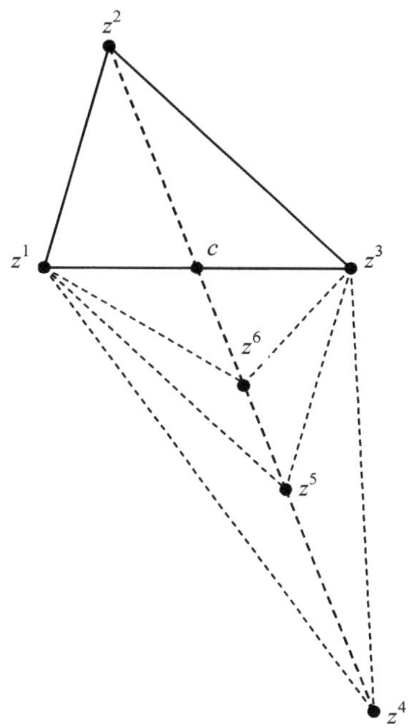

rigorous new analysis and algorithmic improvements to the DIRECT algorithm have been presented. The DIRECT algorithm is a deterministic sampling algorithm developed in the spirit of Lipschitz optimization, and designed to overcome some of the shortcomings of traditional Lipschitzian algorithms (like the algorithm in [59]). One problem of the algorithm in [59] is its reliance on an accurate estimation of the Lipschitz constant. DIRECT solves this problem by replacing the Lipschitz constant with an adaptive internal parameter.

The DIRECT method solves the following mixed-integer nonlinear programming (MINLP) problem [42]:

$$V^*(x_0) = \min_z f(z, x_0) \tag{1.33}$$

subject to :

$$z_{l,i} \leq z_i \leq z_{u,i} \ , \ i = 1, 2, ..., r \tag{1.34}$$

$$z_i \in \mathbb{Z} \ , \ i = r+1, r+2, ..., s \tag{1.35}$$

$$g_j(z, x_0) \leq 0 \ , \ j = 1, 2, ..., q \tag{1.36}$$

where $\mathbb{Z}$ is the set of integer numbers. The vector of optimization variables $z = [z_1, z_2, \ldots, z_r, z_{r+1}, \ldots, z_s]$ includes both real variables $(z_1, z_2, \ldots, z_r)$ and integer variables $(z_{r+1}, z_{r+2}, \ldots, z_s)$. The bounds on the variables limit the search to an s-dimensional hyper-rectangle. DIRECT proceeds by partitioning this rectangle into smaller rectangles, each of which has a sampled point at its center, i.e., a point where the functions have been evaluated [43, 42]. Fig. 1.2 shows the

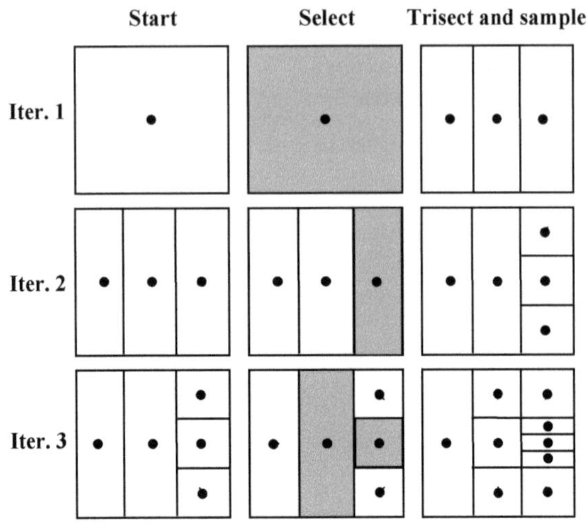

Fig. 1.2 Partitioning of the optimization variables space with DIRECT algorithm [43, 42].

first three iterations of DIRECT on a hypothetical problem with two optimization variables. At the start of each iteration, the space is partitioned into rectangles. DIRECT then selects one or more of these rectangles for further search using a technique described below. Finally, each selected rectangle is trisected along one of its long sides, after which the center points of the new rectangles are sampled. The key step in the algorithm is the selection of rectangles, since this determines how search effort is allocated across the space. The rectangles are selected using all possible relative weightings of *local* versus *global* search [43, 42]. First, it would be necessary to describe how the inequality constraints (1.36) are treated by the DIRECT method. The key to handling constraints in DIRECT is to work with an auxiliary function that combines information on the objective and constraint functions in a special manner [42]. To express this auxiliary function, an additional notation needs to be introduced. Let z_p be the center point of the p-th rectangle. Let $\varphi_1, \varphi_2, \ldots, \varphi_q$ be positive weighting coefficients for the inequality constraints. Let the minimal value of the objective function at the current iteration be $V_{\min}(x_0)$. Let $\widetilde{V}$ be any value that satisfies $\widetilde{V} < V_{\min}(x_0) - \delta$, where $\delta > 0$. The auxiliary function, evaluated at the center of the p-th rectangle, is as follows [42]:

$$V_p^a(\widetilde{V}, x_0) = \max\{f(z_p, x_0) - \widetilde{V}, 0\} + \sum_{j=1}^{q} \varphi_j \max\{g_j(z_p, x_0), 0\} \qquad (1.37)$$

The first term of the auxiliary function represents a penalty for any deviation of the function value $f(z_p, x_0)$ above the value $\widetilde{V}$. The second term is a sum of weighted constraint violations. If $\widetilde{V}$ is the global minimum, the lowest possible value of the auxiliary function is zero and occurs only at the global minimum. At any other point, the auxiliary function is positive either due to suboptimality or infeasibility. For the global minimum to occur in the p-th rectangle, the auxiliary function must fall to zero starting from its positive value at the center point [42]. Moreover, the maximum distance over which this change can occur is the center-vertex distance d_p in the rectangle. Thus, to reach the global minimum in the p-th rectangle, the auxiliary function (1.37) must undergo a minimum rate of change, given by $e_p(\widetilde{V},x_0) = V_p^a(\widetilde{V},x_0)/d_p$ [42]. Since the point x_0 in the MINLP problem (1.33)–(1.36) is fixed, the rate of change function e_p depends only on the argument $\widetilde{V}$. The DIRECT procedure of selecting rectangles for further exploration identifies and selects all rectangles whose rate of change functions $e_p(\widetilde{V},x_0)$ participate in the lower envelope of all curves $V_p^a(\widetilde{V},x_0)/d_p$ for $\widetilde{V} < V_{\min}(x_0) - \delta$ [42]. More details about the DIRECT method can be found in [43, 42].

In this book, the DIRECT algorithm is applied to design explicit model predictive controllers for constrained nonlinear systems with quantized inputs (see Chapter 5).

1.1.4 Sensitivity Results

The solution of a mathematical program can behave in a variety of ways when perturbing the problem parameters. Depending on the problem, the solution may vary smoothly or change drastically for arbitrary small perturbations of parameter values. Let $x_0 \in X$, z_0 satisfy the KKT conditions, and $\mathscr{A}_0$ be the optimal active set at x_0. The Basic Sensitivity Theorem [26] gives local regularity conditions for the optimal solution, Lagrange multipliers and optimal objective function value as functions of x:

Theorem 1.1. *If:*

i). the functions $f(z,x)$ and $g(z,x)$ are twice continuously differentiable in z, and their gradients with respect to z and the constraints are once continuously differentiable in x in a neighborhood of (z_0,x_0),
ii). the second order sufficient condition (1.12) for a local minimum of (1.1)–(1.2) holds at z_0, with associated Lagrange multiplier λ_0,
iii) the active constraint gradients $\nabla_z g_{\mathscr{A}_0}(z_0,x_0)$ are linearly independent,
iv). $(\lambda_0)_i > 0$ when $g_i(z_0,x_0) = 0$ (strict complementary slackness),

Then:

a). z_0 is a local isolated minimizing point with unique associated Lagrange multiplier λ_0,
b). for x in the neighborhood of x_0, there exist unique, once continuously differentiable functions $z^*(x)$ and $\lambda^*(x)$ such that $z^*(x_0) = z_0$ and $z^*(x)$ is a locally unique local minimum of (1.1)–(1.2) with associated Lagrange multiplier $\lambda^*(x)$,
c). in a neighborhood of x_0, the set of active constraints is unchanged, strict complementary slackness holds, and the active constraint gradients at $z^*(x)$ remain linearly independent.

Related results for slightly different conditions, and extensions that show the existence and computation of directional derivatives of the solution with respect to x at x_0 can be found in [45, 50, 62] and others.

For the fixed active set $\mathscr{A}_0$ the KKT conditions (1.7)–(1.8) reduce to the following system of equations parameterized by x:

$$\nabla_z f(z(x),x) + \sum_{i\in\mathscr{A}_0} \lambda_i(x)\nabla_z g_i(z(x),x) = 0 \tag{1.38}$$

$$g_{\mathscr{A}_0}(z(x),x) = 0\,. \tag{1.39}$$

The functions $z(x)$ and $\lambda(x)$ implicitly defined by (1.38)–(1.39) are optimal only for those x where the active set $\mathscr{A}_0$ is optimal. Assuming z and λ are well defined on X, we characterize the critical region $CR_{\mathscr{A}_0}$ where the solution corresponding to the fixed active set $\mathscr{A}_0$ is optimal:

$$CR_{\mathscr{A}_0} \triangleq \{x \in X \mid \lambda(x) \geq 0,\ g(z(x),x) \leq 0\}\,. \tag{1.40}$$

There is a finite number of candidate active sets, so this result suggests a finite partition of X with a piecewise solution to the mp-NLP. Although explicit exact solutions cannot be found in the general nonlinear case, the above result indicates that it is meaningful to search for a continuous approximation to the optimal solution as a function of x. Continuity of the optimal solution depends on several assumptions that may be hard to verify in the general nonlinear case. However, many optimal control problems tend to lead to continuous solution functions.

1.1.5 *Algorithms for Approximate Multi-parametric Nonlinear Programming*

Consider the nonlinear multi-parametric program (1.1)–(1.2) dependent on the parameter x. Let X be a polytopic set of parameters, defined by $X = \{x \in \mathbb{R}^n \mid Ax \leq b\}$. In multi-parametric programming, it is of interest to characterize the solution of the mp-NLP problem (1.1)–(1.2) for the set X. The solution of an mp-NLP problem is a triple $(V^*(x), z^*(x), X_f)$ (see Section 1.1.1), where X_f is the set of feasible parameters, $V^*(x)$ is the optimal value function, and $z^*(x)$ is the optimizer function. It is assumed that X_f is closed and $V^*(x)$ is finite for every $x \in X_f$.

1.1.5.1 Approximate Solution of Convex mp-NLP

1. Convexity results.
The following assumption is made.

Assumption 1.1. *The functions f and g in the nonlinear multi-parametric program (1.1)–(1.2) are jointly convex functions of (z,x).*

The following basic result for convex multi-parametric programming was proved in [52]:

Theorem 1.2. *Consider the nonlinear multi-parametric program (1.1)–(1.2) and suppose that Assumption 1.1 holds. Then, X_f is a convex set and $V^* : X_f \mapsto \mathbb{R}$ is a convex and continuous function of x.*

Convexity of X_f and V^* is a direct consequence of the convexity of f and g, while continuity of V^* can be established under weaker conditions [26].

The main idea is to construct a feasible piecewise approximation to $z^*(x)$ on X, where the constituent functions pieces are defined on hyper-rectangles covering X. The accuracy of approximation is measured by the difference between the optimal and sub-optimal function values rather than the difference between the exact and approximation solutions. Since the optimal function value V^* cannot be assumed known, convexity is exploited to compute simple bounds to be used for constructing the approximate solution, similar to Chapter 9 in [26]. The method is applicable for piecewise linear (PWL) and piecewise nonlinear (PWNL) approximations.

Consider the vertices $\Theta = \{\theta_1, \theta_2, \dots, \theta_{N_\theta}\}$ of any bounded polyhedron $X_0 \subseteq X_f$. Define the affine function $\bar{V}(x) = \bar{V}_0 x + \bar{l}_0$ as the solution to the following linear program (LP) [38]:

$$\min_{\bar{V}_0, \bar{l}_0} (\bar{V}_0 \theta + \bar{l}_0) \tag{1.41}$$

$$\text{subject to } \bar{V}_0 \theta_i + \bar{l}_0 \geq V^*(\theta_i),\ \forall \theta_i \in \Theta\ . \tag{1.42}$$

Likewise, define the convex PWL function [38]:

$$\underline{V}(x) = \max_{\theta_i \in \Theta} (V^*(\theta_i) + \nabla^T V^*(\theta_i)(x - \theta_i)) \tag{1.43}$$

If V^* is not differentiable at θ_i, then $\nabla V^*(\theta_i)$ is taken as any sub-gradient of V^* at θ_i [63]. $\underline{V}$ and $\bar{V}$ have the following properties [26, 39]:

Theorem 1.3. *Consider the nonlinear multi-parametric program (1.1)–(1.2) and suppose that Assumption 1.1 holds. Consider any bounded polyhedron $X_0 \subseteq X_f$. Then $\underline{V}(x) \leq V^*(x) \leq \bar{V}(x)$ for all $x \in X_0$.*

In [38], it is suggested to select a local linear approximation to the solution that minimizes the objective function approximation error subject to feasibility of the solution, similar to [9].

Lemma 1.1. *Consider the nonlinear multi-parametric program (1.1)–(1.2) and suppose that Assumption 1.1 holds. Consider any bounded polyhedron $X_0 \subseteq X_f$ with vertices $\Theta = \{\theta_1, \theta_2, \dots, \theta_{N_\theta}\}$. If K_0 and h_0 solve the convex NLP:*

$$\min_{K_0, h_0} \sum_{i=1}^{N_\theta} (f(K_0 \theta_i + h_0, \theta_i) - V^*(\theta_i) + \mu \| K_0 \theta_i + h_0 - z^*(\theta_i) \|_2^2) \tag{1.44}$$

$$\textit{subject to } g(K_0 \theta_i + h_0, \theta_i) \leq 0,\ \forall \theta_i \in \Theta\ , \tag{1.45}$$

then $\widehat{z}_0(x) = K_0 x + h_0$ is feasible for the mp-NLP (1.1)–(1.2) for all $x \in X_0$.

In (1.44), $\mu > 0$ is a weighting coefficient. In general, the NLP defined in Lemma 1.1 need not have a feasible solution. As a partial remedy, the following result shows that at least for sufficiently small polyhedron X_0, feasibility can be guaranteed [38].

Lemma 1.2. *Consider the nonlinear multi-parametric program (1.1)–(1.2) and suppose that Assumption 1.1 holds. Let $X_0 \subseteq X_f$ be a sufficiently small bounded polyhedron with non-empty interior. Then there exists an affine function $\widetilde{z}(x)$ such that $g(\widetilde{z}(x),x) \leq 0$ for all $x \in X_0$.*

Proof ([38]). Since $X_0 \subseteq X_f$ is small, it follows from [26] that some unique and continuous feasible solution function $z(x)$ exists in a neighborhood that contains X_0. Since g is convex, it is straightforward to construct an affine support $\widetilde{z}(x)$. □

Since $\widehat{z}_0(x)$ defined in Lemma 1.1 is feasible in X_0, it follows that the suboptimal objective function $\widehat{V}(x) = f(\widehat{z}_0(x),x)$ is an upper bound on $V^*(x)$ in X_0 such that for all $x \in X_0$ we have:

$$0 \leq \widehat{V}(x) - V^*(x) \leq \varepsilon_0 \tag{1.46}$$

where:

$$\varepsilon_0 = -\min_{x \in X_0}(-\widehat{V}(x) + \underline{V}(x)) \,. \tag{1.47}$$

Computing ε_0 requires the solution of the NLP (1.47). If $\underline{V}$ is conservatively chosen as affine $\underline{V}(x) = V^*(\theta_i) + \nabla^T V^*(\theta_i)(x - \theta_i)$ (cf. (1.43)), this NLP is concave since $\widehat{V}$ is convex. Hence, the optimization can be done efficiently since X_0 is a polyhedron and it suffices to compare the solution at its vertices due to the concavity [36].

2. Algorithm for approximate explicit solution of convex mp-NLPs.
Consider a hyper-rectangle $X \subset \mathbb{R}^n$ where we seek to approximate the solution function $z^*(x)$ to the mp-NLP (1.1)–(1.2). In many problems of interest the approximate solution function will be evaluated in an embedded computer architecture under strict real-time requirements and with highly limited computational resources. In order to keep the computational complexity at a minimum, we require that the approximating function is PWL with a parameter space partition that is orthogonal and can be represented by a $k-d$ tree [14], [39, 31], such that the real-time search complexity is logarithmic with respect to the number of regions in the partition. The $k-d$ tree (Fig. 1.3) is a hierarchical data structure where a hyper-rectangle can be sub-divided into smaller hyper-rectangles allowing the local resolution to be adapted. When searching the tree, only one scalar comparison is required at each level. Initially the algorithm will consider the whole region $X_0 = X$. Under the convexity Assumption 1.1, the main idea of the approximate mp-NLP algorithm is to compute the solution of problem (1.1)–(1.2) at the 2^n vertices of the hypercube X_0, by solving up to 2^n NLPs. Based on these solutions, assuming they are all feasible, we compute a feasible local linear approximation function $\widehat{z}_0(x)$ to the optimal solution function $z^*(x)$, restricted to the hyper-rectangle X_0, using Lemma 1.1. If such an approximation exists, and the maximal objective function error ε_0 in X_0 is

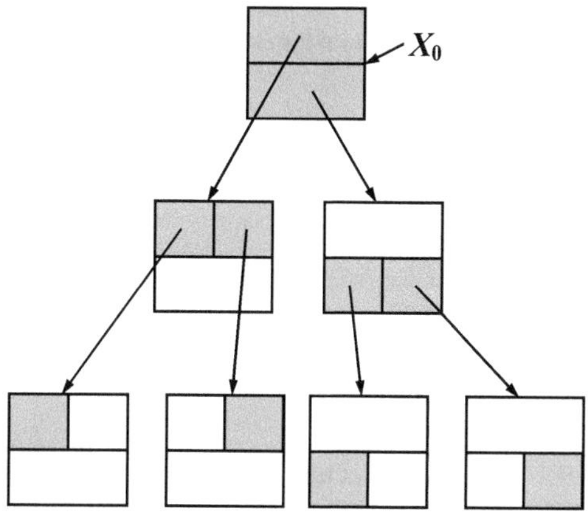

Fig. 1.3 $k-d$ tree partition of a rectangular region.

smaller than some prescribed tolerance $\bar{\varepsilon} > 0$, no further refinement of the region X_0 is needed. Otherwise, we split X_0 into two hyper-rectangles, and repeat the procedure for each of these.

Assume the tolerance $\bar{\varepsilon} > 0$ of the objective function approximation error is given. For simplicity, we consider uniform tolerance in this chapter. In later chapters, sometimes the tolerance will depend on x, which causes no problems. Denote with S_{X_0} the volume of a given hyper-rectangular region $X_0 \subset X \subset \mathbb{R}^n$, i.e. $S_{X_0} = \prod_{i=1}^{n} \Delta x_i$, where Δx_i is the size of X_0 along the axis x_i. Let $S_{\min}$ be the minimal allowed volume of the regions in the partition of X. A nonzero $S_{\min}$ is required to ensure termination of the algorithm in finite time. The following algorithm is proposed to determine an explicit approximate solution of convex mp-NLP (1.1)–(1.2) [38].

Algorithm 1.1. Explicit approximate solution of convex mp-NLP.

Input: Data to problem (1.1)–(1.2), the parameter μ (used in Lemma 1.1), the approximation tolerance $\bar{\varepsilon}$, the minimal allowed volume $S_{\min}$.
Output: Partition $\Pi = \{X_1, X_2, ..., X_{N_X}\}$ and associated PWL solution $\widehat{z}_\Pi = \{\widehat{z}_{X_1}, \widehat{z}_{X_2}, \dots, \widehat{z}_{X_{N_X}}\}$.

1. Initialize the partition to the whole hyper-rectangle, i.e., $\Pi = \{X\}$. Mark the hyper-rectangle X as unexplored, and let $flag = 1$.
2. **while** $flag = 1$ **do**
3.   **while** $\exists$ unexplored hyper-rectangles in Π **do**
4.   Select any unexplored hyper-rectangle $X_0 \in \Pi$ and compute its volume S_{X_0}.
5.   Solve problem (1.1)–(1.2) for x fixed to each of the vertices θ_i, $i = 1, \dots, N_\theta$ of the hyper-rectangle X_0.
6.   **if** (1.1)–(1.2) has a feasible solution at all points θ_i, $i = 1, \dots, N_\theta$ **then**
7.   Compute a linear approximation $\widehat{z}_0(x) = K_0 x + h_0$ using Lemma 1.1, as an approximation to be used in X_0.
8.   **if** a solution $\widehat{z}_0(x)$ was found **then**
9.   Compute the error bound ε_0, using (1.41)–(1.43), and (1.47).

10. **If** $\varepsilon_0 > \bar{\varepsilon}$ and $S_{X_0} > S_{\min}$, mark the hyper-rectangle X_0 to be split.
Otherwise, mark X_0 as explored and feasible.
11. **else**
12. **If** $S_{X_0} < S_{\min}$, mark X_0 infeasible and explored.
Otherwise, mark X_0 to be split.
13. **end if**
14. **else**
15. **If** $S_{X_0} < S_{\min}$, mark X_0 infeasible and explored.
Otherwise, mark X_0 to be split.
16. **end if**
17. **end while**
18. $flag := 0$
19. **if** $\exists$ hyper-rectangles in Π that are marked to be split **then**
20. $flag := 1$
21. **while** $\exists$ hyper-rectangles in Π that are marked to be split **do**
22. Select any hyper-rectangle $X_0 \in \Pi$ marked to be split.
23. Split X_0 into two hyper-rectangles X_1 and X_2 by applying an heuristic splitting rule. Mark X_1 and X_2 unexplored, remove X_0 from Π, and add X_1 and X_2 to Π.
24. **end while**
25. **end if**
26. **end while**

The PWL approximation generated by Algorithm 1.1 is denoted $\widehat{z}_\Pi : \underline{X} \mapsto \mathbb{R}^s$, where $\underline{X}$ is the union of the hyper-rectangles where a feasible solution has been found. It is an inner approximation to X_f and the approximation accuracy is determined by the minimal allowed volume $S_{\min}$ of the regions. The boundary of the feasible region X_f can thus be approximated more closely by allowing smaller infeasible regions by choosing $S_{\min}$ small. We remark that $\widehat{z}_\Pi$ is generally not continuous.

Step 23 needs further specification of how a hyper-rectangle is being partitioned. A hyper-rectangle is split into two equal parts by an axis-orthogonal hyperplane that goes through its center. As in [31], the main idea is to select the hyperplane where the change of error between the solutions on each side of the hyperplane is largest (before splitting). This is implemented by comparing the solutions at the vertices of the hyper-rectangle. It is reasonable to expect that this heuristics may give a significant reduction in the error in both hyper-rectangles after splitting and its effectiveness is observed in a number of examples.

Theorem 1.4. *Consider the nonlinear multi-parametric program (1.1)–(1.2) and suppose that Assumption 1.1 holds and* $S_{\min}$ *is sufficiently small. Assume that the partitioning rule in step 23 guarantees that the error decreases by some minimum amount or factor at each split. Then Algorithm 1.1 terminates with an approximate solution function* $\widehat{z}_\Pi$ *that is feasible and satisfies* $0 \le f(\widehat{z}_\Pi(x), x) - V^*(x) \le \bar{\varepsilon}$ *for all* $x \in \underline{X}$.

Proof ([38]). If the algorithm terminates, the specified tolerance is met because of steps 9, 10, and 23. Since V^* is continuous, it is clear that a $k-d$ tree partition will lead to an approximation with arbitrary uniform accuracy provided the hypercubes are sufficiently small. According to Lemma 1.2, this approximation will be feasible, and since the partitioning rule ensures that the error decreases by some minimum amount or factor at each step, the algorithm will indeed terminate after a finite number of steps. □

In [22], several alternative multi-parametric programming algorithms for explicit approximate solution of convex mp-NLP problems are discussed. Thus, in [24] a multi-parametric outer approximation algorithm for mp-NLP problems and multi-parametric mixed-integer nonlinear programming (mp-MINLP) problems is presented. A multi-parametric quadratic approximation algorithm is proposed in [37] and recently revisited in [23]. An approximate multi-parametric algorithm is proposed in [11], where the parameter space is divided into a set of simplices. Recently, a geometric vertex search algorithm is proposed in [53]. Some of these algorithms are extended to consider the non-convex case (see [58]).

1.1.5.2 Approximate Solution of Non-convex mp-NLP

If convexity does not hold (Assumption 1.1), then global optimization, e.g. [43, 42, 36], is generally needed in several steps of the algorithm to maintain its theoretical properties [38]:

(1) The NLP (1.1)–(1.2) must be solved using global optimization in step 5.
(2) The NLP (1.44)–(1.45) must be reformulated and solved using global optimization in step 7. It is not sufficient to impose the constraints at the vertices of the polyhedron X_0 if g is not convex. In order to resolve this problem, one may use (a conservative) convex underestimation in combination with global optimization as suggested in [26].
(3) The computation of the error bound ε_0 in step 9 assumes the knowledge of a lower bound $\underline{V}$ on the optimal objective function. The bound (1.43) does not necessarily hold if V^* is not convex. Again, convex underestimation and global optimization may be used.

On the other hand, one may argue in the favor of a computationally more efficient ad hoc approach to handle non-convex problems [38]. The reason for this is that an explicit representation of the approximate solution is available, which makes rigorous verification and validation of its properties possible. One heuristic approach is to include some interior points in addition to the set of vertices Θ when used in (1.43)–(1.45). Hence, any non-convexity related error in the computed bounds and approximation of the constraints are likely to be reduced. Moreover, based on the solutions of the associated NLPs one may locally estimate the Hessian of the optimal objective function at the points in Θ and may thus be able to detect if it is locally convex or non-convex, and adjust the number of additional interior points to be added to Θ. The introduction of such additional points does not necessarily lead to additional complexity of the PWL approximate solution, but may only serve to verify its accuracy [38].

Here, practical computational methods for explicit approximate solution of non-convex mp-NLP problems are presented. They don't necessarily lead to guaranteed properties of the explicit approximate solution, but when combined with verification and analysis methods may give a practical tool for explicit approximate solution of non-convex mp-NLPs.

1. Close-to-global solution of non-convex mp-NLPs.

In general, the objective function f can be non-convex with multiple local minima. Therefore, it would be necessary to apply an effective initialization of the mp-NLP problem (1.1)–(1.2) so to find a close-to-global solution. One possible way to obtain this is to find a close-to-global solution at a point $w_0 \in X_0$ by comparing the local minima corresponding to several initial guesses and then to use this solution as an initial guess at the neighboring points $w_i \in X_0$, $i = 1,2,...,N_1$, i.e. to propagate the solution. The following procedure is used to generate a set of points $W_0 = \{w_0, w_1, w_2, \ldots, w_{N_1}\} \subset X_0$ [32].

Procedure 1.1 (generation of set of points). *Consider any hyper-rectangle $X_0 \subseteq X$ with vertices $\Theta^0 = \{\theta_1^0, \theta_2^0, \ldots, \theta_{N_\theta}^0\}$ and center point w_0. Consider also the hyper-rectangles $X_0^j \subset X_0$, $j = 1,2,\ldots,N_0$ with vertices respectively $\Theta^j = \{\theta_1^j, \theta_2^j, \ldots, \theta_{N_\theta}^j\}$, $j = 1,2,\ldots,N_0$. Suppose $X_0^1 \subset X_0^2 \subset \ldots \subset X_0^{N_0}$. For each of the hyper-rectangles X_0 and $X_0^j \subset X_0$, $j = 1,2,\ldots,N_0$, denote the set of its facets centers with $\Psi^j = \{\psi_1^j, \psi_2^j, \ldots, \psi_{N_\psi}^j\}$, $j = 0,1,2,\ldots,N_0$. Define the set of all points*

$$W_0 = \{w_0, w_1, w_2, \ldots, w_{N_1}\}, \text{ where } w_i \in \left\{\bigcup_{j=0}^{N_0} \Theta^j\right\} \cup \left\{\bigcup_{j=0}^{N_0} \Psi^j\right\},\ i = 1,2,\ldots,N_1.$$

For a hyper-rectangle in the n-dimensional parameter space, the number of its vertices is $N_\theta = 2^n$ and the number of its facets centers is $N_\psi = 2n$. Therefore, the number of all points generated with Procedure 1.1 is $1 + (N_0 + 1)(N_\theta + N_\psi)$, where N_0 is the number of interior hyper-rectangles.

The following procedure is applied to search for a close-to-global solution at the points $w_i \in W_0$, $i = 0,1,2,\ldots,N_1$ [32].

Procedure 1.2 (close-to-global solution of mp-NLP). *Consider any hyper-rectangle $X_0 \subseteq X$ with a set of points $W_0 = \{w_0, w_1, w_2, \ldots, w_{N_1}\}$ determined by applying Procedure 1.1. Then:*

a). Suppose local minima of the NLP (1.1)–(1.2) at the center point w_0 of X_0 have been computed. Then, determine a close-to-global solution of (1.1)–(1.2) at w_0 through the following minimization:

$$z^*(w_0) = \arg \min_{z \in \{z_1^{\text{local}}, z_2^{\text{local}}, \ldots, z_{N_z}^{\text{local}}\}} f(z, w_0)\,. \tag{1.48}$$

Here, z_i^{local}, $i = 1,2,\ldots,N_z$ correspond to local minima of the objective function $f(z, w_0)$ obtained for a number of initial guesses z_i^0, $i = 1,2,\ldots,N_z$.

b). Determine a close-to-global solution of the NLP (1.1)–(1.2) at the points $w_i \in W_0$, $i = 1, 2, \ldots, N_1$ in the following way:

1. *Let $z^*(w_0)$ be the close-to-global solution of the NLP (1.1)–(1.2) at the center point w_0, obtained by solving problem (1.48) in step a). Let $i = 1$.*
2. *Let $W^s = \{w_0, w_1, w_2, \ldots, w_{N_2}\} \subset W_0$ be the subset of points at which a feasible solution of the NLP (1.1)–(1.2) has been already determined.*
3. *Find the point $\widetilde{w} \in W^s$ that is most close to the point w_i, i.e., $\widetilde{w} = \arg\min_{w \in W^s} \|w - w_i\|$. Let the solution at $\widetilde{w}$ be denoted $z^*(\widetilde{w})$.*
4. *Solve the NLP (1.1)–(1.2) at the point w_i with initial guess for the optimization variables set to $z^*(\widetilde{w})$.*
5. *If a solution of the NLP (1.1)–(1.2) at the point w_i has been found, mark w_i as feasible and add it to the set W^s. Otherwise, mark w_i as infeasible.*
6. *Let $i = i + 1$. If $i \leq N_1$, go to step 2. Otherwise, terminate.*

With some abuse of notation we do not distinguish between the global solution and the close-to-global solution, and denote both with $z^*(x)$. Procedure 1.2 is illustrated on Fig. 1.4. First, a close-to-global solution to the NLP (1.1)–(1.2) is determined at the center point w_0 of the hyper-rectangle X_0 (the case when no feasible solution at the center point w_0 exists is discussed later). Then, this solution is used as an initial guess when solving the NLP at the points $w_1, w_2, \ldots, w_8$ which represent the vertices and the facets centers of the smallest interior hyper-rectangle X_0^1. Then, the solutions at these points are used as initial guesses when solving the NLP at the points $w_9, w_{10}, \ldots, w_{16}$ which are the vertices and the facets centers of the interior

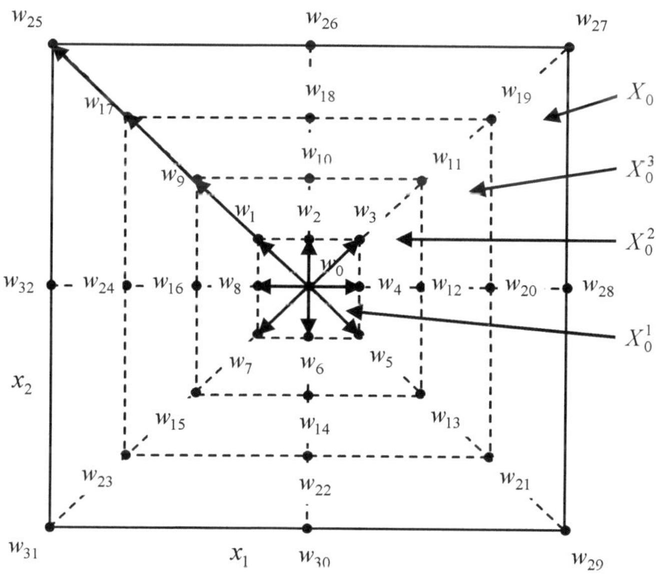

Fig. 1.4 Illustration of Procedure 1.2.

hyper-rectangle X_0^2. Next, the solutions at these points are used as initial guesses when solving the NLP at the points $w_{17}, w_{18}, ..., w_{24}$ which represent the vertices and the facets centers of the interior hyper-rectangle X_0^3. At the end, the solutions at these points are used as initial guesses when solving the NLP at the points $w_{25}, w_{26}, ..., w_{32}$ which are the vertices and the facets centers of the hyper-rectangle X_0.

2. Computation of explicit approximate solution.
We restrict our attention to a hyper-rectangle $X \subset \mathbb{R}^n$ where we seek to approximate the solution $z^*(x)$ to the non-convex mp-NLP (1.1)–(1.2). Like in Section 1.1.5.1, we require that the parameter space partition is orthogonal and can be represented as a $k-d$ tree [14], [39, 31]. The main idea of the approach to explicit approximate solution of non-convex mp-NLPs is to construct a PWL approximation $\widehat{z}(x)$ to the close-to-global solution $z^*(x)$ on X, where the constituent affine functions are defined on hyper-rectangles covering X. It should be noted that sometimes it may be more appropriate to use a piecewise nonlinear (PWNL) approximation. In case of non-convexity, it would not be sufficient to impose the constraints only at the vertices of the hyper-rectangle X_0. This problem is resolved by including some interior points in addition to the set of vertices of X_0 [32]. These additional points represent the vertices and the facets centers of one or more hyper-rectangles contained in the interior of X_0 (see Procedure 1.1). Based on the solutions at all points, a local linear approximation $\widehat{z}_0(x) = K_0x + h_0$ to the close-to-global solution $z^*(x)$, to be used as an approximation in the whole hyper-rectangle X_0, is determined by applying the following procedure [32].

Procedure 1.3 (computation of approximate solution). *Consider any hyper-rectangle $X_0 \subseteq X$ with a set of points $W_0 = \{w_0, w_1, w_2, ..., w_{N_1}\}$ determined by applying Procedure 1.1. Compute K_0 and h_0 by solving the following NLP:*

$$\min_{K_0,h_0} \sum_{i=0}^{N_1} (f(K_0w_i + h_0, w_i) - V^*(w_i) + \mu \|K_0w_i + h_0 - z^*(w_i)\|_2^2) \quad (1.49)$$

$$\textit{subject to } \; g(K_0w_i + h_0, w_i) \leq 0, \; \forall w_i \in W_0 \, . \quad (1.50)$$

In (1.49), the parameter $\mu > 0$ is a weighting coefficient. Note that the linear approximation $\widehat{z}_0(x) = K_0x + h_0$, computed with Procedure 1.3, satisfies the constraints in the mp-NLP problem (1.1)–(1.2) only for the discrete set of points $W_0 \subset X_0$. In order to give an appropriate initialization of the NLP problem (1.49)–(1.50) for the region X_0, the already computed solutions of this problem in some of the neighboring regions can be used as initial guesses.

3. Estimation of error bounds. Suppose that a linear approximation $\widehat{z}_0(x) = K_0x + h_0$ for the region X_0 has been computed by applying Procedure 1.3. Then it follows that the sub-optimal objective function $\widehat{V}(x) = f(\widehat{z}_0(x), x)$ is an approximate upper bound on $V^*(x)$ in X_0, such that for all $x \in X_0$, where $\widehat{z}_0(x)$ is feasible, we have:

$$0 \leq \widehat{V}(x) - V^*(x) \leq \varepsilon_0 \, . \quad (1.51)$$

As already mentioned, the objective function f can be non-convex with multiple local minima. Therefore, (1.51) is only valid if global solutions are found to all sub-problems and feasibility of $\widehat{z}_0(x)$ and $z^*(x)$ holds for all $x \in X_0$. The following procedure can be used to obtain an estimate $\widehat{\varepsilon}_0$ of the maximal approximation error ε_0 in X_0 [32].

Procedure 1.4 (computation of error bound approximation). *Consider any hyper-rectangle $X_0 \subseteq X$ with a set of points $W_0 = \{w_0, w_1, w_2, \ldots, w_{N_1}\}$ determined by applying Procedure 1.1. Compute an estimate $\widehat{\varepsilon}_0$ of the error bound ε_0 through the following maximization:*

$$\widehat{\varepsilon}_0 = \max_{i \in \{0,1,2,\ldots,N_1\}} (\widehat{V}(w_i) - V^*(w_i)) . \tag{1.52}$$

4. Procedure and heuristic rules for splitting a region.
The following procedure is applied to determine the split of a region X_0 for which a local linear approximation $\widehat{z}_0(x) = K_0 x + h_0$ is found, but the required accuracy $\bar{\varepsilon} > 0$ of objective function approximation is not achieved [32].

Procedure 1.5 (Determination of the split of a region). *Consider a hyper-rectangle X_0 and suppose that a local linear approximation $\widehat{z}_0(x) = K_0 x + h_0$ was found by applying Procedure 1.3. Suppose also that the required accuracy $\bar{\varepsilon}$ is not achieved. Then, determine the split of X_0 in the following way:*

1. *Let $j = 1$.*
2. *Consider splitting X_0 by a hyperplane through its center and orthogonal to the axis x_j. Denote the new hyper-rectangles with X_1^j and X_2^j.*
3. *Compute local linear approximations $\widehat{z}_1^j(x)$ and $\widehat{z}_2^j(x)$, candidates for use in X_1^j and X_2^j, respectively, by applying Procedure 1.3.*
4. *Compute estimates $\widehat{\varepsilon}_1^j$ and $\widehat{\varepsilon}_2^j$, respectively of the error bounds ε_1^j in X_1^j and ε_2^j in X_2^j, by applying Procedure 1.4. Let $\widehat{\varepsilon}^j = \widehat{\varepsilon}_1^j + \widehat{\varepsilon}_2^j$.*
5. *Let $j = j + 1$. If $j \leq n$, go to step 2.*
6. *Split X_0 by a hyperplane through its center and orthogonal to the axis x_j where $\widehat{\varepsilon}^j$ is minimal.*

In step 4, the metric $\widehat{\varepsilon}^j = \widehat{\varepsilon}_1^j + \widehat{\varepsilon}_2^j$ could be replaced by other metrics such as $\widehat{\varepsilon}^j = \max(\widehat{\varepsilon}_1^j, \widehat{\varepsilon}_2^j)$.

The following rule is applied when no feasible solution to the NLP problem (1.1)–(1.2) was found at some of the points $w_i \in W_0$, $w_i \neq w_0$ [32]. Here, the set $W_0 = \{w_0, w_1, w_2, \ldots, w_{N_1}\}$ is defined in Procedure 1.1.

Heuristic splitting rule 1.1 (handling infeasibility). *Consider the following two cases:*

1. *The set of the feasible points in X_0 includes the center point w_0 and some non-empty subset of the points $w_i \in W_0$, $w_i \neq w_0$ (the set $W_0 = \{w_0, w_1, w_2, \ldots, w_{N_1}\}$ is defined in Procedure 1.1). Then, split X_0 into two types of hyper-rectangles by hyperplanes containing some of the feasible points $w_i \in W_0$:*

i. Hyper-rectangles $X_1^f, X_2^f, \ldots, X_{N_f}^f$ *containing only feasible points.*

ii. Hyper-rectangles $X_1^{nf}, X_2^{nf}, \ldots, X_{N_{nf}}^{nf}$ *containing some infeasible points.*

2. *The center point* w_0 *of* X_0 *is the only feasible point. Then, split* X_0 *on all parameter space axes by hyperplanes through* w_0.

This rule is illustrated in Fig. 1.5, where the hyper-rectangle X_0 will be split into the hyper-rectangles X_1^f with vertices $\{w_{24}, w_8, w_{17}, \alpha_1\}$, X_2^f with vertices $\{w_{23}, w_{21}, w_{24}, w_{20}\}$ and X_1^{nf} with vertices $\{w_8, w_{20}, \alpha_1, w_{19}\}$.

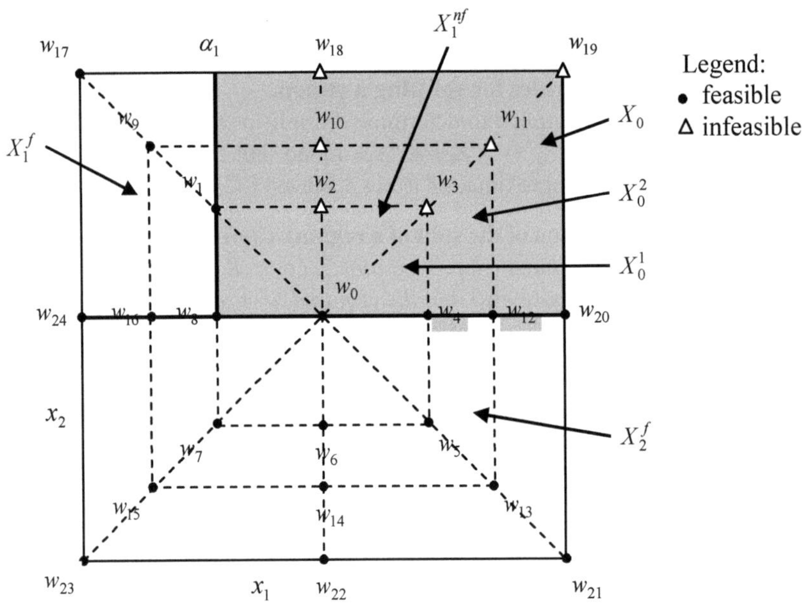

Fig. 1.5 Illustration of Heuristic splitting rule 1.1.

The following rule is applied when there is no feasible solution to the NLP problem (1.1)–(1.2) at the center point w_0 of the hyper-rectangle X_0 [32].

Heuristic splitting rule 1.2 (handling infeasibility). *If there is no feasible solution of the NLP problem (1.1)–(1.2) at the center point* w_0 *of* X_0, *split the hyper-rectangle* X_0 *by a hyperplane through* w_0 *and orthogonal to an arbitrary axis.*

The following rule is used when the NLP problem (1.49)–(1.50) in Procedure 1.3 has no solution [32].

Heuristic splitting rule 1.3 (handling infeasibility). *If the NLP problem (1.49)–(1.50) in Procedure 1.3 is infeasible, split the hyper-rectangle* X_0 *by a hyperplane through its center and orthogonal to an arbitrary axis.*

5. Algorithm for explicit approximate solution of non-convex mp-NLPs. Assume the tolerance $\bar{\varepsilon} > 0$ of the objective function approximation error is given. Denote with S_{X_0} the volume of a given hyper-rectangular region $X_0 \subset X \subset \mathbb{R}^n$, i.e. $S_{X_0} = \prod_{i=1}^{n} \Delta x_i$, where Δx_i is the size of X_0 along the variable x_i. Let $S_{\min} > 0$ be the minimal allowed volume of the regions in the partition of X. The following algorithm is proposed to determine an explicit approximate solution of non-convex mp-NLP (1.1)–(1.2) [32].

Algorithm 1.2. Explicit approximate solution of non-convex mp-NLP.

Input: Data to problem (1.1)–(1.2), the number N_0 of internal regions (used in Procedure 1.1), the parameter μ (used in Procedure 1.3), the approximation tolerance $\bar{\varepsilon}$, the minimal allowed volume $S_{\min}$.
Output: Partition $\Pi = \{X_1, X_2, ..., X_{N_X}\}$ and associated PWL solution function $\widehat{z}_\Pi = \{\widehat{z}_{X_1}, \widehat{z}_{X_2}, \dots, \widehat{z}_{X_{N_X}}\}$.

1. Initialize the partition to the whole hyper-rectangle, i.e., $\Pi = \{X\}$. Mark the hyper-rectangle X as unexplored, $flag := 1$.
2. **while** $flag = 1$ **do**
3.   **while** $\exists$ unexplored hyper-rectangles in Π **do**
4.   Select any unexplored hyper-rectangle $X_0 \in \Pi$ and compute its volume S_{X_0}.
5.   Search for a close-to-global solution to problem (1.1)–(1.2) at the center point w_0 of X_0 by applying Procedure 1.2a.
6.   **if** a feasible solution was found to problem (1.1)–(1.2) at w_0 **then**
7.   Define a set of points $W_0 = \{w_0, w_1, w_2, \dots, w_{N_1}\}$ by applying Procedure 1.1.
8.   Search for a close-to-global solution to problem (1.1)–(1.2) for x fixed to each of the points w_i, $i = 1, 2, \dots, N_1$ by applying Procedure 1.2b.
9.   **if** (1.1)–(1.2) has a feasible solution at all points w_i, $i = 1, \dots, N_1$ **then**
10.   Search for a linear approximation $\widehat{z}_0(x) = K_0 x + h_0$ using Procedure 1.3, as an approximation to be used in X_0.
11.   **if** a solution $\widehat{z}_0(x)$ was found **then**
12.   Compute an estimate $\widehat{\varepsilon}_0$ of the error bound ε_0 in X_0 by applying Procedure 1.4.
13.   **If** $\widehat{\varepsilon}_0 > \bar{\varepsilon}$ and $S_{X_0} > S_{\min}$, mark the hyper-rectangle X_0 to be split. **Otherwise**, mark X_0 as explored and feasible.
14.   **else**
15.   **If** $S_{X_0} < S_{\min}$, mark X_0 infeasible and explored. **Otherwise**, mark X_0 to be split.
16.   **end if**
17.   **else**
18.   **If** $S_{X_0} < S_{\min}$, mark X_0 infeasible and explored. **Otherwise**, mark X_0 to be split.
19.   **end if**
20.   **else**

21. **If** $S_{X_0} < S_{\min}$, mark X_0 infeasible and explored.
    Otherwise, mark X_0 to be split.
22. **end if**
23. **end while**
24. $flag := 0$
25. **if** $\exists$ hyper-rectangles in Π that are marked to be split **then**
26. $flag := 1$
27. **while** $\exists$ hyper-rectangles in Π that are marked to be split **do**
28. Select any hyper-rectangle $X_0 \in \Pi$ marked to be split.
29. Split X_0 into hyper-rectangles $X_1, \ldots, X_{N_s}$ by applying the heuristic splitting rules. Mark $X_1, \ldots, X_{N_s}$ unexplored, remove X_0 from Π, and add $X_1, \ldots, X_{N_s}$ to Π.
30. **end while**
31. **end if**
32. **end while**

1.2 Convex Multi-parametric Quadratic Programming

1.2.1 Problem Formulation

Consider the convex quadratic mathematical program dependent on a parameter x:

$$V^*(x) = \min_{z} \frac{1}{2} z^T H z \tag{1.53}$$

$$\text{s.t.} \;\; Gz \leq W + Sx\,, \tag{1.54}$$

where $z \in \mathbb{R}^s$ is the vector of optimization variables, $x \in \mathbb{R}^n$ is the vector of parameters, and $H \in \mathbb{R}^{s \times s}$, $G \in \mathbb{R}^{q \times s}$, $W \in \mathbb{R}^q$, and $S \in \mathbb{R}^{q \times n}$ are matrices. Here, it is supposed that $H \succ 0$, which leads to a strictly convex multi-parametric quadratic programming (mp-QP) problem (1.53)–(1.54). The case when the multi-parametric programming problem (1.53)–(1.54) is only convex, i.e. $H \succeq 0$, is considered in [73, 69].

Let X be a polytopic set of parameters, defined by $X = \{x \in \mathbb{R}^n \mid Ax \leq b\}$. In parametric programming, it is of interest to characterize the solution of the mp-QP problem (1.53)–(1.54) for the set X. The solution of an mp-QP problem is a triple $(V^*(x), z^*(x), X_f)$ (see Section 1.1.1), where X_f is the set of feasible parameters, $V^*(x)$ is the optimal value function, and $z^*(x)$ is the optimizer function. It is assumed that X_f is closed and $V^*(x)$ is finite for every $x \in X_f$.

In [12, 13], an algorithm has been developed, which expresses the solution $z^*(x)$ and the optimal value $V^*(x)$ of the mp-QP problem (1.53)–(1.54) as an explicit function of the parameters x, and the analytical properties of these functions have been characterized. In particular it has been proved that the solution $z^*(x)$ is a continuous piecewise linear function of x in the following sense [12, 13]:

Definition 1.1. A function $z(x) : X \mapsto \mathbb{R}^s$, where $X \subseteq \mathbb{R}^n$ is a polyhedral set, is piecewise linear if it is possible to partition X into convex polyhedral regions, CR_i, and $z(x) = K_i x + h_i$, $\forall x \in CR_i$.

Piecewise quadraticity is defined analogously by letting $z(x)$ be a quadratic function $x^T Q_i x + K_i x + h_i$.

1.2.2 *Optimality Conditions*

The solution of mp-QP problems can be approached by employing the principles of multi-parametric nonlinear programming and in particular the first-order Karush-Kuhn-Tucker (KKT) optimality conditions, which lead to the Basic Sensitivity Theorem (see Section 1.1). Instead, in [12, 13] a more direct approach has been adopted which exploits the linearity of the constraints and the fact that the function to be minimized is quadratic. The approach [12, 13] is described as follows. In order to start solving the mp-QP problem, an initial vector x_0 inside the polyhedral set X of parameters is needed, such that the QP problem (1.53)–(1.54) is feasible for $x = x_0$. Such a vector can be found for instance by solving the linear program (LP) [12, 13]:

$$\max_{x,z,\varepsilon} \varepsilon \tag{1.55}$$

$$\text{subject to :}$$

$$Gz - Sx + \varepsilon \leq W \tag{1.56}$$

$$\varepsilon \geq 0 \tag{1.57}$$

$$x \in X \; . \tag{1.58}$$

If the LP (1.55)–(1.58) is infeasible, then the QP problem (1.53)–(1.54) is infeasible for all $x \in X$. Otherwise, the QP problem (1.53)–(1.54) is solved with $x = x_0$ in order to obtain the corresponding optimal solution z_0. Such a solution is unique because $H \succ 0$ and therefore uniquely determines a set of active constraints $\tilde{G}z_0 = \tilde{S}x_0 + \tilde{W}$ among the constraints (1.54). Let $\tilde{G}$, $\tilde{S}$ and $\tilde{W}$ denote the rows of G, S and W corresponding to the active constraints. Then, the following theorem is proved [12, 13]:

Theorem 1.5. *Let $H \succ 0$. Consider a combination of active constraints $\tilde{G}$, $\tilde{S}$, $\tilde{W}$ and assume that the rows of $\tilde{G}$ are linearly independent. Let CR_0 be the set of all vectors x for which such a combination is active at the optimum (CR_0 is referred to as critical region). Then, the optimal z and the associated vector of Lagrange multipliers λ are uniquely defined linear functions of x over CR_0.*

Proof ([13]). The first-order KKT conditions for the mp-QP are given by:

$$Hz + G^T \lambda = 0 \; , \; \lambda \in \mathbb{R}^q \tag{1.59}$$

$$\lambda_i (G^i z - W^i - S^i x) = 0 \; , \; i = 1, 2, \ldots, q \tag{1.60}$$

$$\lambda \geq 0 \; , \tag{1.61}$$

where the superscript i denotes the i-th row. Equality (1.59) is solved for z:

$$z = -H^{-1}G^T\lambda \tag{1.62}$$

and the result is substituted into (1.60) to obtain the complementary slackness condition:

$$\lambda_i(-G^iH^{-1}G^{iT}\lambda_i - W^i - S^ix) = 0\,,\ i = 1,2,\dots,q \tag{1.63}$$

Let $\breve{\lambda}$ and $\tilde{\lambda}$ denote the Lagrange multipliers corresponding to inactive and active constraints, respectively. For inactive constraints $\breve{\lambda} = 0$. For active constraints $-\tilde{G}H^{-1}\tilde{G}^T\tilde{\lambda} - \tilde{W} - \tilde{S}x = 0$ and therefore:

$$\tilde{\lambda} = -(\tilde{G}H^{-1}\tilde{G}^T)^{-1}(\tilde{W} + \tilde{S}x)\,, \tag{1.64}$$

where $\tilde{G}$, $\tilde{W}$, $\tilde{S}$ correspond to the set of active constraints and $(\tilde{G}H^{-1}\tilde{G}^T)^{-1}$ exists because the rows of $\tilde{G}$ are linearly independent. Thus λ is a linear function of x for all $x \in CR_0$, where the active constraints set is optimal. By substituting $\tilde{\lambda}$ from (1.64) into (1.62), it is obtained:

$$z^*(x) = H^{-1}\tilde{G}^T(\tilde{G}H^{-1}\tilde{G}^T)^{-1}(\tilde{W} + \tilde{S}x) \tag{1.65}$$

and it is noted that z^* is also a linear function of x in CR_0. □

Theorem 1.5 characterizes the solution only locally in the neighborhood of a specific x_0, as it does not provide the construction of the set CR_0 where this characterization remains valid. On the other hand, this region can be characterized immediately [12, 13]. The variable z from (1.65) must satisfy the constraints (1.54):

$$GH^{-1}\tilde{G}^T(\tilde{G}H^{-1}\tilde{G}^T)^{-1}(\tilde{W} + \tilde{S}x) \le W + Sx \tag{1.66}$$

and by (1.61) the Lagrange multipliers in (1.64) must remain nonnegative:

$$-(\tilde{G}H^{-1}\tilde{G}^T)^{-1}(\tilde{W} + \tilde{S}x) \ge 0 \tag{1.67}$$

as x varies. After removing the redundant inequalities from (1.66) and (1.67), a compact representation of CR_0 is obtained. Obviously, CR_0 is a polyhedron in the x-space and represents a subset of X such that the combination of active constraints at the minimizer remains unchanged (Fig. 1.6(a)). Then, the algorithm in [12, 13] continues with the division of the rest of the parameter space $CR^{rest} = X - CR_0$ as in Fig. 1.6(b) and (c) by reversing one by one the hyperplanes defining the critical region CR_0. Iteratively each new region R_i is subdivided in a similar way as was done with X. An effective approach for partitioning the rest of the space was proposed in [25]. The following theorem justifies such a procedure to characterize the rest of the region CR^{rest} [13].

Theorem 1.6. *Let $Y \subseteq \mathbb{R}^n$ be a polyhedron, and $CR_0 = \{x \in Y \mid Ax \leq b\}$ a polyhedral subset of Y, $CR_0 \neq \emptyset$. Also let:*

$$R_i = \{x \in Y \mid A^i x > b^i, A^j x \leq b^j, \forall j < i\}, \; i = 1, 2, ..., m, \tag{1.68}$$

where $m = \dim(b)$, and let $CR^{rest} = \bigcup_{i=1}^{m} R_i$. Then:
(i) $CR^{rest} \bigcup CR_0 = Y$.
(ii) $CR_0 \bigcap R_i = \emptyset$, $R_i \bigcap R_j = \emptyset$, $\forall i \neq j$, i.e., $\{CR_0, R_1, ..., R_m\}$ is a partition of Y.

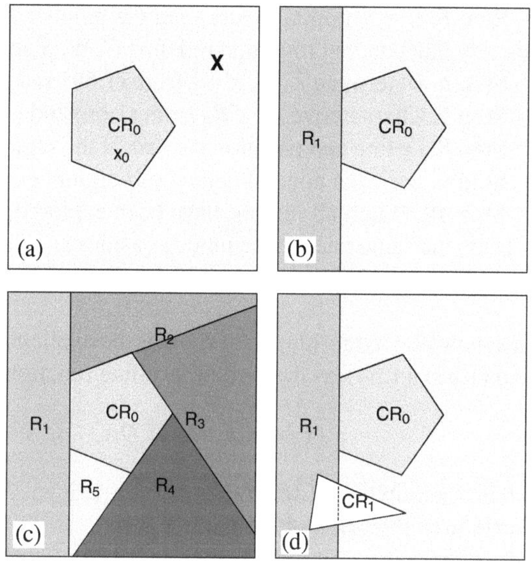

Fig. 1.6 Parameter space exploration strategy in [12, 13].

The properties of the set of feasible parameters $X_f \subseteq X$ (i.e. the set of parameters $x \subset X$ such that a feasible solution $z^*(x)$ exists to the optimization problem (1.53)–(1.54), the value function $V^*(x)$ and the solution $z^*(x)$ are formulated in the following theorem [13]:

Theorem 1.7. *Consider the convex multi-parametric quadratic program (1.53)–(1.54) with $H \succ 0$, X convex. Then the set of feasible parameters $X_f \subseteq X$ is convex, the optimizer $z^*(x) : X_f \mapsto \mathbb{R}^s$ is continuous and piecewise linear and the value function $V^*(x) : X_f \mapsto \mathbb{R}$ is continuous, convex and piecewise quadratic.*

1.2.3 Algorithms for Exact Convex Multi-parametric Quadratic Programming

Based on the above results, the main steps of the off-line mp-QP solver are outlined in the following algorithm [13]:

Algorithm 1.3. Exact mp-QP.

Step 1. Let the current region be the whole polyhedron $X \subseteq \mathbb{R}^n$.
Step 2. Choose a vector x_0 in the current region by solving the linear program (1.55)–(1.58).
Step 3. For $x = x_0$, compute the corresponding optimal solution (z_0, λ_0) by solving a QP.
Step 4. Determine the set of active constraints when $z = z_0$, $x = x_0$, and build $\tilde{G}$, $\tilde{W}$, $\tilde{S}$.
Step 5. If $r = \operatorname{rank}\tilde{G}$ is less than the number l of rows of $\tilde{G}$, take a subset of r linearly independent rows and redefine $\tilde{G}$, $\tilde{W}$, $\tilde{S}$ accordingly.
Step 6. Determine $\tilde{\lambda}(x)$, $z^*(x)$ from (1.64) and (1.65).
Step 7. Characterize the CR_0 from (1.66) and (1.67).
Step 8. Define and partition the rest of the region as illustrated in Fig. 1.6.
Step 9. For each nonempty new sub-region, go to step 2.
Step 10. When all regions have been explored, for all polyhedral regions where $z^*(x)$ is the same and whose union is a convex set, compute such a union.

In conclusion, Algorithm 1.3 provides the explicit solution $z^*(x)$ to the mp-QP problem (1.53)–(1.54), as the piecewise affine function:

$$z^*(x) = K_i x + h_i \ \text{ if } D_i x \le d_i\,,\ i = 1, 2, \ldots, N_r\,, \tag{1.69}$$

where the polyhedral sets $D_i x \le d_i$, $i = 1, 2, \ldots, N_r$ are critical regions that form a partition of the given set of states X.

1.2.3.1 Efficient Implementation of the Exact Approach to Explicit Solution of mp-QP Problems

1. Main theoretical result.
As noted in [72], the main drawback of this algorithm is that the regions R_i are not related to optimality, as they can split some of the critical regions like CR_1 in Fig. 1.6(d). A consequence is that CR_1 will be detected at least twice. The approach in [72] modifies the explicit approach in [12, 13] by analyzing several properties of the geometry of the polyhedral partition and its relation to the combination of active constraints at the optimum of the quadratic program. Based on that, they derive a new exploration strategy for sub-dividing the parameter space, which aims to:

(1) Avoid unnecessary partitioning.
(2) Avoid the solution to LP problems for determining an interior point in each new region of the parameter space.
(3) Avoid the solution to the QP problem for such an interior point.

As a consequence, there is a significant improvement of efficiency with respect to the algorithm in [12, 13]. Before describing the main idea of the approach in [72], some definitions are made [72]:

Definition 1.2. Let $z^*(x)$ be the optimal solution to (1.53)–(1.54) for a given x. We define *active constraints* the constraints with $G^i z^*(x) - W^i - S^i x = 0$ and *inactive constraints* the constraints with $G^i z^*(x) - W^i - S^i x < 0$. The *optimal active set* $\mathscr{A}^*(x)$ is the set of indices of active constraints at the optimum $\mathscr{A}^*(x) = \{i \mid G^i z^*(x) = W^i + S^i x\}$ (a superscript index is used to denote a row of a matrix). We also define as *weakly active constraint* an active constraint with an associated zero Lagrange multiplier λ_i and as *strongly active constraint* an active constraint with a positive Lagrange multiplier λ_i.

Definition 1.3. For an active set, we say that the *linear independence constraint qualification* (LICQ) holds if the set of active constraint gradients are linearly independent, i.e. $\tilde{G}$ has full row rank.

Below, the linear expression of the PWL function $z^*(x)$ over the critical region CR_k is denoted by $z_k^*(x)$.

Definition 1.4. Two polyhedra are called *neighboring* polyhedra if they have a common facet.

Definition 1.5. Let a polyhedron X be represented by $A_0 x \leq b$. We say that $A_0^i x \leq b^i$ is *redundant* if $A_0^j x \leq b^j$, $\forall j \neq i \Rightarrow A_0^i x \leq b^i$ (i.e. it can be removed from the description of the polyhedron). The inequality i is *redundant* with degree h if it is redundant but there exists a h-dimensional subset Y of X such that $A_0^i x = b^i$ for all $x \in Y$.

Let us consider a hyperplane defining the common facet between two polyhedra CR_0, CR_i in the optimal partition of the state space. There are two different kinds of hyperplanes [72]. The first (Type I) are those described by (1.66), which represent a non-active constraint that becomes active at the optimum as x moves from CR_0 to CR_l. This means that if a polyhedron is bounded by a hyperplane which originates from (1.66), the corresponding constraint will be activated on the other side of the facet defined by this hyperplane. In addition, the corresponding Lagrange multiplier may become positive. The other kind (Type II) of hyperplanes which bounds the polyhedra are those described by (1.67). In this case, the corresponding constraint will be non-active on the other side of the facet defined by this hyperplane. This is formulated in the following theorem [72]:

Theorem 1.8. *Consider an optimal active set* $\{i_1, i_2, \ldots, i_k\}$ *and its corresponding n-minimal representation of the critical region* CR_0 *obtained by (1.66)–(1.67) after removing redundant inequalities. Let* CR_i *be a full-dimensional neighboring critical region to* CR_0 *and assume LICQ holds on their common facet* $\Phi = CR_0 \bigcap \Psi$ *where* Ψ *is the separating hyperplane between* CR_0 *and* CR_i. *Moreover, assume that there are no constraints which are weakly active at the optimizer* $z^*(x)$ *for all* $x \in CR_0$. *Then:*

Type I. If Ψ is given by $G^{i_{k+1}} z_0^*(x) = W^{i_{k+1}} + S^{i_{k+1}} x$, then the optimal active set in CR_i is $\{i_1, \ldots, i_k, i_{k+1}\}$.

Type II. If Ψ is given by $\lambda_0^{i_k}(x) = 0$, then the optimal active set in CR_i is $\{i_1, \ldots, i_{k-1}\}$.

In degenerate cases, when the LICQ condition does not hold or there are weakly active constraints, Theorem 1.8 provides no conclusion. In particular, when moving across the facet of one critical region there may not be a single unique critical region that shares the same facet, [67]. As discussed in [72, 73, 67, 69], the method in [12, 13] is effective to handle these special cases.

2. Example.

The example represents a Model Predictive Control (MPC) problem for a double integrator [72, 40], which is transformed into the equivalent mp-QP problem (1.53)–(1.54) with H, G, W, S given by:

$$H = \begin{bmatrix} 1.079 & 0.076 \\ 0.076 & 1.073 \end{bmatrix} \tag{1.70}$$

$$G^T = \begin{bmatrix} 1 & 0 & -1 & 0 & 0.05 & 0.05 & -0.05 & -0.05 \\ 0 & 1 & 0 & -1 & 0 & 0.05 & 0 & -0.05 \end{bmatrix} \tag{1.71}$$

$$W^T = \begin{bmatrix} 1 & 1 & 1 & 1 & 0.5 & 0.5 & 0.5 & 0.5 \end{bmatrix} \tag{1.72}$$

$$S^T = \begin{bmatrix} 1.0 & 0.9 & -1.0 & -0.9 & 0.1 & 0.1 & -0.1 & -0.1 \\ 1.4 & 1.3 & -1.4 & -1.3 & -0.9 & -0.9 & 0.9 & 0.9 \end{bmatrix} \tag{1.73}$$

The partitioning starts with finding the region where no constraints are active. As the mp-QP is created from a feasible MPC problem, the empty active set will be optimal in some full-dimensional region ($\mathscr{A}_0 = \emptyset$ and $\tilde{G}$, $\tilde{W}$ and $\tilde{S}$ are empty matrices, $z^*(x) = 0$). This critical region is then described by $0 \leq W + Sx$ which contains 8 inequalities. Two of these inequalities are redundant with degree 0 (#2 and #4), the remaining 6 hyperplanes are facet inequalities of the polyhedron (see Fig. 1.7(a)). By crossing the facet given by Ψ_1, defined by inequality 1 and of Type I, as predicted by Theorem 1.8 the optimal active set across this facet is $\mathscr{A}_1 = \{1\}$, which leads to the critical region CR_1 (see Fig. 1.7(b)). After removing redundant inequalities we are left with an n-minimal representation of CR_1 containing 4 facets. The first of these is of Type II, $\lambda_1(x) = 0$. The other three are of Type I. These are inequalities #2, #6 and #7. Consider first the other side of the facet which comes from $\lambda_1(x) = 0$, see Fig. 1.7(c). The region should not have constraint 1 active, so the optimal active set is $\mathscr{A}_2 = \emptyset$. This is the same combination of active constraints as $\mathscr{A}_0$, as expected, so $\mathscr{A}_2$ is not pursued. Next, consider crossing the respective facets of inequalities #2, #6 and #7, see Fig. 1.7(d)–Fig. 1.7(f). This results in three different active sets: $\mathscr{A}_3 = \{1, 2\}$, $\mathscr{A}_4 = \{1, 6\}$ and $\mathscr{A}_5 = \{1, 7\}$. The sets $\mathscr{A}_3$ and $\mathscr{A}_4$ lead to new polyhedra as shown in the figures. The combination $\mathscr{A}_5$ leads to an interesting case of "degeneracy". The associated matrix $\tilde{G}$ has linearly dependent rows, which violates the LICQ assumption. In this case, $\mathscr{A}_5$ leads to an infeasible part of the state space.

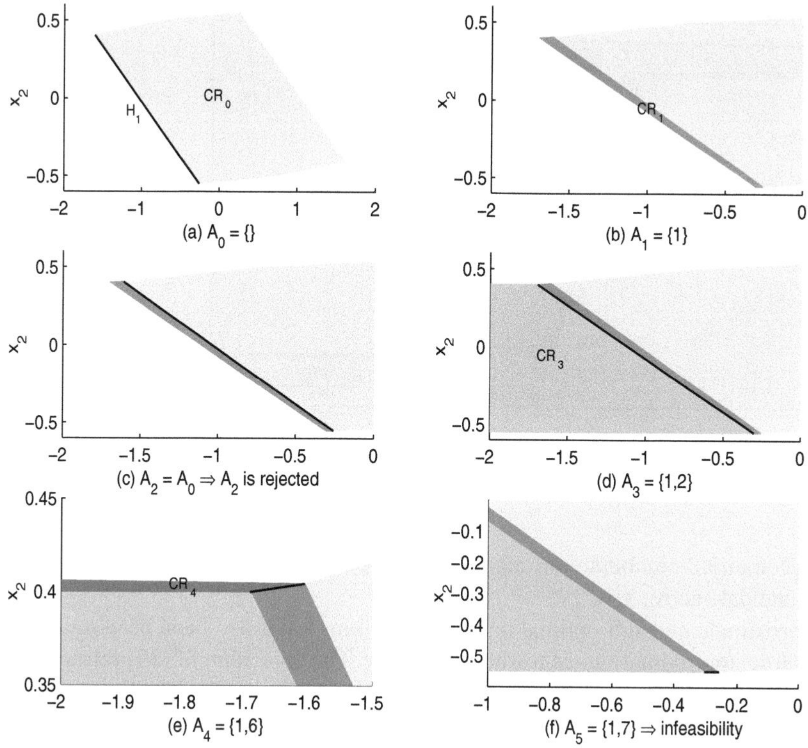

Fig. 1.7 Parameter space exploration strategy in [72].

The parameter space partition of the explicit solution to the mp-QP characterized by (1.70)–(1.73) has 14 polyhedral critical regions and it is given in Fig. 1.8.

1.2.4 Remarks on Alternative mp-QP Algorithms

This section has described in mp-QP algorithms presented in [13] and [72]. A combination of these algorithms, that uses the strategy of [72] to step over facets between neighboring regions, in combination with the QP solution of [13] in order to identify the optimal active set, is given by [4]. Like the algorithm in [72], it depends on the facet-to-facet property and the modifications described in [69] are useful. The mp-QP algorithm of [4] is the primary mp-QP algorithm of the widely used Multi-Parametric Toolbox (MPT), [46].

The combinatorial approach of [65] considers the combinations of potentially optimal active constraints. In many cases, this is not efficient since it tends to lead to many critical regions that are not full dimensional and must therefore be disregarded. An approach that exploits double representation (vertices and hyperplanes) of polyhedrons was given in [57]. It was shown in [34] that a more efficient

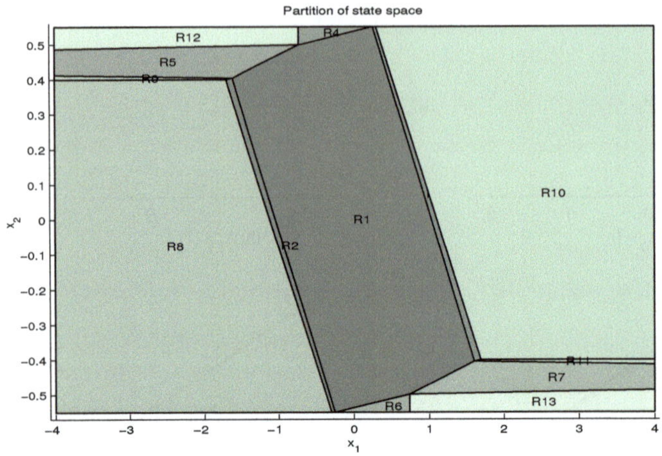

Fig. 1.8 Partition of the explicit solution to the mp-QP characterized by (1.70)–(1.73).

"non-geometric" combinatorial algorithm can be implemented by pruning infeasible candidate active sets.

Approximate and sub-optimal mp-QP algorithms have also been proposed, pursuing close approximations of lower complexity. The algorithm of [39] relies on an orthogonal partitioning of the parameter space that is built recursively to achieve an acceptable guaranteed maximum approximation error when approximating the solution with an affine function within each hyper-rectangle of the parameter space. A similar approach was taken in [9], using a simplex partition instead of hyper-rectangles, and in [64] that exploits the Delaunay tessellation. A reformulation of the MPC problem solving a sequence of simpler explicit MPC problems of horizon $N = 1$ and a nested sequence of terminal sets, a la dynamic programming, was proposed in [33]. Another sub-optimal approach based on short horizons was proposed in [40]. The use of nested invariant sets and interpolation techniques is pursued for approximations in [55]. An approximate solution for explicit MPC using set membership approximation has been introduced in [17]. While all the above approximations lead to PWA representations, [48] considers polynomial approximations.

For further review of existing algorithms, as well as multi-parametric linear programming (mp-LP) algorithms, we refer to [2].

1.3 Evaluating Piecewise Functions

Both mp-NLP and mp-QP algorithms provide the solution as some piecewise function representation, where the function pieces are defined a polyhedral partitioning into regions. All mp-QP algorithms, and most mp-NLP algorithms, return affine function pieces. Since the complexity of representation may be large even for relatively low order systems with constraints on some horizon (often thousands of regions), explicit MPC depend on efficient methods to evaluate piecewise functions.

The direct approach is to evaluate directly for each region if the current parameter (given by the current state, plus other auxiliary variables, perhaps) belongs to that region. This is clearly computationally demanding and may require more computations than solving the corresponding QP using an active-set solver, in particular if a good initialization is available for warm start. Still, it is very simple, and may even be very fast if implemented on a massively parallel computer architecture. The parallel implementation is entirely straightforward since all regions can be evaluated concurrently.

A much more computationally efficient approach was proposed in [71, 74], relying on a binary search tree representation of the polyhedral partitioning where at each level of the search tree one is able to exclude a significant fraction of the remaining candidate regions by evaluating on which side of a given hyperplane the current parameter belongs (typically a reduction of 1/3 is possible to achieve, as a rule of thumb). Hence, due to logarithmic complexity in the number of regions, the search among thousands of regions would amount of evaluating less than 20 hyperplanes. For mp-QP, the weakness of this approach is that extensive off-line computations are needed to construct a balanced binary search tree, and the piecewise function representation may still require extensive on-line computer memory. This is also true for methods that exploit optimal algorithms for selection of the hyperplanes for decisions [29]. To address this issue, the use of a truncated binary search tree in combination with direct search [6] or the lattice representation of piecewise linear functions [5] has been proposed. The realization of such piecewise affine function evaluation algorithms in dedicated hardware is investigated in [60, 41]. It should be mentioned that orthogonal partitions such as [38, 39] builds such a binary search tree as an integral part of the multi-parametric programming strategy.

Data structures other than binary search trees are also useful to support efficient evaluation. Bounding-boxes [19] and hash-tables [7] are proposed as supporting data structures to efficiently narrow down the search for the optimal polyhedral regions.

In MPC, the parameter (state) at one time instant is likely to be close to the parameter at the previous time instant, due to the continuity of trajectories of dynamic systems. Several algorithms have been proposed to build data structures that represent the topology of the polyhedral partitioning in order to quickly identify neighboring regions along the path from one parameter to the next parameter [68, 75].

Complexity of piecewise function representations can also be reduced by joining convex unions of polyhedrons, which share the same affine function piece (for the first control sample) [10]. In particular, the fact that input saturation will typically occur in a large number of regions can be directly exploited, [47]. Several methods for further compression of representation and efficient evaluation are investigated in [70].

References

1. Adams, L., Nazareth, J.L. (eds.): Linear and nonlinear conjugate gradient-related methods. Society for Industrial and Applied Mathematics, Philadelphia (1996)
2. Alessio, A., Bemporad, A.: A Survey on Explicit Model Predictive Control. In: Magni, L., Raimondo, D.M., Allgöwer, F. (eds.) Nonlinear Model Predictive Control: Towards New Challenging Applications. LNCIS, vol. 384, pp. 345–369. Springer, Heidelberg (2009)
3. Armijo, L.: Minimization of functions having Lipschitz-continuous first partial derivatives. Pacific J. Math. 16, 1–3 (1966)
4. Baotič, M.: An efficient algorithm for multi-parametric quadratic programming. Technical Report AUT02-05. Institut für Automatik, ETH Zürich (2002)
5. Bayat, F., Johansen, T.A., Jalali, A.A.: Flexible piecewise function evaluation methods with application to explicit model predictive control. In: Proceedings of the IEEE International Conference on Mechatronics, Istanbul (2011)
6. Bayat, F., Johansen, T.A., Jalali, A.A.: Combining truncated binary search tree and direct search for flexible piecewise function evaluation for explicit MPC in embedded microcontrollers. In: Proceedings of the IFAC World Congress, Milano (2011), www.IFAC-PapersOnLine.net
7. Bayat, F., Johansen, T.A., Jalali, A.A.: Using hash tables to manage time-storage complexity in point location problem: Application to explicit MPC. Automatica 47, 571–577 (2011)
8. Bazaraa, M.S., Sherali, H.D., Shetty, C.M.: Nonlinear programming: Theory and algorithms, 3rd edn. Wiley-Interscience, New Jersey (2006)
9. Bemporad, A., Filippi, C.: Suboptimal explicit MPC via approximate quadratic programming. In: Proceedings of the IEEE Conference on Decision and Control, Orlando, pp. 4851–4856 (2001)
10. Bemporad, A., Fukuda, K., Torrisi, F.D.: Convexity recognition of the union of polyhedra. Computational Geometry: Theory and Applications 18, 141–154 (2001)
11. Bemporad, A., Filippi, C.: An algorithm for approximate multiparametric convex programming. Computational Optimization and Applications 35, 87–108 (2006)
12. Bemporad, A., Morari, M., Dua, V., Pistikopoulos, E.N.: The explicit solution of model predictive control via multiparametric quadratic programming. In: Proceedings of the American Control Conference, Chicago, Illinois, pp. 872–876 (2000)
13. Bemporad, A., Morari, M., Dua, V., Pistikopoulos, E.N.: The explicit linear quadratic regulator for constrained systems. Automatica 38, 3–20 (2002)
14. Bentley, J.L.: Multidimensional binary search trees used for associative searching. Communications of the ACM 18, 509–517 (1975)
15. Box, M.J.: A new method of constrained optimization and a comparison with other methods. Computer J. 8, 42–52 (1965)
16. Boyd, S., Vandenberghe, L.: Convex optimization. University Press, Cambridge (2004)
17. Canale, M., Fagiano, L., Milanese, M.: Set membership approximation theory for fast implementation of model predictive control laws. Automatica 45, 45–54 (2009)
18. Cauchy, A.: Methode generale pour la resolution des systemes d'equations simultanees. Comp. Rend. Acad. Sci. Paris 536–538 (1847)
19. Christophersen, F., Kvasnica, M., Jones, C.N., Morari, M.: Efficient evaluation of piecewise control laws defined over a large number of polyhedra. In: Proceedings of the European Control Conference, pp. 2360–2367 (2007)
20. Deuflhard, P.: Newton methods for nonlinear problems. Springer, New York (2004)

21. Diehl, M., Bock, H.G., Schlöder, J.P.: Newton-type methods for the approximate solution of nonlinear programming problems in real-time. In: Di Pillo, G., Murli, A. (eds.) High Performance Algorithms and Software for Nonlinear Optimization, pp. 177–200. Kluwer Academic Publishers B.V. (2003)
22. Domínguez, L.F., Narciso, D.A., Pistikopoulos, E.N.: Recent advances in multiparametric nonlinear programming. Computers and Chemical Engineering 34, 707–716 (2010)
23. Domínguez, L.F., Pistikopoulos, E.N.: Quadratic approximation algorithm for multiparametric nonlinear programming problems. Technical report. Imperial College London (2009)
24. Dua, V., Pistikopoulos, E.N.: Algorithms for the solution of multi-parametric mixed-integer nonlinear optimization problems. Industrial & Engineering Chemistry Research 38, 3976–3987 (1999)
25. Dua, V., Pistikopoulos, E.N.: An algorithm for the solution of multiparametric mixed integer linear programming problems. Annals of Operations Research 99, 123–139 (2000)
26. Fiacco, A.V.: Introduction to sensitivity and stability analysis in nonlinear programming. Academic Press, Orlando (1983)
27. Finkel, D.E.: Global optimization with the DIRECT algorithm. Ph.D. thesis. North Carolina State University (2005)
28. Fletcher, R., Reeves, C.M.: Function minimization by conjugate gradients. Comput. J. 7, 149–154 (1964)
29. Fuchs, A.N., Jones, C.N., Morari, M.: Optimized decision trees for point location in polytopic data sets - Application to explicit MPC. In: Proceedings of the American Control Conference, Baltimore, pp. 5507–5512 (2010)
30. Gablonsky, J.M.: Modifications of the DIRECT algorithm. Ph.D. thesis. North Carolina State University (2001)
31. Grancharova, A., Johansen, T.A.: Approximate explicit model predictive control incorporating heuristics. In: Proceedings of IEEE International Symposium on Computer Aided Control System Design, Glasgow, Scotland, U.K., pp. 92–97 (2002)
32. Grancharova, A., Johansen, T.A., Tøndel, P.: Computational aspects of approximate explicit nonlinear model predictive control. In: Findeisen, R., Allgöwer, F., Biegler, L. (eds.) Assessment and Future Directions of Nonlinear Model Predictive Control. LNCIS, vol. 358, pp. 181–192. Springer, Heidelberg (2007)
33. Grieder, P., Morari, M.: Complexity reduction of receding horizon control. In: Proceedings of the 42th IEEE Conference on Decision and Control, Maui, Hawaii, USA, pp. 3179–3184 (2003)
34. Gupta, A., Bhartiua, S., Nataraj, P.S.V.: A novel approach to multiparametric quadratic programming. Automatica 47, 2112–2117 (2011)
35. Hooke, R., Jeeves, T.A.: Direct search solution of numerical and statistical problems. J. Assoc. Comput. Mach. 8, 212–229 (1961)
36. Horst, R., Tuy, H.: Global optimization. Springer, Berlin (1995)
37. Johansen, T.A.: On multi-parametric nonlinear programming and explicit nonlinear model predictive control. In: Proceedings of the IEEE Conference on Decision and Control, Las Vegas, NV, vol. 3, pp. 2768–2773 (2002)
38. Johansen, T.A.: Approximate explicit receding horizon control of constrained nonlinear systems. Automatica 40, 293–300 (2004)
39. Johansen, T.A., Grancharova, A.: Approximate explicit constrained linear model predictive control via orthogonal search tree. IEEE Transactions on Automatic Control 48, 810–815 (2003)
40. Johansen, T.A., Petersen, I., Slupphaug, O.: Explicit sub-optimal linear quadratic regulation with state and input constraints. Automatica 38, 1099–1111 (2002)

41. Johansen, T.A., Jackson, W., Schreiber, R., Tøndel, P.: Hardware synthesis of explicit model predictive controllers. IEEE Transactions Control Systems Technology 15, 191–197 (2007)
42. Jones, D.R.: The DIRECT global optimization algorithm. In: Floudas, C.A., Pardalos, P.M. (eds.) Encyclopedia of Optimization, vol. 1, pp. 431–440. Kluwer, Dordrecht (2001)
43. Jones, D.R., Perttunen, C.D., Stuckman, B.E.: Lipschitzian optimization without the Lipschitz constant. Journal of Optimization Theory and Applications 79, 157–181 (1993)
44. Kelley, C.T.: Iterative methods for optimization. Society for Industrial and Applied Mathematics, Philadelphia (1999)
45. Kojima, M.: Strongly stable stationary solutions in nonlinear programs. In: Robinson, S.M. (ed.) Analysis and Computation of Fixed Points, pp. 93–138. Academic Press, New York (1980)
46. Kvasnica, M., Grieder, P., Baotić, M., Morari, M.: Multi-Parametric Toolbox (MPT). In: Alur, R., Pappas, G.J. (eds.) Hybrid Systems: Computation and Control, HSCC 2004. LNCS, vol. 2993, pp. 448–462. Springer, Heidelberg (2004)
47. Kvasnica, M., Fikar, M.: Performance-lossless complexity reduction in explicit MPC. In: Proceedings of the IEEE Conference on Decision and Control, pp. 5270–5275 (2010)
48. Kvasnica, M., Löfberg, J., Fikar, M.: Stabilizing polynomial approximation of explicit MPC. Automatica 47, 2292–2297 (2011)
49. Lau, M.S.K., Yue, S.P., Ling, K.V., Maciejowski, J.M.: A comparison of interior point and active set methods for FPGA implementation of model predictive control. In: Proceedings of the European Control Conference 2009, Budapest, Hungary, pp. 156–161 (2009)
50. Levitin, E.S.: Perturbation theory in mathematical programming. Wiley (1994)
51. Lewis, R.M., Torczon, V., Trosset, M.W.: Direct search methods: then and now. Journal of Computational and Applied Mathematics 124, 191–207 (2000)
52. Mangasarian, O.L., Rosen, J.B.: Inequalities for stochastic nonlinear programming problems. Operations Research 12, 143–154 (1964)
53. Narciso, D.: Developments in nonlinear multiparametric programming and control. PhD thesis, London, U.K. (2009)
54. Nelder, J.A., Mead, R.: A simplex method for function minimization. Computer J. 7, 308–313 (1965)
55. Nguyen, H.N., Gutman, P.O., Olaru, S., Hovd, M.: Explicit constrained control based on interpolation techniques for time-varying and uncertain linear discrete-time systems. In: Proceedings of the IFAC World Congress, Milano (2011), www.IFAC-PapersOnLine.net
56. Nocedal, J., Wright, S.J.: Numerical optimization. Springer, New York (1999)
57. Olaru, S., Dumur, D.: A parameterized polyhedra approach for explicit contrained predictive control. In: Proceedings of the IEEE Conference on Decision and Control, Bahamas, pp. 1580–1585 (2004)
58. Pistikopoulos, E.N., Georgiadis, M.C., Dua, V.: Multi-parametric programming: Theory, algorithms, and applications. Wiley-VCH (2007)
59. Piyawksii, S.A.: An algorithm for finding the absolute extremum of a function. USSR Computational Mathematics and Mathematical Physics 12, 57–67 (1972)
60. Poggi, T., Comaschi, F., Storace, M.: Digital circuit realization of piecewise affine functions with non-uniform resolution: Theory and FPGA implementation. IEEE Trans. Circuits and Systems - II: Express Briefs 57, 131–135 (2010)
61. Powell, M.J.D.: A fast algorithm for nonlinearly constrained optimization calculations. In: Watson, G.A. (ed.) Numerical Analysis, Dundee 1977. Lecture Notes in Mathematics, vol. 630. Springer, Berlin (1978)

62. Ralph, D., Dempe, S.: Directional derivatives of the solution of a parametric nonlinear program. Mathematical Programming 70, 159–172 (1995)
63. Rockafellar, R.T.: Convex analysis. Princeton University Press, New Jersey (1970)
64. Scibilia, F., Olaru, S., Hovd, M.: Approximate explicit linear MPC via Delaunay tessellation. In: Proceedings of the European Control Conference, Budapest, Hungary (2009)
65. Seron, M.M., Goodwin, G.C., De Doná, J.A.: Characterization of receding horizon control for constrained linear systems. Asian J. Control 5, 271–286 (2003)
66. Spendley, W., Hext, G.R., Himsworth, F.R.: Sequential applications of simplex designs in optimisation and evolutionary operation. Technometrics 4, 441 (1962)
67. Spjøtvold, J., Kerrigan, E.C., Jones, C.N., Tøndel, P., Johansen, T.A.: On the facet-to-facet property of solutions to convex parametric quadratic programs. Automatica 42, 2209–2214 (2006)
68. Spjøtvold, J., Rakovic, S.V., Tøndel, P., Johansen, T.A.: Utilizing reachability analysis in point location problems. In: Proceedings of the IEEE Conference on Decision and Control, San Diego (2006)
69. Spjøtvold, J., Tøndel, P., Johansen, T.A.: Continuous selection and unique polyhedral representation of solutions to convex parametric quadratic programs. Journal of Optimization Theory and Applications 134, 177–189 (2007)
70. Szücs, A., Kvasnica, M., Fikar, M.: A memory-efficient representation of explicit MPC solutions. In: Proceedings of the IEEE Conference on Decision and Control, Orlando (2011)
71. Tøndel, P., Johansen, T.A.: Complexity reduiction in explicit model predictive control. In: Proceedings of the IFAC World Congress, Barcelona (2002), `www.IFAC-PapersOnLine.net`
72. Tøndel, P., Johansen, T.A., Bemporad, A.: An algorithm for multi-parametric quadratic programming and explicit MPC solutions. Automatica 39, 489–497 (2003)
73. Tøndel, P., Johansen, T.A., Bemporad, A.: Further results on multi-parametric quadratic programming. In: Proceedings of the IEEE Conference on Decision and Control, Maui (2003)
74. Tøndel, P., Johansen, T.A., Bemporad, A.: Evaluation of piecewise affine control via binary search tree. Automatica 39, 743–749 (2003)
75. Wang, Y., Jones, C.N., Maciejowski, J.: Efficient point location via subdivision walking with application to explicit MPC. In: Proceedings of the European Control Conference, pp. 447–453 (2007)
76. Wright, S.J.: Primal-dual interior-point methods. SIAM Publications, Philadelphia (1997)

Chapter 2
Nonlinear Model Predictive Control

Abstract. A nonlinear model predictive control (NMPC) strategy requires the formulation of an optimization problem. For online NMPC the nonlinear programming problem must be solved numerically at every sampling interval, while explicit NMPC assumes that an explicit representation of the solution can be computed using multi-parametric nonlinear programming. This chapter considers the *formulation* of the optimization problem, which is an essential part of the control design and involves numerous decisions that are important for the control performance, stability, and robustness as well as the computational complexity and the numerical challenges of computing the solution. Key elements are discretization and parameterization procedures that leads to a finite-dimensional numerical optimal control problem that can be addressed using e.g. direct single shooting, direct multiple shooting, or collocation methods. Fundamental properties like feasibility and continuity of solutions are discussed, before various modifications that are needed to explicitly guarantee stability, feasibility, and robustness, are discussed. We then discuss further extensions for handling integer variables, output feedback, decentralized and distributed implementations, before some remarks on numerical and computational challenges are discussed at the end.

2.1 Introduction

Consider the problem of controlling a multi-variable nonlinear system, subject to physical and operational constraints on the input and state. Well known systematic nonlinear control methods such as feedback linearization, [59, 83, 94], and constructive Lyapunov-based methods, [69, 109], lead to very elegant solutions, but they depend on complicated design procedures that does not scale well to large systems and they are not developed in order to handle constraints in a systematic manner. The concept of optimal control, and in particular its practical implementation in terms of Nonlinear Model Predictive Control (NMPC) is an attractive alternative since the complexity of the control design and specification increases moderately with the size and complexity of the system. In particular for systems that can be adequately

A. Grancharova & T.A. Johansen: Explicit Nonlinear Model Predictive Ctrl., LNCIS 429, pp. 39–69.
springerlink.com

modeled with linear models, MPC has become the de-facto standard advanced control method in the process industries, [100]. This is due to its ability to handle large scale multi-variable processes with tens or hundreds of inputs and states that must fulfill physical and operational constraints.

MPC involves the formulation and solution of a numerical optimization problem corresponding to a finite-horizon optimal control problem at each sampling instant. Since the state of the system is updated during each sampling period, a new optimization problem must be solved at each sampling interval. This is known as the receding horizon approach. With linear models the MPC problem is typically a quadratic or linear program, which is known to be convex and for which there exists a variety of numerical methods and software. While the numerical complexity of linear MPC may be a reasonable challenge with powerful computers being available, there is no doubt that NMPC is limited in its industrial impact due to the challenges of guaranteeing a global (or at least sufficiently good) solution to the resulting nonlinear optimization problem within the real-time requirements ([101]). Other limiting factors are the challenges of developing nonlinear dynamic models and state estimators. The nonlinear programming problem may have multiple local minima and will demand a much larger number of computations, even without providing any hard guarantees on the solution. However, it is a powerful approach of great promise that has proven itself in several applications, [101, 45], and with further research in the direction of numerical implementation technology and explicit NMPC methods, it may strengthen its position as the most powerful method available for certain classes of systems.

NMPC has its roots in nonlinear optimal control theory that was developed in the 1950's and 1960's, resulting in powerful characterizations such as the maximum principle, [7] and dynamic programming, [12]. In the direct numerical optimal control literature, [57, 36, 19, 23, 17, 48, 24, 113], numerical methods to compute open loop control trajectories were central research topics. Problem formulations that included constraints on control and state variables were treated using numerical optimization. NMPC involves the repetitive solution of an optimal control problem at each sampling instant in a receding horizon fashion. Unfortunately, there is no guarantee that the receding horizon implementation of a sequences of open loop optimal control solutions will perform well, or even be stable, when considering the closed loop system. This challenge, in combination with the tremendous success of *linear* MPC in the process industries, [100], lead to an increasing academic interest in NMPC research with focus on stability analysis and design modifications that guarantee stability and robustness. The early results [31, 63, 84] boosted a large series of research, including [87, 2, 32, 93, 108, 79, 80, 60, 85]. Industrial applications of NMPC have been reported, and are surveyed in [101, 45].

One of the early contributions of NMPC are given in [73], that uses linearization procedures and Gauss-Newton methods to provide a numerical procedure for NMPC based on SQP that makes only one Newton-iteration at each sampling instant. Theoretical results are also given in [74]. The continuation/GMRES method of [97] is based on a similar philosophy of only one Newton-iteration per sample, while it is based on interior point methods. Recent NMPC research along similar ideas has benefited considerably from progress in numerical optimization, being able to take advantage of structural properties on the NMPC problem and general efficiency improvements, e.g. [18, 39, 114, 124], in addition to important issues such as robustness [81, 78, 76].

2.2 NMPC Optimization Problem Formulation

This section will focus on the *formulation* of the NMPC problem, leading to a formulation such as (1.1)-(1.2), while the detailed issues related to its *numerical solution* are postponed until section 2.3. It is, however, important to have in mind that these two issues are closely linked. While the NMPC problem formulation is driven by the specification of the control objective, constraints and dynamic model formulations, one should also consider potential numerical challenges at this point. In particular, important characteristics of the tradeoff between numerical accuracy, computational complexity and complexity of an approximate explicit representation, are determined already at the point when the NMPC optimization problem is formulation through discretization, choice of parameterizations, and choice of decision variables and constraint formulations in the optimization problem. Some of these relationships are treated also in this section, together with fundamental properties of the optimization problem, including stability, convexity and the link between controllability and well-posedness of the optimization problem.

2.2.1 *Continuous-Time Model, Discretization and Finite Parameterization*

This section will introduce a basic nonlinear optimal control formulation starting from a continuous time model and a finite horizon where the objective is to minimize a cost function

$$J(u[0,T],x[0,T]) \triangleq \int_0^T \ell(x(t),u(t),t)dt + S(x(T),T) \tag{2.1}$$

$$\tag{2.2}$$

subject to the inequality constraints for all $t \in [0,T]$

$$u_{min} \leq u(t) \leq u_{max} \tag{2.3}$$

$$g(x(t),u(t),t) \leq 0 \tag{2.4}$$

and the evolution of the ordinary differential equation (ODE) given by

$$\frac{d}{dt}x(t) = f(x(t),u(t),t) \tag{2.5}$$

with given initial condition $x(0) \in \mathbb{R}^n$. The function ℓ is know as the stage cost, S is the terminal cost, $T > 0$ is the horizon, and together these define the cost function J. The evolution of the state $x(t)$ is given by the function f according to (2.5) and depends on the input signal $u(t) \in \mathbb{R}^m$ and time t, and forms an infinite-dimensional equality constraint to the optimal solution in the formulation above. In addition there is saturation on the input with minimum and maximum thresholds u_{min} and u_{max}, respectively, and general inequality constraints jointly on states and inputs, point-wise in time $t \in [0,T]$, defined by the function g. These constraints may result from both physical and operational constraints of the control system and stability-preserving terminal sets that will be discussed later in section 2.2.3, see also [85].The properties of ℓ and S have consequences for the control performance, including stability, and must be carefully understood and tuned, [85]. We will return to this important issue in section 2.2.3. The explicit time-dependence in f, g, ℓ allows for time-varying reference trajectories, known disturbances and exogenous input signals to be accounted for in the optimal control problem formulation. The solving multi-parametric programs, the time or some parameterization of these exogenous signals, should then be included in the parameter vector. Throughout this chapter we implicitly assume all the functions involved satisfy the necessary regularity assumptions, such as continuity and smoothness.

The above formulation basically defines an infinite-dimensional optimal control problem whose solution can be characterized using classical tools like calculus of variations, Pontryagin's maximum principle ([7]) and dynamic programming, [12]. In these *indirect methods* such characterizations of the solution can help us only in a very limited number of special cases to find an analytic exact representation of the solution.

Although numerical solutions can be found based on the characterizations of the indirect methods, in the context of NMPC we choose to restrict our attention to so-called *direct methods* that seems most promising and popular. They are characterized by discretization and finite parameterization being introduced in the optimal control problem formulation which is then directly solved with numerical methods. The principle of NMPC is to repeatedly solve finite-horizon optimal control problems of the above kind at each sampling instant. This means that the initial state $x(0)$ to (2.5) is viewed as the current state based on the most recent measurements, and the optimal control trajectory $u[0,T]$ solving the above problem is implemented for a short period of time (usually one sampling interval, typically much smaller than T) until the procedure is repeated and an updated optimal control trajectory is available. However, the solution of the above optimal control problem, requires reformulations for the following reasons:

- The solution to the ordinary differential equation (2.5) with given initial conditions must generally be based on discretized to be handled by numerical integration since exact closed-form solutions of the ODE are usually not possible to formulate in the general nonlinear case. Viewed in a different way, the infinite number of equality constraints (2.5) must be represented by a finite approximation.
- The infinite-dimensional unknown solution $u[0,T]$ should be replaced by a finite number of decision variables to be able to define a finite-dimensional optimization problem that can be solved using numerical optimization.

- Measurements are typically sampled data available only at the sampling instants, such that an updated initial state $x(0)$ will normally be available only at defined sampling instants.
- Arbitrary control trajectories cannot be implemented since typically the control command can only be changed at defined sampling instants and is typically assumed to be constant (or some other simple sample-and-hold function such as linear) between the sampling instants.

In order to reformulate the problem into a finite-dimensional and practical setting, we will make the following assumptions that will allow the integral and differentiation operators to be approximated by numerical integration methods.

- The horizon T is finite and given.
- The input signal $u[0,T]$ is assumed to be piecewise constant with a regular sampling interval t_s such that T is an integer multiple of t_s, and parameterized by a vector $U \in \mathbb{R}^p$ such that $u(t) = \mu(t,U) \in \mathbb{R}^r$ is piecewise continuous.
- An (approximate) solution to (2.5) is assumed to be defined in the form $x(t) = \phi(t,U,x(0))$ at N discrete time instants $T_d = \{t_1,t_2,...,t_N\} \subset [0,T]$ for some ODE solution function $\phi(\cdot)$. The discrete set of time instants T_d results from discretization of the ODEs and its time instants may not be equidistant. A simulation of the ODEs embedded in the function $\phi(\cdot)$ may incorporate additional intermediate time-steps not included in T_d, since the purpose of T_d is primarily to discretize the inequality constraints (2.3)-(2.4) at a finite number of representative points in time and to approximate the integral in (2.1) with a finite sum. In general, the time instants T_d need not coincide with sampling instants.

The assumption of given horizon T is typical for many NMPC problems, but there are important exceptions such as minimum-time formulations in e.g. robotics, [112], batch process control ([46, 91, 92]), and other problems such as diving decompression ([41]), where the horizon T may be considered a free variable. The resulting modifications to the problem formulations may lead to additional challenges related to the time discretization and may make the optimization problem more challenging.

The basis for the NMPC is the nominal model (2.5), and we remark that model uncertainty, unknown disturbances and measurement errors are not accounted for in this formulation of the NMPC problem. Various extensions and variations that can relax many of the assumptions above can be made relatively easy as straightforward modifications to the basic problem formulation. For simplicity of presentation, we stick to the formulation above and return to some alternatives and opportunities that will be discussed in later sections.

The parameterization of the input signal $\mu(t,U)$ on the horizon $t \in [0,T]$ is important and will influence both the control performance and computational performance. In general, it should satisfy the following objectives

- Be sufficiently flexible in order to allow for a solution of the reformulated optimal control problem close to the solution original problem (2.1)-(2.5).
- Be parsimonos in the sense that it does not contain unnecessary parameters that will lead to unnecessary computational complexity, complexity of an approximate explicit representation, and numerical sensitivity.

- Be implementable within the capabilities of the control system hardware and software, meaning that particular consideration may be needed for any parameterization beyond a piecewise constant input trajectory that is restricted to change its value only at the sampling instants.

Based on the last very practical point, a general choice is the piecewise constant control input $\mu(t,U) = U_k$ for $t_k \leq t < t_{k+1}$ parameterized by the vector $U = col(U_0,...,U_{N-1}) \in \mathbb{R}^{mN}$. Practical experience shows that the receding horizon implementation offers considerable flexibility for a NMPC to recover performance due to sub-optimality at each step. Consequently, it is common practice to implement move-blocking strategies such that a smaller number of parameters is required by restricted the input from change at every sampling instant on the horizon, in particular towards the end of the horizon. For example, MPC has been successfully implemented for stable plants based on linear models by optimizing a constant input on the whole horizon, [100].

2.2.2 *Numerical Optimal Control*

In this section the basic optimal control formulation in section 2.2.1 is reformulated into a form suitable for numeric solution by a nonlinear optimization solver.

As classified in [39] there are two main avenues to direct numerical optimal control

- **The sequential approach.** The ODE constraint (2.5) is solved via numeric simulation when evaluating the cost and constraint functions. This means that the intermediate states $x(t_1),....,x(t_N)$ disappear from the problem formulation by substitution into the cost and constraint functions, while the control trajectory parameters U are treated as unknowns. This leads to a sequence of simulate-optimize iterations, often known as *Direct Single Shooting*, [57, 105, 68].
- **The simultaneous approach.** The ODE constraints (2.5) are discretized in time and the resulting finite set of nonlinear algebraic equations are treated as nonlinear equality constraints. The intermediate states $x(t_1),....,x(t_N)$ are treated as unknown variables together with the control trajectory parameters U, and the cost function is evaluated simply by replacing the integral (2.1) by a finite sum. This leads to simultaneous solution of the ODEs and the optimization problem with a larger number of constraints and variables. The most well known methods of this type are *Direct Multiple Shooting*, [36, 23, 24, 72], and *Collocation methods*, [116, 19, 113].

It is fair to say that all the above mentioned approaches have advantages that could make them the method of choice when considering a specific problem:

- The simultaneous approach involves a larger number of constraints and variables and therefore leads to "bigger problems". On the other hand, the cost and constraint function evaluation is much simpler and there are structural properties of the equations and numerical advantages that can be exploited in some cases. This will be discussed in section 2.3.

- Neglecting errors due to discretization and numerical approximations, all methods results in the same optimal control trajectory. Hence, one may expect the main difference between these alternatives to be related to numerical properties and complexity of computation and explicit representation. Numerical accuracy of the solution is a consequence of discretization, round-off errors, sensitivity to initial conditions and input, differences in linear algebraic methods, etc. and must be balanced against computational cost. These aspects will be treated in more detail in section 2.3.
- Nonlinear optimization problems are generally non-convex, and the convergence and success of a given optimization algorithm depend largely on the initial guess provided for the solution. The sequential and simultaneous approach are in this sense fundamentally different, since the simultaneous approach not only requires an initial control trajectory guess, but also one for the state trajectory. The availability of a good initial guess for the state trajectory is an advantage that can be exploited by the simultaneous approach. On the other hand, the presence of nonlinear equality constraints (which by definition are non-convex) in the simultaneous approach, one cannot expect feasible initial guesses, which has consequences for the choice of numerical methods, and will be further discussed in section 2.3.
- The sequential approach may use more or less arbitrary and separate ODE and optimization solvers, which may in some cases be simple and convenient when compared to the simultaneous approach that tend to require more specialized and integrated numeric software combining these tasks. This may be a particularly important issue for industrial users that must use software tools based on an extensive set of requirements and constraints.

2.2.2.1 Direct Single Shooting

In direct single shooting ([57, 105, 68]), the ODE constraint (2.5) is eliminated by substituting its discretized numerical solution $x(t_k) = \phi(t_k, U, x(0))$ into the cost and constraints;

$$
\begin{aligned}
V^*(x(0)) = \min_{U \in \mathbb{R}^p} V(U; x(0)) \triangleq \sum_{k=1}^{N} \ell(\phi(t_k, U, x(0)), \mu(t_k, U), t_k)(t_k - t_{k-1}) \\
+ S(\phi(T, U, x(0)), T) \qquad (2.6)
\end{aligned}
$$

subject to

$$u_{min} \leq \mu(t_k, U) \leq u_{max}, \quad t_k \in T_d \tag{2.7}$$

$$g(\phi(t_k, U, x(0)), \mu(t_k, U), t_k) \leq 0, \quad t_k \in T_d \tag{2.8}$$

and the ODE solution function $\phi(\cdot)$ is the result of a numerical integration scheme. In its simplest form, an explicit integration scheme may be used

$$x(t_{k+1}) = F(x(t_k), \mu(t_k, U), t_k), \; x(t_0) = x(0) \text{ given,} \tag{2.9}$$

for $k = 0, ..., N-1$, leading to

$$\phi(t_k, U, x(0)) = F(\dots F(F(x(0), \mu(t_0, U), t_0), \mu(t_1, U), t_1), ..., \mu(t_{k-1}, U), t_{k-1}) \tag{2.10}$$

However, $\phi(t_k, U, x(0))$ may also be computed using any other (implicit) discretization scheme in the simulation.

The problem (2.6) - (2.8) is a nonlinear program in U parameterized by the initial state vector $x(0)$ and time, and hence in the multi-parametric nonlinear programming form (1.1)-(1.2). Dependence on time-varying external signals such as references and known disturbances are left implicit in order to keep the notation simple. They can be included in the parameter vector in a straightforward manner. The receding horizon MPC strategy will therefore re-optimize U when new state or external input information appears, typically periodically at each sample. We assume the solution exists, and let it be denoted U^*.

We note that the introduction of common modifications such as terminal constraints and infeasibility relaxations still gives a nonlinear multi-parametric program, but with additional decision variables and constraints.

2.2.2.2 Direct Collocation

In direct collocation ([116, 19, 113]) the numerical solution for $x(t_k)$ is not substituted into the cost and constraint functions, but the associated nonlinear algebraic equations resulting of an ODE discretization scheme are kept. Hence, the variables $x(t_k),\ k = 1, ...N$ are treated as unknown decision variables:

$$\begin{aligned} V^*(x(0)) = & \min_{U \in \mathbb{R}^p, x(t_1) \in \mathbb{R}^n, ..., x(t_N) \in \mathbb{R}^n} V(U, x(t_1), ..., x(t_N); x(0)) \\ & \triangleq \sum_{k=1}^{N} \ell(x(t_k), \mu(t_k, U), t_k)(t_k - t_{k-1}) + S(x(t_N), T) \end{aligned} \tag{2.11}$$

subject to

$$u_{min} \leq \mu(t_k, U) \leq u_{max},\ \ t_k \in T_d \tag{2.12}$$
$$g(x(t_k), \mu(t_k, U), t_k) \leq 0,\ \ t_k \in T_d \tag{2.13}$$
$$F(x(t_{k+1}), x(t_k), \mu(t_k, U), t_k) = 0,\ \ k = 0, ..., N-1 \tag{2.14}$$
$$x(t_0) = x(0) \text{ given} \tag{2.15}$$

where F is a function defined by the discretization scheme of the ODE (2.5). We observe from (2.14) that it directly allows for implicit numerical integration methods to be used, and that the algebraic equations resulting from the implicit integration scheme will be solved simultaneously with the optimization.

The problem (2.11) - (2.13) is a nonlinear program in the variables $U, x(t_1), ...,$ $x(t_N)$ parameterized by the initial state vector $x(0)$. It is therefor in the multi-parametric nonlinear programming form (1.1)-(1.2), except that equality constraints are present. They pose no difficulty, as they should always be included in the optimal active set, and can be treated by standard nonlinear programming solvers. In addition, dependence on time-varying external signals such as references and known disturbances are left implicit in order to keep the notation simple. The receding horizon MPC strategy will therefore re-optimize U when new state or external input information appears, typically periodically at each sample. We assume the solution exists, and let it be denoted $U^*, x^*(t_1), ..., x^*(t_N)$.

2.2.2.3 Direct Multiple Shooting

Direct multiple shooting ([36, 23, 24, 72]) combines elements of both direct single shooting and direct collocation. It is a simultaneous approach in the sense it reformulates the ODE (2.5) to a set of nonlinear algebraic equality constraints that are solved simultaneously with the optimization. It differs from the direct collocation method since an ODE solver is used to simulate the ODE (2.5) in each time interval $t_k \leq t \leq t_{k+1}$ for $k = 0, ..., N-1$:

$$
\begin{aligned}
V^*(x(0)) &= \min_{U \in \mathbb{R}^p, (x(t_1), ..., x(t_N))^T \in \mathbb{R}^{nN}} V(U, x(t_1), ..., x(t_N); x(0)) \\
&\triangleq \sum_{k=1}^{N} \ell(x(t_k), \mu(t_k, U), t_k)(t_k - t_{k-1}) + S(x(t_N), T) \qquad (2.16)
\end{aligned}
$$

subject to

$$
\begin{aligned}
u_{min} \leq \mu(t_k, U) &\leq u_{max}, \quad t_k \in T_d &(2.17)\\
g(x(t_k), \mu(t_k, U), t_k) &\leq 0, \quad t_k \in T_d &(2.18)\\
x(t_{k+1}) &= \phi(x(t_k), \mu(t_k, U), t_k), \quad k = 0, ..., N-1 &(2.19)\\
x(t_0) &= x(0) \text{ given}, &(2.20)
\end{aligned}
$$

where ϕ is a function defined by the simulation of the ODE (2.5). The main difference between direct multiple shooting and direct collocation is due to the use of an arbitrary ODE solver between the time-instants in T_d. Direct multiple shooting may have advantages when adaptive discretization schemes are needed (due to stiff dynamics, for example) since they might require a varying number of grid points for each iteration of the solver. With multiple shooting this can in principle be "hidden" within the direct single shooting solver used between each time-instant in T_d, while it directly leads to a change in the dimensions of the optimization problem at each iteration with a direct collocation method. Direct multiple shooting decouples the grids required for the point-wise discretization of the constraints (2.18) and the discretization grid required to integrated the ODE. In a sense, direct multiple shooting provides additional flexibility compared to both direct single shooting and direct collocation. On the other hand, direct collocation leads to a more sparse structure that can be exploited by the numerical optimization solver.

2.2.2.4 The Nonlinear Program – Feasibility and Continuity

This section summarizes some features of the numeric optimization problem resulting from the direct approach to numerical optimal control in NMPC. Important issues related to the well-posedness of the problem are reviewed. They are related to existence and uniqueness of the solution and continuous dependence of the solution on data such as the initial state $x(0)$. These are again related to regularity properties and fundamental properties such as controllability.

In summary, all formulations reviewed in this section lead to a nonlinear optimization problem of the form

$$V^*(\theta) = \min_z V(z,\theta) \tag{2.21}$$

subject to

$$G(z,\theta) \leq 0 \tag{2.22}$$
$$H(z,\theta) = 0 \tag{2.23}$$

where z is a vector of decision variables (control trajectory parameters, intermediate states, slack variables, etc.) while θ is a vector of parameters to the problem (initial states, parameters of reference trajectories, exogenous inputs, etc.).

Existence of a solution corresponds to feasibility of the optimization problem. We define the feasible set of parameters Θ_F as the set that contains all θ for which the optimization problem (2.21)-(2.23) has a solution $z^*(\theta)$

$$\Theta_F = \{z \mid \text{there exists a } z \text{ such that } G(z,\theta) \leq 0,\ H(z,\theta) = 0\} \tag{2.24}$$

The feasible set is a result of the dynamics of the systems and basically all design parameters of the NMPC problem. Generally speaking, it is desired to make this set as large as possible while fulfilling the physical and operational constraints of the control system. We will return to this design issue in section 2.2.3.

The concept of controllability of nonlinear systems can be defined in several ways. One may take a pragmatic point of view, and focus on conditions that leads to feasibility of the solution, and continuity of the value function or solution as a function of the time-varying data θ that includes the initial conditions (see section 1.1.4). In the context of numerical optimal control, issues related to lack of controllability or inappropriate design choices will typically manifest themselves in terms of infeasibility (no solution exists), indefiniteness of the Hessian (a global solution is not found), or singularity or poor conditioning of the Hessian (the solution is not unique and continuously dependent on the input data, or is highly sensitive to changes in decision variables). The latter case means that small changes in the state may require very large control actions to compensate. Since the above sufficient optimality conditions are practically impossible to verify a priori, these are important issues to be analysed in the solution of the approximate multi-parametric programming algorithms, based on output from the numerical solver and other verification methods, in order to asses the quality of the NMPC design and identify problems related to lack of controllability or inappropriate design or tuning of the NMPC criterion and constraints. It is a great advantage of explicit NMPC that this can be analyzed during the design phase. The simplest special case for which strong properties can be guarantees a priori is the case of joint convexity (see section 1.1.5.1).

2.2.3 *Tuning and Stability*

The specification of the NMPC control functionality and dynamic performance is essentially provided through the cost function and the constraints. We will not go into details on the practical tuning tradeoffs and the types of physical and operational constraints, but note that one may typically choose l_2 or l_1 type cost function

$$\ell(x,u,t) = ||x - r_x(t))|_Q^2 + ||u - r_u(t)||_R^2 \qquad (2.25)$$

$$\ell(x,u,t) = ||Q(x - r_x(t))||_1 + ||R(u - r_u(t)||_1 \qquad (2.26)$$

where the properties of the weight matrices $Q \succeq 0$ and $R \succeq 0$ are essential for performance, and in some cases also stability. In the simplest case when there exists an $\varepsilon > 0$ such that

$$\ell(x,u,t) \geq \varepsilon(||x||^2 + ||u||^2) \qquad (2.27)$$

it is clear that all states and control actions are directly observable through the cost function, an it follows intuitively that minimization of the cost function will influence all states that are controllable. Based on the similar arguments, it is in fact sufficient for stabilization that only the unstable modes of the system are observable through the cost function, such that $Q \succeq 0$ may be sufficient if weights are given on the unstable modes, [85]. In order to ensure uniqueness of the control trajectory it is generally recommended that $R \succ 0$. In summary, conventional LQR tuning guidelines (e.g. [7]) are very helpful as a starting point also for NMPC. We note that parameterization of the reference trajectories $r_x(t)$ and $r_u(t)$ will enter the parameter vector of the multi-parametric nonlinear program, and may significantly contribute to complexity of the representation of an explicit solution approximation.

Although the effect of modeling errors and disturbances will be discussed in section 2.2.4.2, we remark that incorrect choice of the reference $r_u(t)$ for the control input may lead to a steady-state error that will be important to compensate for in many applications.

NMPC is based on the receding horizon control principle, where a finite horizon open loop optimal control solution is implemented until a new optimized control trajectory is available at the next sampling instant. This leads to closed-loop control since each new optimized control trajectory is based on the most recent state information. However, the numerical optimal control problem updated at each sampling instant provides essentially an open-loop control trajectory. The finite-horizon cost function imposes in principle no stability requirement by itself, and with an unfortunate choice of design parameters (horizon T, weight matrices Q and R, terminal cost S, and certain constraints) the closed loop NMPC may be unstable. In particular for open loop unstable systems, it is important to understand how these design parameters should be chosen to avoid an unstable NMPC.

2.2.3.1 Stability Preserving Constraints and Cost-to-Go

This section discusses stability of the NMPC in more depth, and how this property is related to design parameters in the cost function and constraints. The description will be fairly informal, and we avoid the technical details in order to focus on the most important concepts. For simplicity we assume that the objective is regulation to a constant set-point r. Further details and a more rigorous treatment of the topic are found in [32, 85, 87, 63, 84], and we remark that the concepts relevant for NMPC are essentially the same as for linear MPC.

The following principles are generally useful to ensure stability of an NMPC [85]:

- The control trajectory parameterization $\mu(t,U)$ must be "sufficiently rich" - most theoretical work assume piecewise constant control input trajectory that is allowed to move at each sampling instant.
- From the optimality principle of dynamic programming, [12], an infinite horizon cost may be expected to have a stabilizing property. Theoretically, this leads to an infinite dimensional problem (except in simple special cases), so more practical approaches are
  - Sufficiently large horizon T. However, it is not obvious to know what is large enough, in particular for an open loop unstable system and when the constrained outputs are non-minimum phase (see [104] for results on the importance of the zero-dynamics of the constrained outputs for the linear case).
  - A terminal cost chosen to approximate the cost-to-go, i.e. $S(x(T),T) \approx \int_{t=T}^{\infty} \ell(x(t),u(t),t)dt$ such that the total cost function approximates an infinite horizon cost. Unfortunately, the cost-to-go is generally hard to compute and simple (quadratic) approximations are usually chosen.
- Terminal set constraints of the type $x(t_N) \in \Omega$ that ensures that the state is regulated "close enough" to the set-point such that after T it is a priori known that there exists a feasible and stabilizing controller that will ensure that $x(t), t \geq T$ never leaves Ω and eventually goes asymptotically to the set-point. There are many algorithms based on this philosophy, some of them are defined as dual mode NMPC ([87]) since they switch to a stabilizing simpler (non-NMPC) control law once Ω is reached, while others continue to use NMPC also in Ω with the confidence that there exist an (explicit or implicit) stabilizing control law that the NMPC may improve upon.
- Terminal equality constraints of the type $x(t_N) = r$, [63], that ensures convergence in finite time. This basically implies that the cost after time T is zero, and is therefore related to both an infinite-cost strategy and a stability-preserving-constraint strategy. However, this may be a demanding constraint, both due to numerical issues of nonlinear equality constraints, and feasibility issues with a short horizon T.

- Finally, the idea of choosing the cost-to-go to approximate an infinite-horizon cost and the use of a terminal set may be combined. With the use of a terminal set it will be sufficient to approximate the cost-to-go for states that are within the terminal set, and simple tools like local linearization can be applied to make this a fairly practical approach; [32].

A formal treatment of these issues are found in the references, see [85] for additional references. The main tools are the use of either the value function $V^*(x)$ as a Lyapunov function, or investigating monotony of a sequences of function values. Instead, we provide an example that is similar to the method in [32].

Example. Consider the discrete-time non-linear system

$$x(t_{k+1}) = F(x(t_k), u(t_k)) \tag{2.28}$$

where $x \in \mathbb{R}^n$ is the state, and $u \in \mathbb{R}^m$ is the input. We assume the control objective is regulation to the origin. For the current $x(t_k)$, we formulate the optimization problem

$$V^*(x(t_k)) = \min_U J(U, x(t_k)) \tag{2.29}$$

subject to $x_{k|k} = x(t_k)$ and

$$\begin{aligned}
&y_{\min} \leq y_{k+i|k} \leq y_{\max}, \quad i = 1, ..., N \\
&u_{\min} \leq u_{k+i} \leq u_{\max}, \quad i = 0, 1, ..., N-1, \\
&x_{k+N|k} \in \Omega \\
&x_{k+i+1|k} = F(x_{k+i|k}, u_{k+i}), \quad i = 0, 1, ..., N-1 \\
&y_{k+i|k} = Cx_{k+i|k}, \quad i = 1, 2, ..., N
\end{aligned} \tag{2.30}$$

with $U = \{u_k, u_{k+1}, ..., u_{k+N-1}\}$ and the cost function given by

$$J(U, x(t_k)) = \sum_{i=0}^{N-1} \left(||x_{k+i|k}||_Q^2 + ||u_{k+i}||_R^2\right) + ||x_{k+N|k}||_P^2 \tag{2.31}$$

The compact and convex terminal set Ω is defined by

$$\Omega = \{x \in \mathbb{R}^n \mid x^T P x \leq \alpha\} \tag{2.32}$$

where $P = P^T \succ 0$ and $\alpha > 0$ will be specified shortly. An optimal solution to the problem (2.29)-(2.30) is denoted $U^* = \{u_t^*, u_{t+1}^*, ..., u_{t+N-1}^*\}$, and the control input is chosen according to the receding horizon policy $u(t_k) = u_t^*$. This and similar optimization problems can be formulated in a concise form

$$V^*(x) = \min_U J(U, x) \quad \text{subject to} \quad G(U, x) \leq 0 \tag{2.33}$$

Define the set of N-step feasible initial states as follows

$$X_F = \{x \in \mathbb{R}^n \mid G(U,x) \le 0 \text{ for some } U \in \mathbb{R}^{Nm}\} \tag{2.34}$$

Suppose Ω is a control invariant set, such that X_F is a subset of the N-step stabilizable set, [64]. Notice that the origin is an equilibrium and interior point in X_F. It remains to specify $P \succ 0$ and $\alpha > 0$ such that Ω is a control invariant set. For this purpose, we use the ideas of [32], where one simultaneously determine a linear feedback such that Ω is positively invariant under this feedback. Define the local linearization at the origin

$$A = \frac{\partial f}{\partial x}(0,0), \quad B = \frac{\partial F}{\partial u}(0,0) \tag{2.35}$$

Now, the following assumptions are made:

- (A,B) is stabilizable.
- $P,Q,R \succ 0$.
- $y_{min} < 0 < y_{max}$ and $u_{min} < 0 < u_{max}$.
- The function f is twice continuously differentiable, with $f(0,0) = 0$.

Since (A,B) is stabilizable, let K denote the associated LQ optimal gain matrix, such that $A_0 = A - BK$ is strictly Hurwitz. A discrete-time reformulation of Lemma 1 in [32] can be made, [61]:

Lemma 2.1. *If $P \succ 0$ satisfies the Lyapunov-equation*

$$A_0^T P A_0 - P = -\kappa P - Q - K^T R K \tag{2.36}$$

for some $\kappa > 0$, there exists a constant $\alpha > 0$ such that Ω defined in (2.32) satisfies

1. *$\Omega \subset \mathscr{C} = \{x \in \mathbb{R}^n \mid u_{min} \le -Kx \le u_{max}, y_{min} \le Cx \le y_{max}\}$.*
2. *The autonomous nonlinear system*

$$x(t_{k+1}) = F(x(t_k), -Kx(t_k)) \tag{2.37}$$

is asymptotically stable for all $x(0) \in \Omega$, i.e. Ω is positively invariant.

3. *The infinite-horizon cost for the system (2.37)*

$$J_\infty(x(t_k)) = \sum_{i=0}^{\infty} \left(||x_{k+i|k}||_Q^2 + ||Kx_{k+i|k}||_R^2\right) \tag{2.38}$$

satisfies $J_\infty(x) \le x^T P x$ for all $x \in \Omega$.

Proof. In order to prove this result we first remark that the Lyapunov-equation (2.36) is generally satisfied for sufficiently small $\kappa > 0$ because A_0 is strictly Hurwitz and the right-hand side is negative definite. One may define a set of the form

$$\Omega_{\alpha_1} = \{x \in \mathbb{R}^n \mid x^T P x \le \alpha_1\} \tag{2.39}$$

with $\alpha_1 > 0$, such that $\Omega_{\alpha_1} \subseteq \mathscr{C}$, i.e. an ellipsoidal inner approximation Ω_{α_1} to the polyhedron $\mathscr{C}$ where the input and state constraints are satisfied. Hence, the first claim holds for all $\alpha \in (0, \alpha_1]$.

Define the positive definite function $W(x) = x^T Px$. Along trajectories of the autonomous system (2.37) we have

$$\begin{aligned} W(x(t_{k+1})) - W(x(t_k)) &= (A_0 x(t_k) + \phi(x(t_k)))^T P (A_0 x(t_k) + \phi(x(t_k))) \\ &\quad - x^T(t_k) P x(t_k) \\ &= x^T(t_k) \left(A_0^T P A_0 - P\right) x(t_k) + \phi^T(x(t_k)) P \phi(x(t_k)) \\ &\quad + x^T(t_k)(A_0^T P + P A_0)\phi(x(t_k)) \end{aligned}$$

where $\phi(x) = F(x, -Kx) - A_0 x$ satisfies $\phi(0) = 0$. From (2.36)

$$\begin{aligned} W(x(t_{k+1})) - W(x(t_k)) &= -x^T(t_k) \left(Q + K^T RK + \kappa P\right) x(t_k) \\ &\quad + x^T(t_k)(A_0^T P + P A_0)\phi(x(t_k)) + \phi^T(x(t_k)) P \phi(x(t_k)) \end{aligned}$$

Let L_ϕ be a Lipschitz constant for ϕ in Ω_α (which must exist because f is differentiable). Since $\partial\phi/\partial x(0) = 0$ and ϕ is twice differentiable we can choose $L_\phi > 0$ as close to zero as desired by selecting $\alpha > 0$ sufficiently small. Hence, there exist $\alpha \in (0, \alpha_1]$ such that

$$W(x(t_{k+1})) - W(x(t_k)) \leq -x^T(t_k) \left(\frac{\kappa}{2} P + Q + K^T RK\right) x(t_k) \qquad (2.40)$$

for all $x(t_k) \in \Omega$ and positive invariance of Ω follows since Ω is a level set of W.

Notice that from (2.40) we have

$$W(x(\infty)) - W(x(0)) \leq -J_\infty(x(0)) - \frac{\kappa}{2} \sum_{k=0}^{\infty} ||x(t_k)||_P^2 \qquad (2.41)$$

and the third claim holds because $W(x(\infty)) = 0$ for all $x(0) \in \Omega$.

Hence, the result is proven, and it follows from [85, 32] that the RHC makes the origin asymptotically stable with region of attraction equal to the feasible set X_F.

A procedure for selecting P, κ and α can be adapted from [32].

2.2.4 *Extensions and Variations of the Problem Formulation*

2.2.4.1 Infeasibility Handling and Slack Variables

Feasibility of the NMPC optimization problem is an essential requirement for any meaningful state and reference command, and it is importance in practice that the NMPC optimization problem is formulated such that feasibility is ensured as far as possible by relaxing the constraints when needed and when possible. Obviously, physical constraints like input saturation can never be relaxed, but operational constraints can generally be relaxed according to certain priorities under the additional requirement that safety constraints are fulfilled by a separate system (like an emergency shutdown system, pressure relief valves, or by functions in a decentralized control system). Stability-enforcing terminal constraints may also be relaxed in practice, or even skipped completely, since they tend to be conservative and often not needed when the NMPC is otherwise carefully designed, in particular for open loop stable systems.

A general way to reformulate an optimization problem to guarantee feasibility is to use slack variables (e.g. [118]). Taking the fairly general NLP formulation (2.21)-(2.23) as the starting point, we reformulate it in the following way

$$V_s^*(\theta) = \min_{z,s,q} V(z,\theta) + ||W_s s||_1 + ||W_q q||_1 \tag{2.42}$$

subject to

$$G(z,\theta) \leq s \tag{2.43}$$

$$H(z,\theta) = q \tag{2.44}$$

$$s \geq 0 \tag{2.45}$$

where $W_s \succeq 0$ and $W_q \succeq 0$ are weight matrices of appropriate dimension. They are usually chosen such that the two latter penalty terms of (2.42) dominates the first term in order to ensure that the feasibility constraints are not relaxed when not needed. We observe that this formulation is still in the multi-parametric non-linear programming form, although with an augmentation of the vector of decision parameters.

2.2.4.2 Robustness

Practical industrial experience shows that MPC tend to be inherently robust, [100, 101], even without any particular consideration in the design phase beyond ensuring the accuracy of dynamic models and formulating realistic specifications in terms of operational constraints, prediction horizon, and cost function weights. In addition, mechanisms to handle steady state model errors (integral action like mechanisms) are usually implemented, and the processes tend to have a low-level conventional control system that stabilizes and counteracts uncertainties in the plant.

As a contrast to this practical experience, it is shown by examples, [55], that when the NMPC problem involves state constraints, or terminal constraints in combination with short prediction horizons, the asymptotic stability of the closed-loop may have not be robust. A necessary condition for lack of robustness is that the value function and state feedback law are discontinuous, [55], while at the same time lack of continuity does not necessarily lead to lack of robustness, [70].

There exist a wide range of NMPC formulations that include robustness into the formulation of the optimization problem. One can mainly distinguish between three types of approaches; stochastic NMPC, min-max NMPC, and mechanisms to avoid steady-state errors.

There are two formulations of min-max NMPC: the open-loop and the closed-loop formulation (see [78] for review of the min-max NMPC approaches). The open-loop min-max NMPC ([87, 75, 78]) guarantees the robust stability and the robust feasibility of the system, but it may be very conservative since the control sequence has to ensure constraints fulfillment for all possible uncertainty scenarios without considering the fact that future measurements of the state contain information about past uncertainty values. As a result, the open-loop min-max NMPC controllers may have a small feasible set and a poor performance because they do not include the effect of feedback provided by the receding horizon strategy of MPC.

Most min-max MPC robustness approaches assume a fairly simple additive uncertainty model of the form

$$x_{k+1} = F(x_k, u_k) + w_k \tag{2.46}$$

where some bound on the unknown uncertainty w_k is assumed. The conservativeness of the open-loop approaches is overcome by the closed-loop min-max NMPC ([81, 78, 76]), where the optimization is performed over a sequence of feedback control policies. With the closed-loop approach, the min-max NMPC problem represents a differential game where the controller is the minimizing player and the disturbance is the output of the maximizing player. The controller chooses the control input as a function of the current state so as to ensure that the effect of the disturbance on the system output is sufficiently small for any choice made by the maximizing player. In this way, the closed-loop min-max NMPC would guarantee a larger feasible set and a higher level of performance compared to the open-loop min-max NMPC ([81]).

Stochastic NMPC formulations are based on a probabilistic description of uncertainty, and can also be characterized as open-loop [28, 62] and closed-loop [50, 6] similarly to min-max robust NMPC as described above. They also share similar challenges due to significantly increased computational complexity when compared to nominal NMPC formulations.

The reformulation of nonlinear models as Linear Parameter Varying (LPV) models allows for the use of linear and bi-linear matrix inequality formulations of robust NMPC, [5, 29, 119]. The embedding of nonlinear systems into the class of LPV models

$$x_{k+1} = A(p_k)x_k + B(p_k)u_k + w(p_k) \tag{2.47}$$

causes loss of information in the model that leads to more conservative robust control. However, using tools of semi-definite and convex programming, [26], the LPV re-formulation allows for the computational complexity to be significantly reduced in many cases. In (2.47), p_k is a parameter whose value is known to belong to some bounded set, and some approaches also assume that its time-derivative has a known bound, and the LPV re-formulation clearly allows a richer class of uncertainty to be modeled, compared to (2.46).

Steady-state control errors may result if there are steady-state model errors. While linear control design offers several tools to deal with this problem (including integral action, integrating models in linear MPC, and others), not all of them are directly transferable to nonlinear systems. The commonly used cure for steady-state errors in MPC, which can be directly transferred to NMPC, appears to be the use of a state estimator or observer that estimates an input or output disturbance for direct compensation in the NMPC cost function, [90, 99, 98, 25].

2.2.4.3 Observers and Output Feedback

Most formulations of nonlinear MPC assume state feedback. They are usually based on state space models, e.g. [8, 45], although certain black-box using discrete-time nonlinear input/output models have also been proposed [95, 1]. Since all states are usually not measured, implementation of NMPC based on a state space model may require a state estimator, which is often a critical component of an NMPC. State space models have the advantage that they are most conveniently based on first principles.

Although practical rules of thumb for observer design such as separation of time-scales (typically one order of magnitude faster state estimator relative to the control loop response time) tend to be applicable in practical implementations also for NMPC, there also exist a number of rigorous theoretical results on the stability of the combination of observers with NMPC, see [43] for an overview. Although a general separation principles does not exists for NMPC, there are some results in this direction, [44, 86, 103].

2.2.4.4 Mixed-Integer MPC

General NMPC formulations based on nonlinear models suffer from the fact that it is hard to verify whether the underlying optimization problem is convex or not, such that in general it must be assumed to be non-convex. At the same time, all practical optimization solvers will assume some form of local convexity and guarantee convergence only with good initial guesses for the solution. This challenge will be further discussed in section 2.3. On the other hand, NMPC based on piecewise linear (PWL) models and cost functions will in general lead to mixed-integer linear programs (MI-LP) for which there exists solvers that guarantee global convergence, [117, 15, 14]. The equivalence between a wide class of hybrid systems models, mixed logic models and PWL models, [56], makes this approach attractive in many practical applications. Despite its applicability and importance, we only remark that the MI-LP theory and software are well developed, and refer to the references above and the large literature on MI-LP, [120].

Mixed-integer programming also allows for general classes of models that are formulated using logic and integer variables, including on/off type of inputs and system modes in combination with nonlinear dynamics. More generally, mixed-integer nonlinear programming (MI-NLP) solvers are available for problems that cannot be converted into PWL or equivalent form.

2.2.4.5 Decentralized and Distributed NMPC

Recently, several approaches for decentralized and distributed implementation of NMPC algorithms have been developed. A review of architectures for distributed and hierarchical MPC can be found in [106]. The possibility to use MPC in a decentralized fashion has the advantage to reduce the original, large size, optimization problem into a number of smaller and more tractable ones.

In [77], a stabilizing decentralized MPC algorithm for nonlinear systems consisting of several interconnected local subsystems is developed. It is derived under the main assumptions that no information can be exchanged between local control laws, i.e. the coupling between the subsystems is ignored, and only input constraints are imposed on the system. In [40], it is supposed that the dynamics and constraints of the nonlinear subsystems are decoupled, but their state vectors are coupled in a single cost function of a finite horizon optimal control problem. In [65], an optimal control problem for a set of dynamically decoupled nonlinear systems, where the cost function and constraints couple the dynamical behavior of the systems, is solved.

2.3 Numerical Optimization

The topic of this book is on *explicit* NMPC, with the goal that numerical optimization can be avoided in the online real-time computer. Still, nonlinear programming solutions of NMPC problems is the work-horse in the offline computations that are necessary to construct the data structures needed for the online explicit NMPC computations. As a complement to the introduction to numerical optimization methods in Chapter 1, we selectively review some additional aspects of numerical optimization methods that are particularly important for NMPC, both online optimization and in offline multi-parametric nonlinear programming to construct explicit NMPC solutions.

2.3.1 Problem Structure

The choice of numerical optimization solver strategy will have significant impact on both the need for computational resources and the quality of the solution in online NMPC as well as for the offline computations in explicit NMPC. In this context, computational resources usually means the CPU time required for the solution to converge to meet the tolerance requirements, while quality of solution is related to lack of convergence or high sensitivity to initial guesses.

There are several features of NMPC problems that are relevant to consider

- Formulation of the numerical optimal control, e.g. sequential or simultaneous approaches. The sequential approach leads to a smaller, denser problem with a computationally complex cost function usually without nonlinear equality constraints, while the simultaneous approach leads to a larger, more structured, sparse

problem with nonlinear equality constrains and relatively simple cost and constraint functions to evaluate.
- Online NMPC solves a sequence of numerical optimal control or estimation problems, where the parameters of the problem are usually subject to fairly small changes from one run to the next. There is usually benefits of warm starting the next optimization run using the solution and other internal data from the previous run as initial guesses, data or conditions. Likewise, neighboring solutions provide useful warm start information also in multi-parametric nonlinear programming for use in explicit NMPC.
- In online NMPC, the optimization will be repeated at the next sample, and the optimization problem is formulated using uncertain data, it may not always be essential that the solver has converged (or equivalently that the tolerances may not need to be very strict) due to the forgiving effect of feedback in asymptotically stable systems. However, a feasible solution is generally required at each run in order to operate the control and monitoring systems. This means that problems tend to be re-formulated using slack variables with some prioritization of constraints that can be relaxed, and that is it generally desirable to start the next optimization run with a feasible initial guess generated from the previous run such that even with a limited number of iterations one can guarantee feasibility. Explicit NMPC will in general not give exact solutions, and various forms of approximations are essential to limit the offline computations.
- Safety and reliability are essential features of most control and monitoring systems, which means that post-optimal analysis and checks on the quality of the solution must usually be implemented. Issues such as non-convexity and non-smoothness of models and constraints are essential to understand and take into account. In explicit NMPC, such verification can be made offline, which means that more rigorous testing can be made compared to online NMPC.

Although all NMPC problems have certain features in common, they may also differ considerably with respect to size, models, cost functions and constraints. This means that there will not be a single numerical method that will be the best, in general. In section 1.1.3, we briefly outlined some commonly used numerical methods and remark that sequential quadratic programming and interior point methods are most commonly used in NMPC. Below, we discuss some important details that result from the structure of NMPC problems.

2.3.1.1 Numerical Linear Algebra

At heart of both the QP sub-problems of SQP and the Newton-step of IP methods are the solution of a set of linear algebraic equations resulting from the Karush-Kuhn-Tucker conditions. Efficiency of the numerical optimization solver heavily depends on the efficiency of solving this problem, since it will be repeated many times towards the solution of the NLP at each sampling instant of an NMPC. Exploiting structural properties is essential.

Sequential approaches such as direct single shooting formulation will in general lead to a dense set of linear algebraic equations, for which there is not much structure to exploit. Simultaneous approaches such as direct multiple shooting and collocation methods retain a large set of nonlinear algebraic equations that links groups of variables, where each group can be associated with the selected time-intervals. Hence, the links between groups may be limited to the previous or next time-interval only, leading to a sparseness structure that can be exploited in the numerical linear algebra.

Generally, useful structural properties are often related to symmetry and positive definiteness of the Hessian (approximation), sparseness and block-diagonal structure of the linear systems of equations, and what information from the previous optimization run can be used to initialize the next run. Using factorization methods one may solve triangular equation systems efficiently, eliminate algebraic variables, and operate in reduced spaces to save computations. Being able to efficiently maintain and update factorized matrices between the various iterations is usually essential to implement this. We refer to excellent and comprehensive treatments in [96, 39, 48, 47] and the references therein.

2.3.1.2 Initialization

A NLP problem may be closely related to the NLP problem at either the previous sampling instant in online NMPC, since the sampling interval is usually short compared to the dynamics of the plant and the controller, or the NLP problem at neighboring states (or parameters, more generally) in explicit NMPC. Assuming the reference signals and other input to the controller changes slowly, this means that the solution in terms of future input and state trajectories (for MPC problems) can be time shifted one sampling period and still provide a reasonably accurate solution to the next NLP. Assuming no uncertainty in MPC problems, this is a perfectly valid assumption and is commonly used to guarantee feasibility at the next step in stability arguments, e.g [108, 85]. Even without time-shifting, the previous solution itself may also provide a good initialization for warm start purposes in NMPC, [24, 38].

Unlike SQP methods, IP methods can usually not make effective use of initial guesses of the solution due to the reformulation of the KKT conditions that follows the parameterized center path controlled by the parameter $\tau > 0$ that is sequentially reduced towards zero. This does not necessarily imply that IP methods are less suited for NMPC and NMHE problems, in particular for large scale problems where IP methods have advantages that may compensate for this shortcoming. Modified IP methods that can efficiently incorporate warm start is a current research topic, [49, 111].

Warm start is potentially most efficient when including data beyond just the solution point, but also consider the internal data of the optimization algorithm such as initial estimates of the Hessian approximation (in case exact Hessians are not computed), or initial estimates of factorizations of the Hessian (approximation), initial estimates of optimal active sets, and other data. This is in particular a challenge when the dimensions and structure of these internal data will change from one sample to the next. This may for example be the case in the simultaneous formulations (in particular direct collocation) of numerical optimal control (see section 2.2.2), since the discretization may be changed from one sample to the next, in general. One must also have in mind that simultaneous formulations require that both state and control trajectories are initialized, while sequential formulations only require the control trajectory initialization. What is most beneficial will depend on the accuracy of the available information for initialization, amongst other things. We refer to [39, 58] and the references therein for a deeper treatment of this topic.

2.3.2 *Computation of Jacobian and Hessian Matrices*

The computation of the Jacobians of the cost and constraint functions is often the main computational cost of numerical optimization methods, and even fairly small inaccuracies in the calculation of the Jacobians due to may lead to severe convergence problems.

Simultaneous approaches offer advantages over sequential approaches with respect to Jacobian computations:

- The prediction horizon is broken up into several intervals where ODE solutions are computed from given initial conditions. Since these intervals will be shorter than the single interval of a single shooting approach, numerical errors due to the ODE solver tend to accumulate less.
- Implicit ODE solvers, which generally have more stable numerical properties than explicit solvers, can in general be used in simultaneous approach.
- Simultaneous approaches are characterized by simpler cost and constraint functions, where automatic differentiation is more easily exploited to avoid numerical Jacobian computation errors, see section 2.3.2.2.

The numerical challenges are in particular important to consider for plants that are unstable or marginally stable. Like in linear MPC, there may be advantages of prestabilizing an open-loop unstable plant model with a feedback compensator before used in NMPC, [27].

2.3.2.1 Finite Difference

The finite difference method approximates the (i, j)-th element of the Jacobian of a vector function $f(z)$ as

$$(\nabla_z f(z))_{i,j} \approx \frac{f_i(z_j+\delta) - f_i(z_j)}{\delta} \tag{2.48}$$

for some small $\delta > 0$. If δ is too large there will be inaccuracies due to the nonlinearity of f_i, since the method computes the average slope between two points. If the two points are not infinitely close and the function is not linear, there will be a "nonlinearity error". If δ is too small, any finite numerical error ε_1 in the computation of $f_i(z_j+\delta)$ and ε_2 in the computation of $f_i(z_j)$ will lead to an error $\varepsilon = (\varepsilon_1 - \varepsilon_2)/\delta$ in the computation of the derivative. Obviously, this error may be large when $\delta \to 0$, since $\varepsilon_1 - \varepsilon_2$ may not go to zero, so a tradeoff between these errors must be made.

It should be noticed that the finite difference approximation error ε depends on the difference between the errors in the two point-wise evaluations of f_i. This means that systematic errors (i.e. the same error in both ε_1 and ε_2) will cancel each other and have a much smaller effect than a random error of the same magnitude. Practical experience shows that the use of variable-step (adaptive) ODE solvers tend to give a small random numerical error, while the use of fixed-step ODE solvers tend to give a larger systematic error, but even smaller random error. For the reasons described above, one may find that a fixed-step ODE solver leads to considerably smaller error in finite difference Jacobian computations and performs better with less convergence problems in many numerical methods for NMPC.

It is also worthwhile to emphasize that scaling of all variables involved in the optimization problem to the same order of magnitude is in many cases a pre-requisite for numerical nonlinear optimization methods to work satisfactorily. This is evident in the context of finite difference Jacobian computations, but also relevant for other numeric computations.

As a final remark, it is possible to exploit square-root factorizations (like Cholesky factorization) for improved numerical accuracy and computational complexity in finite difference computations, [107].

2.3.2.2 Symbolic and Automatic Differentiation

The most accurate result and computationally most efficient approach is to calculate gradients by symbolically differentiating the cost and constraint functions. Doing this by hand, or even using symbolic computations in Matlab, Maple or Mathematica, may easily become intractable for NMPC problems that may contain a large number of variables, equations and inequalities. A more convenient solution is to rely on so-called *automatic differentiation* software ([53]) that achieved this objective either by overlaying operators in object oriented languages such as C++ ([54]), or automatically generates source code for gradient functions based on source code of the original function, [20].

2.4 Motivation for Explicit Nonlinear Model Predictive Control

The NMPC problem is formulated in (2.21)-(2.23) as a multi-parametric nonlinear program, in a form that is suited to use the computational algorithms outlined in Chapter 1. The solution of the multi-parametric nonlinear program leads to an explicit representation of the approximate solution that can be directly exploited in an explicit NMPC. The explicit solution tend to have a quite extensive representation, often in the form of a piecewise affine function that consists of a binary search tree representation of a large number of polyhedral regions with associated affine functions. This requires online computations that include traversal of binary search trees, [115], and sometimes with the use of additional data structures such as hash tables, [9], lattice representations, [11], bounding boxes, [33], and combination with direct search [10]. The same type of representations and computations can in some cases be used in piecewise nonlinear function approximation and evaluation that results from some explicit NMPC approaches, e.g. [52].

This has several benefits that serves as the main motivation behind the idea of explicit MPC:

- Dramatical reduction in online computations, since online numerical optimization is avoided and replaced by piecewise function evaluation. This may lead to significant reduction in the requirements to real-time embedded computer hardware. In particular, since recursive numerical computations are avoided there will be no accumulation of round-off errors such that fixed point arithmetic or single precision floating point arithmetic is generally sufficient.
- NMPC optimization depend on appropriate initialization in order to avoid local minima, and appropriate formulation of constraints in order to avoid infeasibility. With explicit NMPC the validation of initialization procedures and infeasibility handling can be conducted based on a complete and explicit solution, which is much simpler than evaluating the input and result of a nonlinear numerical optimization solver. In effect, this greatly enhances the reliability and validity of the computations which is essential in a safety-critical application.
- Significant reduction in online software complexity since the code for piecewise function evaluation is much simpler than a nonlinear numerical optimization solver. This may lead to formal software verification being a feasible practical tool, and the need to detailed tuning of numerical parameters such as thresholds, tolerances, iteration limits, and finite difference perturbation steps can be avoided. In effect, this greatly enhances the reliability and validity of the computations which is essential in a safety-critical application.
- Approximate explicit solutions with reduced complexity, and with guaranteed levels of sub-optimality, may be computed offline. Formal analysis of performance, sub-optimality and stability may be possible since an explicit representation of the controller is known.
- Formulations such as stochastic NMPC and robust NMPC may not lead to increased online computations in an explicit NMPC approach, compared to a nominal NMPC formulation, although they will require more offline computations.

Certainly, there are also drawbacks and challenges, which is why explicit MPC is an active research topic that still has not found its way into a wide number of commercial products and applications yet:

- Extensive offline computations in order to solve multi-parametric nonlinear programs and generate the associated data structures for online implementation of explicit NMPC. This can in some cases be prohibitive for NMPC problems of high order or large prediction horizons.
- The computer memory required to store the data structures required for the online computations, may be prohibitive for NMPC problems of high order.
- Analysis of sub-optimality and stability for approximate solutions may be computationally complex, and computationally prohibitive for NMPC problems of high complexity.

- Introduction of flexibility in terms of time-varying model parameters, reference trajectories, setpoints, time-varying constraint limits, and other auxiliary variables, will add to the number of parameters to the multi-parametric nonlinear program and lead to multi-parametric nonlinear programs that may be prohibitive from an offline computational point of view.

While linear MPC may have exact solutions that can be computed explicitly, the explicit NMPC will in general be an approximation. A feature of such approximations, based on our extensive experience with the methods described in this book, is that the main driver behind complexity is the number of parameters to the problem, i.e. the number of states and auxiliary parameters such as reference trajectory parameters, setpoints, or variable constraint limits. This implies that the prediction horizon, the number of constraints, and the fact that the system is nonlinear, are not the main contributions to complexity of the solution representation in approximate explicit MPC based on the main methods in this book. However, it may significantly contribute to offline computational complexity. This is distinct from exact explicit linear MPC, where the prediction horizon and number of constraints are the main drivers for complexity.

The experience that the online solution complexity of explicit NMPC is typically not more than that of a *linear* MPC for a system of the same order (e.g. a linearization of the nonlinear system) is a strong encouragement for future research into explicit NMPC since it may indicate that explicit MPC will have stronger motivation for nonlinear systems than for linear systems.

References

1. Åkesson, B.M., Toivonen, H.T.: A neural network model predictive controller. Journal of Process Control 16, 937–946 (2006)
2. Alamir, M., Bornard, G.: Stability of a truncated infinite constrained receding horizon scheme: The general discrete nonlinear case. Automatica 31, 1353–1356 (1995)
3. Alessandri, A., Baglietto, M., Parisini, T., Zoppoli, R.: A neural state estimator with bounded errors for nonlinear systems. IEEE Transactions on Automatic Control 44, 2028–2042 (1999)
4. Allgöwer, F., Badgwell, T.A., Qin, J.S., Rawlings, J.B., Wright, S.J.: Nonlinear predictive control and moving horizon estimation – An introductory overview. In: Frank, P.M. (ed.) Advances in Control, Highlights of ECC 1999, pp. 391–449. Springer (1999)
5. Angeli, D., Casavola, A., Mosca, E.: Constrained predictive control of nonlinear plants via polytopic linear system embedding. Int. J. Robust Nonlinear Control 10, 1091–1103 (2000)
6. Arellano-Garcia, H., Wendt, M., Barz, T., Wozny, G.: Closed-loop stochastic dynamic optimization under probabilistic output constraints. In: Findeisen, R., Allgöwer, F., Biegler, L.T. (eds.) Assessment and Future Directions of Nonlinear Model Predictive Control. LNCIS, vol. 358, pp. 305–315. Springer (2007)
7. Athans, M., Falb, P.L.: Optimal control. An introduction to the theory and its applications. McGraw Hill Ltd. (1966)
8. Balchen, J.G., Ljungquist, D., Strand, S.: State-space predictive control. Chemical Engineering Science 47, 787–807 (1992)

9. Bayat, F., Johansen, T.A., Jalali, A.A.: Using hash tables to manage time-storage complexity in point location problem: Application to Explicit MPC. Automatica 47, 571–577 (2011)
10. Bayat, F., Johansen, T.A., Jalali, A.A.: Combining truncated binary search tree and direct search for flexible piecewise function evaluation for explicit MPC in embedded microcontrollers. In: Proceedings of the 18th IFAC World Congress, Milano (2011), www.IFAC-PapersOnLine.net
11. Bayat, F., Johansen, T.A., Jalali, A.A.: Flexible piecewise function evaluation methods with application to explicit model predictive control. In: IEEE Int. Conf. Mechatronics, Istanbul (2011)
12. Bellman, R.: Dynamic programming. Princeton University Press, New Jersey (1957)
13. Bemporad, A., Filippi, C.: Suboptimal explicit RHC via approximate multiparametric quadratic programming. J. Optimization Theory and Applications 117, 9–38 (2003)
14. Bemporad, A., Morari, M.: Control of systems integrating logic, dynamics, and constraints. Automatica 35, 407–427 (1999)
15. Bemporad, A., Borrelli, F., Morari, M.: Optimal controllers for hybrid systems: Stability and piecewise linear explicit form. In: Proceedings of the IEEE Conference on Decision and Control, Sydney (2000)
16. Bemporad, A., Morari, M., Dua, V., Pistikopoulos, E.N.: The explicit linear quadratic regulator for constrained systems. Automatica 38, 3–20 (2002)
17. Betts, J.T.: Practical methods for optimal control using nonlinear programming. SIAM, Philadelphia (2001)
18. Biegler, L.: Efficient solution of dynamic optimization and NMPC problems. In: Allgöwer, F., Zheng, A. (eds.) Nonlinear Predictive Control, pp. 219–244. Birkhäuser (2000)
19. Biegler, L.T.: Solution of dynamic optimization problems by successive quadratic programming and orthogonal collocation. Computers and Chemical Engineering 8, 243–248 (1984)
20. Bischof, C., Carle, A., Kadhemi, P., Mauer, A.: ADIFOR2.0: Automatic differentiation of Fortran 77 programs. IEEE Computational Science and Engineering 3, 18–32 (1996)
21. Biyik, E., Arcak, M.: A hybrid redesign of Newton observers in the absence of an exact discrete time model. Systems and Control Letters 55, 429–436 (2006)
22. Bølviken, E., Acklam, P.J., Christophersen, N., Størdal, J.M.: Monte Carlo filters for non-linear state estimation. Automatica 37, 177–183 (2001)
23. Bock, H.G., Plitt, K.J.: A multiple shooting algorithm for direct solution of optimal control problems. In: Proceedings of the 9th IFAC World Congress, pp. 243–247. Pergamon Press, Budapest (1984)
24. Bock, H.G., Diehl, M., Leineweber, D.B., Schlöder, J.P.: Efficient direct multiple shooting in nonlinear model predictive control. In: Keil, F., Mackens, W., Voss, H., Werther, J. (eds.) Scientific Computing in Chemical Engineering II, vol. 2, pp. 218–227. Springer, Berlin (1999)
25. Borrelli, F., Morari, M.: Offset free model predictive control. In: Proceedings of the IEEE Conference on Decision and Control, pp. 1245–1250 (2007)
26. Boyd, S., Ghaoui, L.E., Feron, E., Balahrishnan, V.: Linear matrix inequalities in system and control theory. SIAM, Philadelphia (1994)
27. Cannon, M., Kouvaritakis, B.: Optimizing prediction dynamics for robust MPC. IEEE Trans. Automatic Control 50, 1892–1897 (2005)

28. Cannon, M., Ng, D., Kouvaritakis, B.: Successive Linearization NMPC for a Class of Stochastic Nonlinear Systems. In: Magni, L., Raimondo, D.M., Allgöwer, F. (eds.) Nonlinear Model Predictive Control: Towards New Challenging Applications. LNCIS, vol. 384, pp. 249–262. Springer, Heidelberg (2009)
29. Casavola, A., Famularo, D., Franze, G.: Predictive control of constrained nonlinear systems via LPV linear embeddings. Int. J. Robust Nonlinear Control 13, 281–294 (2003)
30. Cervantes, A., Biegler, L.: Large-scale DAE optimization using a simultaneous NLP formulation. AIChE Journal 44, 1038–1050 (1998)
31. Chen, C.C., Shaw, L.: On receding horizon feedback control. Automatica 18, 349–352 (1982)
32. Chen, H., Allgöwer, F.: A quasi-infinite horizon nonlinear model predictive control scheme with guaranteed stability. Automatica 34, 1205–1217 (1998)
33. Christophersen, F., Kvasnica, M., Jones, C.N., Morari, M.: Efficient evaluation of piecewise control laws defined over a large number of polyhedra. In: Proceedings of the European Control Conference, pp. 2360–2367 (2007)
34. Conn, A., Scheinberg, K., Vicente, L.N.: Introduction to derivative-free optimization. SIAM (2009)
35. Conn, A.R., Gould, N.I.M., Toint, P.L.: Trust region methods. SIAM (2000)
36. Deuflhard, P.: A modified Newton method for the solution of ill-conditioned systems of nonlinear equations with application to multiple shooting. Numer. Math. 22, 289–315 (1974)
37. Diehl, M., Bock, H.G., Schlöder, J.P., Findeisen, R., Nagy, Z., Allgöwer, F.: Real-time optimization and nonlinear model predictive control of processes governed by differential-algebraic equations. J. Process Control 12, 577–585 (2002)
38. Diehl, M., Magni, L., de Nicolao, G.: Online NMPC of unstable periodic systems using approximate infinite horizon closed loop costing. Annual Reviews in Control 28, 37–45 (2004)
39. Diehl, M., Ferreau, H.J., Haverbeke, N.: Efficient Numerical Methods for Nonlinear MPC and Moving Horizon Estimation. In: Magni, L., Raimondo, D.M., Allgöwer, F. (eds.) Nonlinear Model Predictive Control: Towards New Challenging Applications. LNCIS, vol. 384, pp. 391–417. Springer, Heidelberg (2009)
40. Dunbar, W.B., Murray, R.M.: Distributed receding horizon control for multi-vehicle formation stabilization. Automatica 42, 549–558 (2006)
41. Feng, L., Gutvik, C.R., Johansen, T.A., Sui, D.: Barrier function nonlinear optimiza tion for optimal decompression of divers. In: Proceedings of the IEEE Conference on Decision and Control, Shanghai (2009)
42. Fiacco, A.V.: Introduction to sensitivity and stability analysis in nonlinear programming. Academic Press, Orlando (1983)
43. Findeisen, R., Imsland, L., Allgöwer, F., Foss, B.A.: Output feedback stabilization for constrained systems with nonlinear model predictive control. Int. J. Robust and Nonlinear Control 13, 211–227 (2003a)
44. Findeisen, R., Imsland, L., Allgöwer, F., Foss, B.A.: State and output feedback nonlinear model predictive control: An overview. European J. Control 9, 179–195 (2003b)
45. Foss, B.A., Schei, T.S.: Putting nonlinear model predictive control into use. In: Findeisen, R., Allgöwer, F., Biegler, L. (eds.) Assessment and Future Directions of Nonlinear Model Predictive Control. LNCIS, vol. 358, pp. 407–417. Springer, Heidelberg (2007)
46. Foss, B.A., Johansen, T.A., Sørensen, A.V.: Nonlinear predictive control using local models - applied to a batch fermentation process. Control Engineering Practice 3, 389–396 (1995)

47. Gill, P., Murray, W., Wright, M.: Practical optimization. Academic Press, Inc. (1981)
48. Gill, P., Barclay, A., Rosen, J.B.: SQP methods and their application to numerical optimal control. Tech. Rep. NA 97–3, Department of Mathematics, University of California, San Diego (1997)
49. Gondzio, J., Grothey, A.: A new unblocking technique to warmstart interior point methods based on sensitivity analysis. SIAM J. Optimization 19, 1184–1210 (2008)
50. Goodwin, G.C., Østergaard, J., Quevedo, D.E., Feuer, A.: A Vector Quantization Approach to Scenario Generation for Stochastic NMPC. In: Magni, L., Raimondo, D.M., Allgöwer, F. (eds.) Nonlinear Model Predictive Control: Towards New Challenging Applications. LNCIS, vol. 384, pp. 235–248. Springer, Heidelberg (2009)
51. Grancharova, A., Johansen, T.A., Tøndel, P.: Computational aspects of approximate explicit nonlinear model predictive control. In: Findeisen, R., Allgöwer, F., Biegler, L. (eds.) Assessment and Future Directions of Nonlinear Model Predictive Control. LNCIS, vol. 358, pp. 181–192. Springer, Heidelberg (2007)
52. Grancharova, A., Johansen, T.A.: Computation, approximation and stability of explicit feedback min-max nonlinear model predictive control. Automatica 45, 1134–1143 (2009)
53. Griewank, A., Walther, A.: Evaluating derivatives, 2nd edn. SIAM (2008)
54. Griewank, A., Juedes, D., Utke, J.: ADOL-C, A package for the automatic differentiation of algorithms written in C/C++. ACM Trans. Mathematical Software 22, 131–167 (1996)
55. Grimm, G., Messina, M.J., Tuna, S.E., Teel, A.R.: Examples when nonlinear model predictive control is nonrobust. Automatica 40, 1729–1738 (2004)
56. Heemels, W.P.M.H., Schutter, B.D., Bemporad, A.: Equivalence of hybrid dynamical models. Automatica 37, 1085–1091 (2001)
57. Hicks, G.A., Ray, W.H.: Approximation methods for optimal control synthesis. Can. J. Chem. Engng. 49, 522–528 (1971)
58. Houska, B., Ferreau, H.J., Diehl, M.: ACADO toolkit - an open-source framework for automatic control and dynamic optimization. Optimal Control Applications and Methods (2010)
59. Isidori, A.: Nonlinear control systems, 2nd edn. Springer, Berlin (1989)
60. Jadbabaie, A., Yu, J., Hauser, J.: Unconstrained receding-horizon control of nonlinear systems. IEEE Trans. Automatic Control 46, 776–783 (2001)
61. Johansen, T.A.: Approximate explicit receding horizon control of constrained nonlinear systems. Automatica 40, 293–300 (2004)
62. Kantas, N., Maciejowski, J.M., Lecchini-Visintini, A.: Sequential Monte Carlo for Model Predictive Control. In: Magni, L., Raimondo, D.M., Allgöwer, F. (eds.) Nonlinear Model Predictive Control: Towards New Challenging Applications. LNCIS, vol. 384, pp. 263–274. Springer, Heidelberg (2009)
63. Keerthi, S.S., Gilbert, E.G.: Optimal infinite horizon feedback laws for a general class of constrained discrete-time systems: Stability and moving horizon approximations. J. Optimization Theory and Applications 57, 265–293 (1988)
64. Kerrigan, E., Maciejowski, J.M.: Invariant sets for constrained nonlinear discrete-time systems with application to feasibility in model predictive control. In: Proceedings of the IEEE Conference on Decision and Control, Sydney (2000)
65. Keviczky, T., Borrelli, F., Balas, G.J.: Decentralized receding horizon control for large scale dynamically decoupled systems. Automatica 42, 2105–2115 (2006)
66. Kim, I., Liebman, M., Edgar, T.: A sequential error-in-variables method for nonlinear dynamic systems. Comp. Chem. Engr. 15, 663–670 (1991)

67. Kojima, M.: Strongly stable stationary solutions in nonlinear programs. In: Robinson, S.M. (ed.) Analysis and Computation of Fixed Points, pp. 93–138. Academic Press, New York (1980)
68. Kraft, D.: On converting optimal control problems into nonlinear programming problems. In: Schittkowski, K. (ed.) Computational Mathematical Programming. NATO ASI Series, vol. F15, pp. 261–280. Springer (1985)
69. Krstic, M., Kanellakopoulos, I., Kokotovic, P.: Nonlinear and adaptive control design. Wiley and Sons (1995)
70. Lazar, M., Heemels, W., Bemporad, A., Weiland, S.: Discrete-time non-smooth nonlinear MPC: Stability and robustness. In: Findeisen, R., Allgöwer, F., Biegler, L.T. (eds.) Assessment and Future Directions of Nonlinear Model Predictive Control. LNCIS, vol. 358, pp. 93–103. Springer, Heidelberg (2007)
71. Lazar, M., Heemels, W.P.M.H., Roset, B.J.P., Nijmeijer, H., van den Bosch, P.P.J.: Input-to- state stabilizing sub-optimal NMPC with an application to DC-DC converters. International Journal of Robust and Nonlinear Control 18, 890–904 (2008)
72. Leineweber, D.B., Bauer, I., Bock, H.G., Schlöder, J.P.: An efficient multiple shooting based reduced SQP strategy for large-scale dynamic process optimization. part i: Theoretical aspects. Comp. Chem. Eng. 27, 157–166 (2003)
73. Li, W.C., Biegler, L.T.: Multistep Newton-type control strategies for constrained nonlinear processes. Chem. Eng. Res. Des. 67, 562–577 (1989)
74. Li, W.C., Biegler, L.T.: Newton-type controllers for constrained nonlinear processes with uncertainty. Ind. Engr. Chem. Res. 29, 1647–1657 (1990)
75. Limon, D., Alamo, T., Camacho, E.F.: Input-to-state stable MPC for constrained discrete-time nonlinear systems with bounded additive uncertainties. In: Proceedings of the IEEE Conference on Decision and Control, Las Vegas, NV, pp. 4619–4624 (2002)
76. Limon, D., Alamo, T., Salas, F., Camacho, E.F.: Input-to-state stability of min-max MPC controllers for nonlinear systems with bounded uncertainties. Automatica 42, 797–803 (2006)
77. Magni, L., Scattolini, R.: Stabilizing decentralized model predictive control of nonlinear systems. Automatica 42, 1231–1236 (2006)
78. Magni, L., Scattolini, R.: Robustness and robust design of MPC for nonlinear discrete-time systems. In: Findeisen, R., Allgöwer, F., Biegler, L. (eds.) Assessment and Future Directions of Nonlinear Model Predictive Control. LNCIS, vol. 358, pp. 239–254. Springer, Heidelberg (2007)
79. Magni, L., Nicolao, G.D., Scattolini, R.: Output feedback and tracking of nonlinear systems with model predictive control. Automatica 37, 1601–1607 (2001a)
80. Magni, L., Nicolao, G.D., Scattolini, R.: A stabilizing model-based predicitve control algorithm for nonlinear systems. Automatica 37, 1351–1362 (2001b)
81. Magni, L., De Nicolao, G., Scattolini, R., Allgöwer, F.: Robust model predictive control for nonlinear discrete-time systems. International Journal of Robust and Nonlinear Control 13, 229–246 (2003)
82. Mangasarian, O.L., Rosen, J.B.: Inequalities for stochastic nonlinear programming problems. Operations Research 12, 143–154 (1964)
83. Marino, R., Tomei, P.: Nonlinear control design: Geometric, adaptive and robust. Prentice-Hall, UK (1995)
84. Mayne, D.Q., Michalska, H.: Receding horizon control of nonlinear systems. IEEE Trans. Automatic Control 35, 814–824 (1990)
85. Mayne, D.Q., Rawlings, J.B., Rao, C.V., Scokaert, P.O.M.: Constrained model predictive control: Stability and optimality. Automatica 36, 789–814 (2000)

86. Messina, M.J., Tuna, S.E., Teel, A.R.: Discrete-time certainty equivalence output feedback: allowing discontinuous control laws including those from model predictive control. Automatica 41, 617–628 (2005)
87. Michalska, H., Mayne, D.Q.: Robust receding horizon control of constrained nonlinear systems. IEEE Trans. Automatic Control 38, 1623–1633 (1993)
88. Michalska, H., Mayne, D.Q.: Moving horizon observers and observer-based control. IEEE Transactions Automatic Control 40, 995–1006 (1995)
89. Morari, M., Lee, J.: Model predicitve control: Past, present and future. Comp. and Chem. Eng. 23, 667–682 (1999)
90. Muske, K.R., Badgwell, T.A.: Disturbance modeling for offset-free linear model predictive control. J. Process Control 12, 617–632 (2002)
91. Nagy, Z., Braatz, R.D.: Robust nonlinear model predictive control of batch processes. AIChE Journal 49, 1776–1786 (2003)
92. Nagy, Z., Mahn, B., Franke, R., Allgöwer, F.: Real-time implementation of nonlinear model predictive control of batch processes in an industrial framework. In: Findeisen, R., Allgöwer, F., Biegler, L.T. (eds.) Assessment and Future Directions of Nonlinear Model Predictive Control. LNCIS, vol. 358, pp. 465–472. Springer, Heidelberg (2007)
93. Nicolao, G.D., Magni, L., Scattolini, R.: Stability and robustness of nonlinear receding horizon control. In: Allgöwer, F., Zheng, A. (eds.) Nonlinear Predictive Control, pp. 3–23. Birkhäuser (2000)
94. Nijmeijer, H., van der Schaft, A.J.: Nonlinear dynamical control systems. Springer, New York (1990)
95. Nørgaard, M., Ravn, O., Poulsen, N.K., Hansen, L.K.: Neural networks for modelling and control of dynamic systems. Springer, London (2000)
96. Nocedal, J., Wright, S.J.: Numerical optimization. Springer, New York (1999)
97. Ohtsuka, T.: A continuation/GMRES method for fast computation of nonlinear receding horizon control. Automatica 40, 563–574 (2004)
98. Pannocchia, G., Bemporad, A.: Combined design of disturbance model and observer for offset-free model predictive control. IEEE Trans. Automatic Control 52, 1048–1053 (2007)
99. Pannocchia, G., Rawlings, J.B.: Disturbance models for offset-free model predictive control. AIChE J. 49, 426–437 (2003)
100. Qin, S.J., Badgwell, T.A.: An overview of industrial model predictive control technology, preprint CPC-V, Lake Tahoe (1996)
101. Qin, S.J., Badgwell, T.A.: An overview of nonlinear model predictive control applications. In: Nonlinear Predictive Control, pp. 369–392. Springer (2000)
102. Rawlings, J.B.: Tutorial overview of model predictive control. IEEE Contr. Syst. Magazine 20, 38–52 (2000)
103. Roset, B.J.P., Lazar, M., Heemels, W.P.M.H., Nijmeijer, H.: A stabilizing output based nonlinear model predictive control scheme. In: Proceedings of the 45th IEEE Conference on Decision and Control, San Diego, USA, pp. 4627–4632 (2006)
104. Saberi, A., Han, J., Stoorvogel, A.A.: Constrained stabilization problems for linear plants. Automatica 38, 639–654 (2002)
105. Sargent, R.W.H., Sullivan, G.R.: The development of an efficient optimal control package. In: Proceedings of the 8th IFIP Conference on Optimization Techniques. Springer, Heidelberg (1977)
106. Scattolini, R.: Architectures for distributed and hierarchical model predictive control - a review. Journal of Process Control 19, 723–731 (2009)
107. Schei, T.S.: A finite-difference method for linearization in nonlinear estimation algorithms. Automatica 33, 2053–2058 (1997)

108. Scokaert, P.O.M., Mayne, D.Q., Rawlings, J.B.: Suboptimal model predictive control (feasibility implies stability). IEEE Trans. Automatic Control 44, 648–654 (1999)
109. Sepulchre, R., Jankovic, M., Kokotovic, P.: Constructive nonlinear control. Springer, London (1997)
110. Sequeira, S.E., Graells, M., Luis, P.: Real-time evolution for on-line optimization of continuous processes. Industrial and Engineering Chemistry Research, 1815–1825 (2002)
111. Shahzad, A., Kerrigan, E.C., Constantinides, G.A.: A warm-start interior-point method for predictive control. In: Proceedings of UKACC (2010)
112. Shin, K.G., McKay, N.D.: Minimum-time control of robotic manipulators with geometric path constraints. IEEE Transactions on Automatic Control 30, 531–541 (1985)
113. von Stryk, O.: Numerical solution of optimal control problems by direct collocation. In: Optimal Control. International Series in Numerical Mathematics, vol. 111, pp. 129–143 (1993)
114. Tenny, M.J., Rawlings, J.B., Wright, S.J.: Closed-loop behavior of nonlinear model predictive control. AIChE Journal 50, 2142–2154 (2004)
115. Tøndel, P., Johansen, T.A., Bemporad, A.: Evaluation of piecewise affine control via binary search tree. Automatica 39, 743–749 (2003)
116. Tsang, T.H., Himmelblau, D.M., Edgar, T.F.: Optimal control via collocation and nonlinear programming. Int. J. Control 21, 763–768 (1975)
117. Tyler, M.L., Morari, M.: Propositional logic in control and monitoring problems. Automatica 35, 565–582 (1999)
118. Vada, J., Slupphaug, O., Foss, B.A.: Infeasibility handling in linear MPC subject to prioritized constraints. In: Proceedings of IFAC World Congress, Beijing, vol. D, pp. 163–168 (1999)
119. Wan, Z.Y., Kothare, M.V.: Efficient scheduled stabilizing output feedback model predictive control for constrained nonlinear systems. IEEE Trans. Automatic Control 49, 1172–1177 (2004)
120. Williams, H.P.: Model building in mathematical programming (1999)
121. Wright, S.J.: Primal-dual interior-point methods. SIAM, Philadelphia (1997)
122. Wright, S.J., Tenny, M.J.: A feasible trust-region sequential quadratic programming algorithm. SIAM Journal on Optimization 14, 1074–1105 (2004)
123. Xiong, Q., Jutan, A.: Continuous optimization using a dynamic simplex method. Chemical Engineering Science, 3817–3828 (2003)
124. Zavala, V.M., Biegler, L.T.: The advanced step NMPC controller: Optimality, stability and robustness. Automatica 45, 86–93 (2009)

Chapter 3
Explicit NMPC Using mp-QP Approximations of mp-NLP

Abstract. A numerical algorithm for approximate multi-parametric nonlinear programming (mp-NLP) is developed. The algorithm locally approximates the mp-NLP with a multi-parametric quadratic program (mp-QP). This leads to an approximate mp-NLP solution that is composed from the solution of a number of mp-QP solutions. The method allows approximate solutions to nonlinear optimization problems to be computed as explicit piecewise linear functions of the problem parameters. In control applications such as nonlinear constrained model predictive control this allows efficient online implementation in terms of an explicit piecewise linear state feedback without any real-time optimization.

3.1 Introduction

For multi-parametric nonlinear programs (mp-NLPs) one cannot expect to find exact solutions, in general. There is a large body of theory that develops local regularity conditions and local sensitivity results [7, 17], and algorithms for non-local parameter variations are derived for single-parametric problems [12]. Here we describe an approximate mp-NLP algorithm utilizing NLP and mp-QP algorithms to solve local sub-problems, first proposed in [13].

Before we describe the main idea behind the algorithm, we recall that a widely used family of algorithms for the numerical solution of nonlinear programs (NLPs) is Sequential Quadratic Programming (SQP) algorithms, e.g. [20]. They are iterative algorithms, where at each iteration the nonlinear program is locally approximated by a convex quadratic program (QP) at the current candidate solution point. This means that the nonlinear cost function is locally approximated by a positive definite quadratic function, and the nonlinear constraints are locally approximated by linear constraints. The QP is then solved to find a search direction towards a better point, a step in this direction is made, and the procedure is repeated and will eventually converge to a locally optimal solution for the NLP.

In the approximate mp-NLP algorithm described in this chapter, the idea is to locally approximate mp-NLPs with mp-QPs, similar to the use of QPs within SQP.

A. Grancharova & T.A. Johansen: Explicit Nonlinear Model Predictive Ctrl., LNCIS 429, pp. 71–85.
springerlink.com

An iterative (recursive) partitioning of the parameter space is used to control the accuracy of the approximation. It refines the partition in order to improve the accuracy of the local mp-QP approximation in the parts of the parameter space where this is needed in order to meet accuracy specifications in terms of sub-optimality bounds on the cost.

The proposed method is different from the approximate mp-NLP algorithm in Section 1.1.5, and the function approximation methods for non-linear optimal control are described in [1, 21, 22, 4, 23, 15, 2, 19]. While these references approximate the mp-NLP solution based on solution points computed for an extensive number of parameter values using an NLP algorithm, in the present chapter the mp-NLP is approximated by a number of mp-QPs that are solved using the mp-QP algorithm [28]. In [5], several alternative multi-parametric programming algorithms for explicit approximate solution of convex mp-NLP problems are compared, and a modification of the algorithm described in this section was found to be efficient. The main modifications is a different approach for the partitioning outside the mp-QP solutions.

The mp-NLP problem is formulated as follows:

$$\min_{z} V(z,x) \tag{3.1}$$

subject to

$$G(z,x) \leq 0 \tag{3.2}$$

for all $x \in X$, where X is some parameter set. Eqs. (3.1) - (3.2) define an mp-NLP, since it is an NLP in z parameterized by the parameter vector x. Assume the solution exists, and let it be denoted $z^*(x)$. In the special case when V and G are quadratic and linear, respectively, in both z and x, a solution can be found explicitly and exactly as a continuous PWL mapping $z^*(x)$ using mp-QP.

In [13] it is suggested to utilize an mp-QP algorithm to approximately solve the mp-NLP (3.1)-(3.2). In the mp-QP case, this algorithm will iteratively build a polyhedral partition of the state-space with an exact solution corresponding to a fixed active set within each polyhedral critical region. This leads to a PWL solution $z^*(x)$ since a fixed active set leads to a solution that is linear in x, [3]. In the mp-NLP case we keep the PWL structure of the solution, but in each polyhedral region we approximate the (exact) nonlinear solution by a PWL approximate solution found by solving a mp-QP constructed as a locally accurate quadratic approximation to V and linear approximation to G. Under regularity assumptions on V and G, one may expect that the approximation error and constraint violations will be small if each of the regions are sufficiently small. We therefore suggest to analyze the approximation error within each region and introduce a sub-partitioning of some regions when needed in order to keep the approximation error and constraint violations within specified bounds.

3.2 Local mp-QP Approximation to mp-NLP

In this section we study how the cost function and constraints can be locally approximated by mp-QP problems, based on [13]. Let $x_0 \in X$ be arbitrary and denote the corresponding optimal solution $z_0 = z^*(x_0)$. Taylor series expansions of V and G about the point (z_0, x_0) leads to the following locally approximate mp-QP problem:

$$V_0(z,x) \triangleq \frac{1}{2}(z-z_0)^T H_0(z-z_0) + (D_0 + F_0(x-x_0))(z-z_0) + Y_0(x) \quad (3.3)$$

subject to

$$G_0(z-z_0) \leq E_0(x-x_0) + T_0 \quad (3.4)$$

The cost and constraints are defined by the matrices

$$\begin{aligned} H_0 &\triangleq \nabla^2_{zz} V(z_0,x_0), \quad F_0 \triangleq \nabla^2_{xz} V(z_0,x_0) \\ D_0 &\triangleq \nabla_z V(z_0,x_0), \quad G_0 \triangleq \nabla_z G(z_0,x_0) \\ E_0 &\triangleq -\nabla_x G(z_0,x_0), \quad T_0 \triangleq -G(z_0,x_0) \\ Y_0(x) &\triangleq V(z_0,x_0) + \nabla_x V(z_0 z x_0)(x-x_0) + \tfrac{1}{2}(x-x_0)^T \nabla^2_{xx} V(z_0,x_0)(x-x_0) \end{aligned}$$

Let the PWL solution to the mp-QP (3.3) - (3.4) be denoted $z_{QP}(x)$ with associated Lagrange multipliers $\lambda_{QP}(x)$. This solution satisfies the following KKT conditions

$$\begin{aligned} H_0\,(z_{QP}(x)-z_0) + F_0(x-x_0) + D_0 + G_0^T \lambda_{QP}(x) &= 0 \quad (3.5) \\ \mathrm{diag}(\lambda_{QP}(x))\,(G_0(z_{QP}(x)-z_0) - E_0(x-x_0) - T_0) &= 0 \quad (3.6) \\ \lambda_{QP}(x) &\geq 0 \quad (3.7) \\ G_0\,(z_{QP}(x)-z_0) - E_0(x-x_0) - T_0 &\leq 0 \quad (3.8) \end{aligned}$$

Consider the optimal active set $\mathscr{A}$ of the QP (3.3) - (3.4) at a given $x \in X$, and let $G_{0,\mathscr{A}}$ and $\lambda_{QP,\mathscr{A}}$ denote the rows of G_0 and λ_{QP}, respectively, with indices in $\mathscr{A}$. Eqs. (3.5) - (3.6) define the following linear equations

$$\begin{pmatrix} H_0 & G_{0,\mathscr{A}}^T \\ G_{0,\mathscr{A}} & 0 \end{pmatrix} \begin{pmatrix} z_{QP,\mathscr{A}}(x) - z_0 \\ \lambda_{QP,\mathscr{A}}(x) \end{pmatrix} = \begin{pmatrix} F_0(x-x_0) + D_0 \\ E_0(x-x_0) + T_0 \end{pmatrix} \quad (3.9)$$

The following results is an extension of Theorem 2 in [3] (where $H_0 > 0$ was assumed in addition to LICQ).

Assumption 3.1. *V and G are twice continuously differentiable in a neighborhood of* (z_0, x_0).

Assumption 3.2. *The sufficient conditions (1.7)-(1.10) and (1.12) for a local minimum at* z_0 *hold.*

Assumption 3.3. *Linear independence constraint qualification (LICQ) holds, i.e. the active constraint gradients $\nabla_z G_{\mathscr{A}_0}(z_0, x_0)$ are linearly independent.*

Assumption 3.4. *Strict complementary slackness holds, i.e. $(\lambda_0)_{\mathscr{A}_0} > 0$.*

Assumption 3.5. *For an optimal active set $\mathscr{A}$, the matrix $G_{0,\mathscr{A}}$ has full row rank (LICQ) and $Z_{0,\mathscr{A}}^T H_0 Z_{0,\mathscr{A}} > 0$, where the columns of $Z_{0,\mathscr{A}}$ is a basis for null$(G_{0,\mathscr{A}})$.*

Theorem 3.1. *Consider the problem (3.3)-(3.4), and let X be a polyhedral set with $x_0 \in X$. The system of linear equations (3.9) has a unique solution*

$$\begin{pmatrix} z_{QP,\mathscr{A}}(x) - z_0 \\ \lambda_{QP,\mathscr{A}}(x) \end{pmatrix} = \begin{pmatrix} H_0 & G_{0,\mathscr{A}}^T \\ G_{0,\mathscr{A}} & 0 \end{pmatrix}^{-1} \begin{pmatrix} F_0^T(x - x_0) + D_0 \\ E_0(x - x_0) + T_0 \end{pmatrix} \tag{3.10}$$

and the critical region where the solution is optimal is given by the polyhedral set

$$\mathscr{X}_{0,\mathscr{A}} \triangleq \left\{ x \in X \mid \lambda_{QP,\mathscr{A}}(x) \geq 0, G_0(z_{QP,\mathscr{A}}(x) - z_0) \leq E_0(x - x_0) + T_0 \right\}$$

Hence, $z_{QP}(x) = z_{QP,\mathscr{A}}(x)$ and $\lambda_{QP}(x) = \lambda_{QP,\mathscr{A}}(x)$ if $x \in \mathscr{X}_{0,\mathscr{A}}$, and the solution z_{QP} is a continuous, PWL function of x defined on a polyhedral partition of X.

Proof ([13]). Non-singularity of the matrix on the left-hand-side of (3.9) follows from standard 2nd order considerations such as Lemma 16.1 in [20], due to Assumption 3.5. The rest of the proof is similar to [3]. □

Algorithms for solving such an mp-QP (with straightforward modifications to account for the relaxed second-order condition of Assumption 3.5) are given in Section 1.2. The following result compares the primal and dual local QP solution with the global NLP solution.

Theorem 3.2. *Consider the problem (3.1)-(3.2). Let $x_0 \in X$ and suppose there exists a z_0 satisfying the above assumptions. Then for x in a neighborhood of x_0*

$$z_{QP}(x) - z^*(x) = \mathscr{O}(||x - x_0||_2^2) \tag{3.11}$$

$$\lambda_{QP}(x) - \lambda^*(x) = \mathscr{O}(||x - x_0||_2^2) \tag{3.12}$$

Proof ([13]). Let the neighborhood of x_0 under consideration be restricted to $\mathscr{X}_{0,\mathscr{A}_0}$, where $\mathscr{A}_0$ is the optimal active set at x_0. This is without loss of generality since the assumptions imply that x_0 is an interior point in $\mathscr{X}_{0,\mathscr{A}_0}$. The first KKT condition for the QP is

$$H_0\left(z_{QP}(x) - z_0\right) + F_0(x - x_0) + \left(D_0 + G_0^T \lambda_{QP}(x)\right) = 0 \tag{3.13}$$

Since $z_0 = z^*(x_0)$ we have $z^*(x) - z_0 = \mathscr{O}(||x - x_0||_2)$, and the first KKT condition (1.7) for the NLP can be rewritten as follows using a Taylor series expansion

$$0 = \nabla_z V(z^*(x),x) + \nabla_z^T G(z^*(x),x)\lambda^*(x) \tag{3.14}$$
$$= \nabla_z V(z_0,x_0) + \nabla_{zz}^2 V(z_0,x_0)(z^*(x)-z_0) + \nabla_{xz}^2 V(z_0,x_0)(x-x_0) + \left(\nabla_z^T G(z_0,x_0) + \mathscr{O}(||x-x_0||_2)\right)\lambda^*(x) + \mathscr{O}(||x-x_0||_2^2) \tag{3.15}$$
$$= D_0 + H_0(z^*(x)-z_0) + F_0(x-x_0) + G_0^T\lambda_{QP}(x) + G_0^T(\lambda^*(x)-\lambda_{QP}(x)) + \mathscr{O}(||x-x_0||_2^2) + \mathscr{O}(||x-x_0||_2)(\lambda^*(x)-\lambda_{QP}(x)) \tag{3.16}$$

Comparing (3.13) and (3.16) we get

$$H_0\left(z_{QP}(x)-z^*(x)\right) + G_0^T\left(\lambda_{QP}(x)-\lambda^*(x)\right) = \mathscr{O}(||x-x_0||_2^2) \tag{3.17}$$

From Theorem 1.1, part 3, it is known that the set of active constraints is unchanged in a neighborhood of x_0. Hence, for the QP we have

$$G_0\left(z_{QP}(x)-z_0\right) = E_0(x-x_0) + T_0 \tag{3.18}$$

When x is in a neighborhood of x_0, Taylor expanding the NLP constraints gives

$$0 = G(z^*(x),x) \tag{3.19}$$
$$= G(z_0,x_0) + \nabla_z G(z_0,x_0)(z^*(x)-z_0) + \nabla_x G(z_0,x_0)(x-x_0) + \mathscr{O}(||x-x_0||_2^2)$$
$$= G_0(z^*(x)-z_0) - E_0(x-x_0) - T_0 + \mathscr{O}(||x-x_0||_2^2)$$

Comparing (3.18) and (3.19) it follows that

$$G_0\left(z_{QP}(x)-z^*(x)\right) = \mathscr{O}(||x-x_0||^2) \tag{3.20}$$

and the result follows by inverting the system (3.17) and (3.20). This system is indeed invertible: Due to Assumption 3.4 it follows that $\nabla_z G_{\mathscr{A}_0}(z_0,x_0)\zeta = 0$ for all $\zeta \in \mathscr{F}$. Since $G_{0,\mathscr{A}_0} = \nabla_z G_{\mathscr{A}_0}(z_0,x_0)$, it is clear that $\mathscr{F} = \mathrm{null}(G_{0,\mathscr{A}_0})$ and Assumptions 3.2 and 3.3 (and in particular eq. (1.12)) ensures that Assumption 3.5 holds and non-singularity of

$$\begin{pmatrix} H_0 & G_0^T \\ G_0 & 0 \end{pmatrix}$$

follows from Lemma 16.1 in [20]. □

Theorem 3.2 concerns only a small neighborhood of x_0 and is therefore of limited computational use. However, it provides a qualitative indication that the mp-QP approximation of the mp-NLP is locally accurate, under some assumptions. We therefore proceed by deriving some quantitative estimates and bounds on the cost and solution errors, as well as the maximum constraint violation. The solution error bound is defined as

$$\rho \triangleq \max_{x \in X_0} |w^T(\mu(0, z_{QP}(x)) - \mu(0, z^*(x)))| \tag{3.21}$$

where $X_0 \subset X$ is arbitrary, and w is a vector with positive weights. Likewise, we define the cost error bound

$$\varepsilon \triangleq \max_{x \in X_0} |V(z_{QP}(x), x) - V^*(x)| \tag{3.22}$$

where $V^*(x) \triangleq V(z^*(x); x)$. In addition, one may compute the maximum constraint violation

$$\delta \triangleq \max_{x \in X_0} \omega^T G(z_{QP}(x), x) \tag{3.23}$$

where ω is a vector of non-negative weights. Typically, the elements of w corresponding to the first sample of the trajectory will be positive, while the remaining will be zero since in receding horizon control the primary interest is the first sample of the trajectory. The maximum constraint violation (3.23) can be computed by solving an NLP, while the solution and cost error bounds (3.21) and (3.22) are not easily computed without introducing additional assumptions or allowing underestimation. A further problem is that they require computation of the exact $z^*(x)$ for several x, which relies on the solution of several NLPs and is therefore expensive. Obvious estimation techniques for ρ and ε is to take the maximum over a finite number of points X_0, such as extreme points (vertices), points generated by Monte Carlo methods, or combinations. It should be emphasized that these methods can underestimate the bounds, in general.

3.3 Convexity

For the case when V and G are convex functions, it is possible to derive a guaranteed bound on ε from knowledge of $z^*(x)$ only at all the vertices $\mathscr{V} = \{v_1, v_2, ..., v_M\}$ of the bounded polyhedron X_0, see section 1.1.5.1. This immediately gives the following bounds on the cost function error $-\varepsilon_1 \leq V^*(x) - V(z_{QP}(x), x) \leq \varepsilon_2$, where

$$\varepsilon_1 = \max_{x \in X_0} (V(z_{QP}(x), x) - \underline{V}(x)) \tag{3.24}$$

$$\varepsilon_2 = \max_{x \in X_0} \left(\overline{V}(x) - V(z_{QP}(x), x)\right) \tag{3.25}$$

Hence, the cost error bound $\tilde{\varepsilon} \triangleq \max(\varepsilon_1, \varepsilon_2) \geq \varepsilon$ can be computed by solving two NLPs. A solution error bound can be shown to exist as in Chapter 9.7 of [7].

3.4 Algorithm

So far it has been established that under some regularity conditions, local mp-QP solutions give accurate approximation to the mp-NLP solution when restricted to a sufficiently small subset $X_0 \subset X$. It remains to determine a sub-partition of the polyhedral region X such that the local mp-QP solutions associated with each region are sufficiently accurate. In [13] the following algorithm was suggested to approximate the mp-NLP solution, based on recursive sub-partitioning guided by the approximation errors discussed above.

Algorithm 3.1. Approximate mp-NLP.

Step 1. Let $X_0 := X$.
Step 2. Select x_0 as the Chebychev center of X_0, by solving an LP.
Step 3. Compute $z_0 = z^*(x_0)$ by solving the NLP (3.1)-(3.2) with $x(0) = x_0$.
Step 4. Compute the local mp-QP problem (3.3) - (3.4) at (z_0, x_0). If H_0 is not positive definite, then modify H_0 such that it is positive definite (e.g. by an eigenvalue decomposition where negative eigenvalues are replaced by small positive numbers) and the mp-QP is convex.
Step 5. Estimate the approximation errors ε, ρ and δ on X_0.
Step 6. If $\varepsilon > \overline{\varepsilon}$, $\rho > \overline{\rho}$, or $\delta > \overline{\delta}$, then sub-partition X_0 into polyhedral regions using the heuristic rules described in Section 1.1.5.2.
Step 7. Select a new X_0 from the partition. If no further sub-partitioning is needed, go to step 8. Otherwise, repeat Steps 2-7 until the tolerances $\overline{\varepsilon}$, $\overline{\rho}$ and $\overline{\delta}$ are respected in all polyhedral regions in the partition of X.
Step 8. For all sub-partitions X_0, solve the mp-QP (3.3) - (3.4) using the mp-QP solver [28, 26].

Computation of the approximation errors in Step 5 are carried out based on the results in Section 3.3 if the cost function and constraints are known to be convex. If not, we suggest to estimate error bounds by solving NLPs at a number of points in X_0, typically the vertices and possibly other points, as in [10]. If the convexity assumption does not hold, this seems to be a fairly robust strategy. The sub-partitioning in Step 6 is based on heuristic criteria, where the purpose is to select one axis-orthogonal hyperplane to split X_0 such that the approximation error after splitting is minimized (as described in Section 1.1.5.2). Alternatively, the hyperplane is selected such that the change of error at the vertices (before splitting) across the hyperplane is maximal (as used in [9]).

3.5 Example: Compressor Surge Control

3.5.1 NMPC Formulation

Consider the following 2nd-order compressor model [11, 8] with x_1 being normalized mass flow, x_2 normalized pressure and u normalized mass flow through a close coupled valve in series with the compressor

$$\dot{x}_1 = B(\Psi_e(x_1) - x_2 - u) \tag{3.26}$$
$$\dot{x}_2 = \frac{1}{B}(x_1 - \Phi(x_2)) \tag{3.27}$$

The following compressor and valve characteristics are used

$$\Psi_e(x_1) = \psi_{c0} + H\left(1 + 1.5\left(\frac{x_1}{W} - 1\right) - 0.5\left(\frac{x_1}{W} - 1\right)^3\right)$$
$$\Phi(x_2) = \gamma \mathrm{sign}(x_2)\sqrt{|x_2|}$$

with $\gamma = 0.5$, $B = 1$, $H = 0.18$, $\psi_{c0} = 0.3$ and $W = 0.25$. The control objective is to avoid surge, i.e. stabilize the system. This may be formulated as

$$J(u[0,T], x[0,T]) \triangleq \int_0^T l(x(t), u(t), t)dt + S(x(T), T) + Rv^2 \tag{3.28}$$

where

$$l(x,u) = \alpha(x - x^*)^T(x - x^*) + \kappa u^2$$
$$S(x) = \beta(x - x^*)^T(x - x^*)$$

with $\alpha, \beta, \kappa, \rho \geq 0$ and the setpoint $x_1^* = 0.40$, $x_2^* = 0.60$ corresponds to an unstable equilibrium point, subject to the inequality constraints for $t \in [0,T]$

$$u_{min} \leq u(t) \leq u_{max} \tag{3.29}$$
$$-x_2 + 0.4 \leq v \tag{3.30}$$
$$-v \leq 0 \tag{3.31}$$

and the ordinary differential equation (ODE) given by

$$\frac{d}{dt}x(t) = f(x(t), u(t)) \tag{3.32}$$

with given initial condition $x(0) \in X \subset \mathbb{R}^n$. Valve capacity requires the constraint $0 \leq u(t) \leq 0.3$ to hold, and the pressure constraint $x_2 \geq 0.4 - v$ avoids operation too far left of the operating point. The variable $v \geq 0$ is a slack variable introduced in order to avoid infeasibility and $R = 8$ is a large weight. The input signal $u[0,T]$ is assumed to be piecewise constant and parameterized by a vector $U \in \mathbb{R}^p$ such that $u(t) = \mu(t, U) \in \mathbb{R}^r$ is piecewise constant. The solution to (3.32) is assumed in the form $x(t) = \phi(t, U, x(0))$ for $t \in [0,T]$ and some piecewise continuous function ϕ. Relaxing the inequality constraints (3.30) to hold only at N time instants $\{t_1, t_2, ..., t_N\} \subset [0,T]$, we can rewrite the optimization problem in the following standard parametric form (direct single shooting, Section 2.2.2.1) where the ODE constraint (3.32) has been eliminated by substituting its solution ϕ into the cost and constraints; minimize with respect to $z = (U, v)$ the cost

$$V(z,x(0)) \triangleq \int_0^T l(\phi(t,U,x(0)),\mu(t,U),t)dt + S(\phi(T,U,x(0)),T) + Rv^2 \tag{3.33}$$

subject to

$$G(z,x(0)) \triangleq \begin{pmatrix} \tilde{G}(U;x(0)) \\ U - U_{max} \\ U_{min} - U \\ -v \end{pmatrix} \leq 0 \tag{3.34}$$

with blocks $\tilde{G}_i(U;x(0)) \triangleq \hat{g}(\phi(t_i,U,x(0)),\mu(t_i,U))$ as defined by (3.30). Eqs. (3.33) - (3.34) define an mp-NLP, since it is an NLP in z parameterized by the initial state vector $x(0)$.

3.5.2 *Tuning and Settings*

We have chosen $\alpha = 1$, $\beta = 0$, and $\kappa = 0.08$. The horizon is chosen as $T = 12$, which is split into $N = p = 15$ equal-sized intervals, leading to a piecewise constant control input parameterization. Numerical analysis of the cost function shows that it is non-convex. It should be remarked that the constraints on u and v are linear, such that any mp-QP solution is feasible for the mp-NLP. The bounds ε and ρ are estimates by computing the errors at the vertices only, and the tolerances $\overline{\varepsilon} = 0.5$ and $\overline{\rho} = 0.03$ were applied.

3.5.3 *Results*

The mp-NLP contains 16 free variables, 47 constraints and 2 parameters. The partition contains 379 regions, resulting from 45 mp-QPs, cf. Fig. 3.1. This can be reduced to 101 polyhedral regions without loss of accuracy in a postprocessing step, where regions with the same solution at the first sample are joined whenever their union remains polyhedral, as in [3]. The computed approximate PWL feedback is shown in Fig. 3.2, together with the exact feedback computed by solving the NLP on a dense grid. The corresponding optimal costs are shown in Fig. 3.3, and simulation results are shown in Fig. 3.4, where the controller is switched on after $t = 20$. We note that it quickly stabilizes the deep surge oscillations. Euler integration with step size 0.02 is applied to solve the ODE.

By generating a search tree using the method of [27], the PWL mapping with 379 regions can be represented as a binary search tree with 329 nodes, of depth 9. Real-time evaluation of the controller therefore requires 49 arithmetic operations, in the worst case, and 1367 numbers needs to be stored in real-time computer memory.

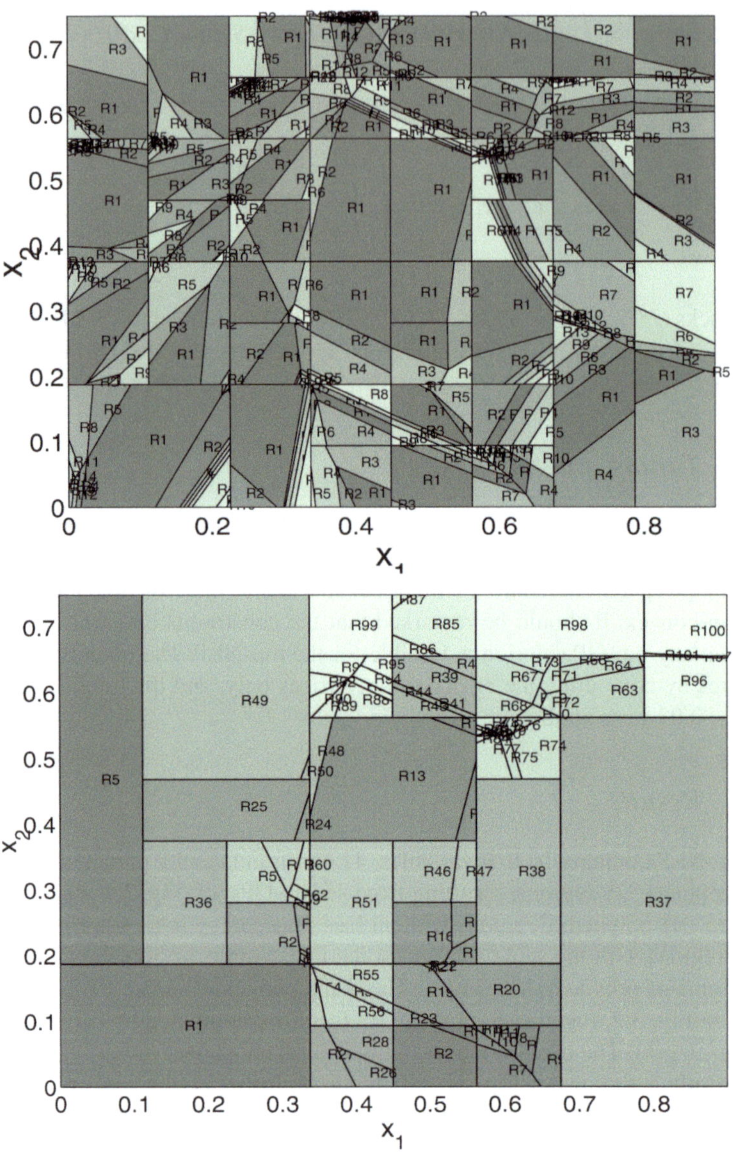

Fig. 3.1 State space partition (top), and after reduction (bottom).

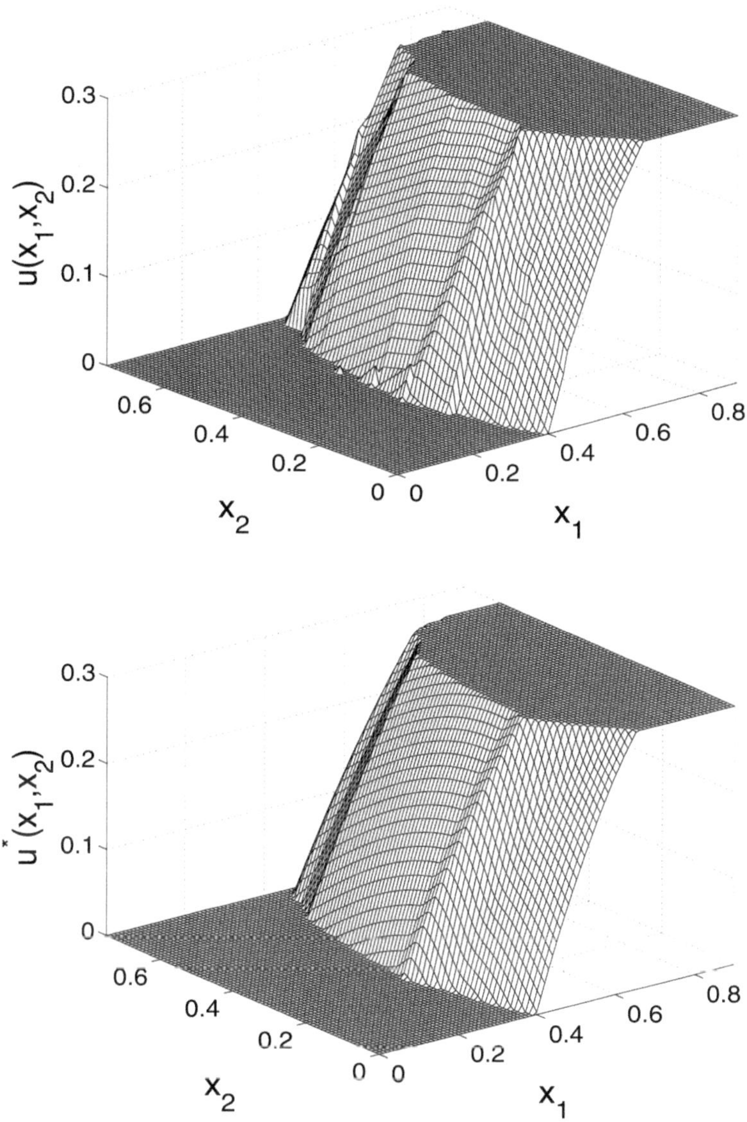

Fig. 3.2 Piecewise linear approximate feedback control law (top) and exact feedback control law (bottom).

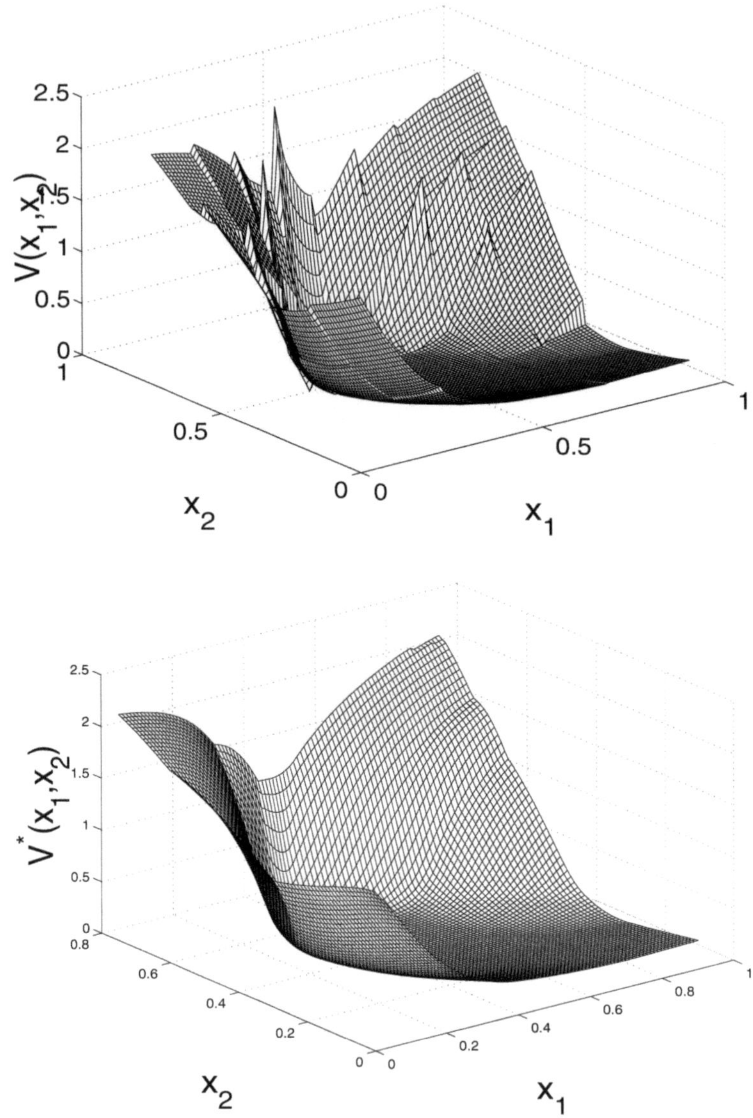

Fig. 3.3 Optimal costs of the approximate feedback control law (top) and exact feedback control law (bottom).

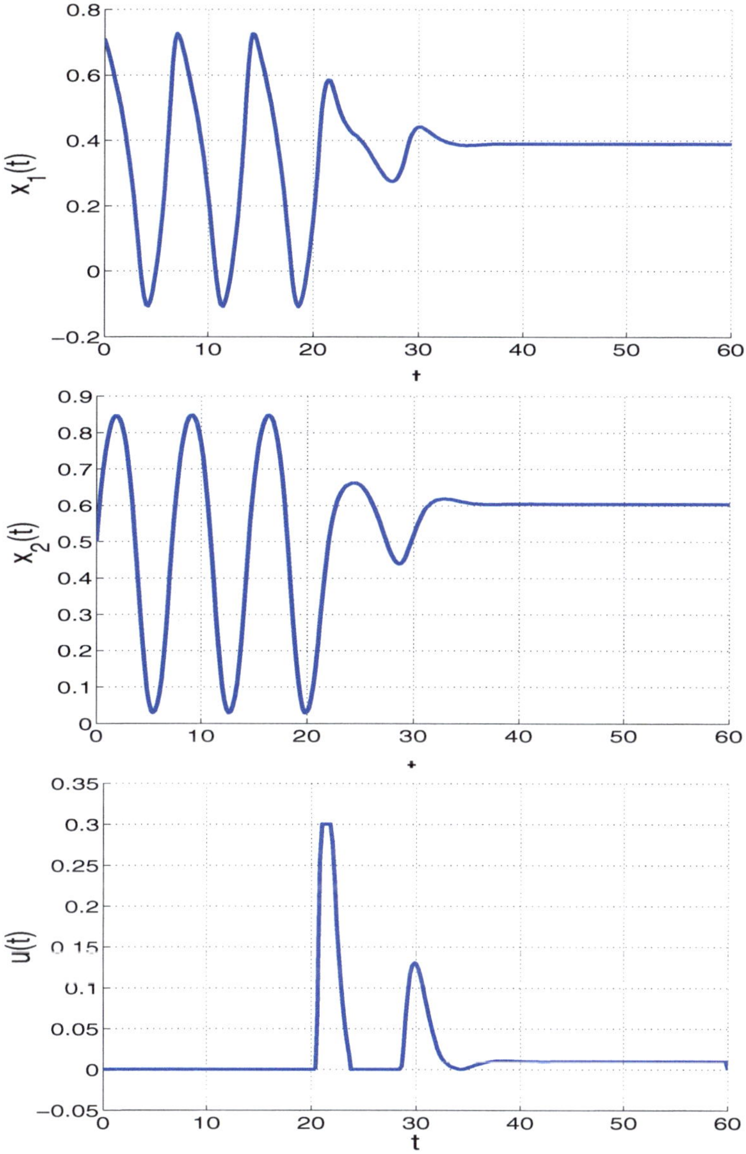

Fig. 3.4 Simulation of compressor with approximate explicit nonlinear MPC. The solution with the exact explicit MPC cannot be distinguished graphically.

References

1. Bemporad, A., Filippi, C.: Suboptimal explicit RHC via approximate multiparametric quadratic programming. J. Optimization Theory and Applications 117, 9–38 (2003)
2. Bemporad, A., Filippi, C.: An algorithm for approximate multiparametric convex programming. Computational Optimization and Applications 35, 87–108 (2006)
3. Bemporad, A., Morari, M., Dua, V., Pistikopoulos, E.N.: The explicit linear quadratic regulator for constrained systems. Automatica 38, 3–20 (2002)
4. Bertsekas, D.P., Tsitsiklis, J.N.: Neuro-dynamic programming. Athena Scientific, Belmont (1998)
5. Domínguez, L.F., Narciso, D.A., Pistikopoulos, E.N.: Recent advances in multiparametric nonlinear programming. Computers and Chemical Engineering 34, 707–716 (2010)
6. Domínguez, L.F., Pistikopoulos, E.N.: Quadratic approximation algorithm for multiparametric nonlinear programming problems. Technical report. Imperial College London (2009)
7. Fiacco, A.V.: Introduction to sensitivity and stability analysis in nonlinear programming. Academic Press, Orlando (1983)
8. Gravdahl, J.T., Egeland, O.: Compressor surge control using a close-coupled valve and backstepping. In: Proceedings of the American Control Conference, Albuquerque, NM, vol. 2, pp. 982–986 (1997)
9. Grancharova, A., Johansen, T.A.: Approximate explicit model predictive control incorporating heuristics. In: Proceedings of IEEE International Symposium on Computer Aided Control System Design, Glasgow, Scotland, U.K., pp. 92–97 (2002)
10. Grancharova, A., Johansen, T.A., Tøndel, P.: Computational aspects of approximate explicit nonlinear model predictive control. In: Findeisen, R., Allgöwer, F., Biegler, L. (eds.) Assessment and Future Directions of Nonlinear Model Predictive Control. LNCIS, vol. 358, pp. 181–192. Springer, Heidelberg (2007)
11. Greitzer, E.M.: Surge and rotating stall in axial flow compressors, part i: Theoretical compression system model. J. Engineering for Power 98, 190–198 (1976)
12. Guddat, J., Guerra Vazquez, F., Jongen, H.T.: Parametric optimization: Singularities, pathfollowing and jumps. Wiley (1990)
13. Johansen, T.A.: On multi-parametric nonlinear programming and explicit nonlinear model predictive control. In: Proceedings of the IEEE Conference on Decision and Control, Las Vegas, NV, vol. 3, pp. 2768–2773 (2002)
14. Johansen, T.A., Grancharova, A.: Approximate explicit constrained linear model predictive control via orthogonal search tree. IEEE Trans. Automatic Control 48, 810–815 (2003)
15. Johansen, T.A.: Approximate explicit receding horizon control of constrained nonlinear systems. Automatica 40, 293–300 (2004)
16. Kojima, M.: Strongly stable stationary solutions in nonlinear programs. In: Robinson, S.M. (ed.) Analysis and Computation of Fixed Points, pp. 93–138. Academic Press, London (1980)
17. Levitin, E.S.: Perturbation theory in mathematical programming. Wiley (1994)
18. Mangasarian, O.L., Rosen, J.B.: Inequalities for stochastic nonlinear programming problems. Operations Research 12, 143–154 (1964)
19. Narciso, D.: Developments in nonlinear multiparametric programming and control. PhD thesis. London, U.K. (2009)
20. Nocedal, J., Wright, S.J.: Numerical optimization. Springer, New York (1999)
21. Parisini, T., Zoppoli, R.: A receding-horizon regulator for nonlinear systems and a neural approximation. Automatica 31, 1443–1451 (1995)

22. Parisini, T., Zoppoli, R.: Neural approximations for multistage optimal control of nonlinear stochastic systems. IEEE Trans. on Automatic Control 41, 889–895 (1996)
23. Pistikopoulos, E.N., Georgiadis, M.C., Dua, V.: Multi-parametric programming: Theory, algorithms, and applications. Wiley-VCH (2007)
24. Ralph, D., Dempe, S.: Directional derivatives of the solution of a parametric nonlinear program. Mathematical Programming 70, 159–172 (1995)
25. Rockafellar, R.T.: Convex analysis. Princeton University Press, New Jersey (1970)
26. Spjøtvold, J., Kerrigan, E.C., Jones, C.N., Tøndel, P., Johansen, T.A.: On the facet-to-facet property of solutions to convex parametric quadratic programs. Automatica 42, 2204–2209 (2006)
27. Tøndel, P., Johansen, T.A., Bemporad, A.: Evaluation of piecewise affine control via binary search tree. Automatica 39, 743–749 (2003)
28. Tøndel, P., Johansen, T.A., Bemporad, A.: An algorithm for multi-parametric quadratic programming and explicit MPC solutions. Automatica 39, 489–497 (2003)

Chapter 4
Explicit NMPC via Approximate mp-NLP

Abstract. This chapter considers the design of explicit MPC controllers for constrained nonlinear systems by applying the approximate multi-parametric Nonlinear Programming (mp-NLP) approaches, described in Section 1.1. It is organized as follows. In Section 4.2, general regulation and reference tracking NMPC problems are formulated and represented as an mp-NLP problem. In Section 4.3, it is shown that bounding the approximation error of the approximate explicit solution to convex regulation NMPC problems ensures the asymptotic stability of the suboptimal closed-loop system. In Section 4.4, an explicit approximate NMPC for compressor surge regulation is designed. In Section 4.5, the approximate mp-NLP approach is applied to design an explicit reference tracking NMPC controller for position control of an electropneumatic clutch actuator. The performance of the explicit NMPC controller is compared with that of a sliding mode controller and of a PID controller. Section 4.6 briefly discusses an explicit NMPC application to time-optimal decompression of divers for implementation in low cost diving computers.

4.1 Introduction

The main objective of this chapter is to present some examples on how to use the approximate mp-NLP algorithms of Section 1.1.5 together with various NMPC formulations, as described in Chapter 2. Formulations and conditions that guarantee stability of the closed loop system may require some attention, since the control strategy will be sub-optimal due to the approximation errors of the mp-NLP algorithms.

The main emphasis in this chapter is on case studies that present mathematical models, NMPC formulations, mp-NLP computational results, and closed loop simulations. The case studies are taken from diverse areas such as automotive mechatronics, compressor control, and diving computers. They share common features such as safety-critical requirements in combination with very limited computational resources per updated control computation. While the mechatronics and compressor control examples are relatively fast dynamics that require fast sampling, the diving

A. Grancharova & T.A. Johansen: Explicit Nonlinear Model Predictive Ctrl., LNCIS 429, pp. 87–110.
springerlink.com

computers are extremely inexpensive devices with very low computational capabilities although the update frequencies are not very high.

4.2 Formulation of the NMPC Problem as an mp-NLP Problem

Consider the discrete-time nonlinear system:

$$x(t+1) = f(x(t), u(t)) \tag{4.1}$$

$$y(t) = Cx(t) \tag{4.2}$$

where $x(t) \in \mathbb{R}^n$, $u(t) \in \mathbb{R}^m$, and $y(t) \in \mathbb{R}^p$ are the state, input and output variable, and $f : \mathbb{R}^n \times \mathbb{R}^m \to \mathbb{R}^n$ is a nonlinear function. It is supposed that a full measurement of the state $x(t)$ is available at the current time t.

First, consider the optimal *regulation* problem where the goal is to steer the system state to the origin. For the current $x(t)$, the regulation NMPC solves the optimization problem:

Problem 4.1:

$$V^*(x(t)) = \min_U J(U, x(t)) \tag{4.3}$$

subject to $x_{t|t} = x(t)$ and:

$$y_{\min} \leq y_{t+k|t} \leq y_{\max},\ k = 1, \ldots, N \tag{4.4}$$

$$u_{\min} \leq u_{t+k} \leq u_{\max},\ k = 0, 1, \ldots, N-1 \tag{4.5}$$

$$x_{t+N|t} \in \Omega \tag{4.6}$$

$$x_{t+k+1|t} = f(x_{t+k|t}, u_{t+k}),\ k \geq 0 \tag{4.7}$$

$$y_{t+k|t} = Cx_{t+k|t},\ k \geq 0 \tag{4.8}$$

with $U = [u_t, u_{t+1}, \ldots, u_{t+N-1}]$ and the cost function given by:

$$J(U, x(t)) = \sum_{k=0}^{N-1} \left[\|x_{t+k|t}\|_{Q_x}^2 + \|u_{t+k}\|_R^2 \right] + \|x_{t+N|t}\|_{P_x}^2 \tag{4.9}$$

Here, N is a finite horizon, P_x, Q_x, R are weighting matrices, and $\Omega \subset \mathbb{R}^n$ is a terminal set. The following assumptions are made:

Assumption 4.1. $P_x, Q_x, R \succ 0$.

Assumption 4.2. $y_{\min} < 0 < y_{\max}$ *and* $u_{\min} < 0 < u_{\max}$.

Assumption 4.3. *The function* $f : \mathbb{R}^n \times \mathbb{R}^m \to \mathbb{R}^n$ *is twice continuously differentiable, with* $f(0,0) = 0$.

The compact and convex terminal set Ω is defined by:

$$\Omega = \{x \in \mathbb{R}^n |\ x^T P_x x \leq \alpha\} \tag{4.10}$$

where $\alpha > 0$ is specified in Section 4.3.

Now, consider the optimal *reference tracking* problem where the goal is to have the output vector $y(t)$ track the reference signal $r(t) \in \mathbb{R}^p$. For the current $x(t)$, the reference tracking NMPC solves the optimization problem:

Problem 4.2:

$$V^*(x(t), r(t), u(t-1)) = \min_U J(U, x(t), r(t), u(t-1)) \tag{4.11}$$

subject to $x_{t|t} = x(t)$ and:

$$y_{\min} \leq y_{t+k|t} \leq y_{\max},\ k = 1, \ldots, N \tag{4.12}$$
$$u_{\min} \leq u_{t+k} \leq u_{\max},\ k = 0, 1, \ldots, N-1 \tag{4.13}$$
$$\Delta u_{\min} \leq \Delta u_{t+k} \leq \Delta u_{\max},\ k = 0, 1, \ldots, N-1 \tag{4.14}$$
$$\|y_{t+N|t} - r(t)\| \leq \delta \tag{4.15}$$
$$\Delta u_{t+k} = u_{t+k} - u_{t+k-1},\ k = 0, 1, \ldots, N-1 \tag{4.16}$$
$$x_{t+k+1|t} = f(x_{t+k|t}, u_{t+k}),\ k \geq 0 \tag{4.17}$$
$$y_{t+k|t} = Cx_{t+k|t},\ k \geq 0 \tag{4.18}$$

with $U = [u_t, u_{t+1}, \ldots, u_{t+N-1}]$ and the cost function:

$$J(U, x(t), r(t), u(t-1)) = \sum_{k=0}^{N-1} \left[\|y_{t+k|t} - r(t)\|_{Q_y}^2 + \|\Delta u_{t+k}\|_R^2 \right] + \|y_{t+N|t} - r(t)\|_{P_y}^2 \tag{4.19}$$

Similar to above, N is a finite horizon and P_y, Q_y, R are weighting matrices. This formulation is also somewhat extended since it includes input-rate constraints and cost. The following assumptions are made:

Assumption 4.4. P_y, Q_y, $R \succ 0$.

Assumption 4.5. $\Delta u_{\min} < 0 < \Delta u_{\max}$.

From a stability point of view it is desirable to choose Ω in (4.6) or δ in (4.15) as small as possible [24]. However, the feasibility of Problems 4.1 and 4.2 will rely on either Ω, δ or N being sufficiently large. A part of the NMPC design will be to address this tradeoff.

The optimization Problems 4.1 and 4.2 can be formulated in a compact form as follows using direct single shooting formulation (see Section 2.2.2):

Problem 4.3:

$$V^*(\tilde{x}(t)) = \min_U J(U, \tilde{x}(t)) \text{ subject to } G(U, \tilde{x}(t)) \leq 0 \tag{4.20}$$

Here $\tilde{x}(t) \in \mathbb{R}^{\tilde{n}}$ is the parameter vector. For the regulation Problem 4.1 it is given by:

$$\tilde{x}(t) = x(t),\ \tilde{n} = n \tag{4.21}$$

while for the reference tracking Problem 4.2 it is [3]:

$$\tilde{x}(t) = [x(t), r(t), u(t-1)] \in \mathbb{R}^{\tilde{n}},\ \tilde{n} = n + p + m \tag{4.22}$$

Problem 4.3 defines an mp-NLP, since it is an NLP in U parameterized by $\tilde{x}(t)$. An optimal solution to this problem is denoted $U^* = [u_t^*, u_{t+1}^*, \ldots, u_{t+N-1}^*]$ and the control input is chosen according to the receding horizon policy $u(t) = u_t^*$. Define the set of feasible parameters as follows:

$$X_f = \{\tilde{x} \in \mathbb{R}^{\tilde{n}} |\ G(U, \tilde{x}) \leq 0 \text{ for some } U \in \mathbb{R}^{Nm}\} \tag{4.23}$$

For Problem 4.1, X_f is the set of N-step feasible initial states. If Ω, δ and N are such that the Problem 4.1 or 4.2 is feasible, then X_f is a non-empty set. In parametric programming problems one seeks the solution $U^*(\tilde{x})$ as an explicit function of the parameters $\tilde{x}$ in a set $X \subseteq X_f \subseteq \mathbb{R}^{\tilde{n}}$ [10]. In case the Problem 4.3 is convex, its approximate solution can be found by applying the approximate mp-NLP approach, described in Section 1.1.5.1. Otherwise, the approximate mp-NLP approach from Section 1.1.5.2 should be used, where in addition to the set of vertices of a given hyper-rectangle in the parameter space, the optimal solution is also computed at several interior points and global optimization methods are applied.

4.3 Stability of Regulation NMPC

Consider the regulation NMPC Problem 4.1. As in [18], it is supposed that Ω is a control invariant set, such that X_f is a subset of the N-step stabilizable set [20]. The Assumptions 4.2 and 4.3 imply that the origin is an equilibrium and interior point in X_f. Then, it remains to specify $P_x \succ 0$ and $\alpha > 0$ such that Ω is a control invariant set. For this purpose in [18], the ideas in [7] are used, where one simultaneously determines a linear feedback such that Ω is positively invariant under this feedback. Define the local linearization at the origin:

$$A = \frac{\partial f}{\partial x}(0,0),\ B = \frac{\partial f}{\partial u}(0,0) \tag{4.24}$$

and make the following assumption:

Assumption 4.6. (A, B) *is stabilizable.*

Let K denote the associated LQ optimal gain matrix, such that $A_0 = A - BK$ is strictly Hurwitz. The following lemmas from [18] are discrete-time versions of Lemma 1 in [7]:

Lemma 4.1. *If $\kappa > 0$ is such that $A_0 + \kappa I$ is strictly Hurwitz, the Lyapunov equation:*

$$(A_0 + \kappa I)^T P_x (A_0 + \kappa I) - P_x = -Q_x - K^T RK \tag{4.25}$$

has a unique solution $P_x \succ 0$.

Proof ([18]). The result is trivial since $Q_x + K^T RK \succ 0$. □

Lemma 4.2. *Let κ and P_x satisfy the conditions in Lemma 4.1. Then there exists a constant $\alpha > 0$ such that Ω defined in (4.10) satisfies:*
(1) $\Omega \subset \mathscr{C} = \{x \in \mathbb{R}^n | \; u_{\min} \leq -Kx \leq u_{\max}, \; y_{\min} \leq Cx \leq y_{\max}\}$.
(2) The autonomous nonlinear system:

$$x(t+1) = f(x(t), -Kx(t)) \tag{4.26}$$

is asymptotically stable for all $x(0) \in \Omega$, i.e. Ω is positively invariant.
(3) The infinite-horizon cost for the system (4.26):

$$J_\infty(x(t)) = \sum_{k=0}^{\infty} \left[\|x_{t+k|t}\|^2_{Q_x} + \|Kx_{t+k|t}\|^2_R \right] \tag{4.27}$$

satisfies $J_\infty(x) \leq x^T P_x x$ for all $x \in \Omega$.

The proof of this lemma can be found in Section 2.2.3.

It follows from [24, 7] that the exact NMPC makes the origin asymptotically stable with region of attraction X_f. A procedure for selecting P_x, κ and α is given in [7]. It is mentioned in [18] that one feature of the explicit approach to NMPC is that it is not generally desirable to select Ω as large as possible since this may lead to loss of performance and robustness. Moreover, any computational advantages of choosing Ω large are less important since the optimization is carried out entirely off-line.

Consider the mp-NLP Problem 4.3 and recall that for the regulation NMPC $\tilde{x} \equiv x$ and $\tilde{n} \equiv n$, and that the explicit solution is sought in a set $X \subset X_f \subset \mathbb{R}^n$. In [18], it is shown that the asymptotic stability is inherited by the approximate explicit NMPC under the following additional assumptions.

Assumption 4.7. *J and G in Problem 4.3 are jointly convex for all $(U, x) \in \mathbb{U} \times X$, where $\mathbb{U} = [u_{\min}, u_{\max}]^N$ is the set of admissible inputs.*

Assumption 4.8. *The explicit approximate NMPC is obtained by applying Algorithm 1.1 in Chapter 1 with the following additional steps, which are applied immediately after step 6:*
6a. If $0 \in X_0$, let $\hat{U}_0(x) = -Kx$ and go to step 6b. Otherwise, go to step 7.
6b. If $X_0 \subseteq \Omega$, mark X_0 as explored and feasible and go to step 17. Otherwise, mark X_0 to be split and go to step 17.

This assumption requires that in a neighborhood of the origin the LQ optimal gain matrix is used, as in dual-mode MPC [25].

Assumption 4.9. *Assume the approximation tolerances are set such that the partition Π generated by Algorithm 1.1, extended with Assumption 4.8, has the property that for any hypercube $X_0 \in \Pi$ that does not contain the origin:*

$$\bar{\varepsilon} \leq \gamma \min_{x \in X_0} \|x\|^2_{Qx} \tag{4.28}$$

where $\gamma \in (0, 1)$ is given.

The PWL approximation generated by Algorithm 1.1, extended with Assumption 4.8, is denoted $\hat{U} : X' \to \mathbb{R}^{Nm}$, where X' is the union of the hyper-rectangles of the partition where a feasible solution has been found. Let the associated sub-optimal cost function be denoted $\hat{V} : X' \to \mathbb{R}$, i.e. $\hat{V}(x) = J(\hat{U}(x), x)$. The set X' is an inner approximation to X_f and the approximation accuracy is determined by the minimal allowed regions volume $S_{\min}$ (used as an input parameter to Algorithm 1.1). The boundary of the feasible region X_f can thus be approximated more closely by allowing smaller regions.

Then, the following theorem is formulated in [18].

Theorem 4.1. *Suppose that Assumptions 4.1, 4.2, 4.3, 4.6, 4.7, and 4.9 hold. Then, the origin is an asymptotically stable equilibrium point for the system (4.1)–(4.2) in closed loop with the explicit approximate NMPC given by Algorithm 1.1, extended with Assumption 4.8, for all $x(0) \in X'$.*

Proof ([18]). Let $x(t) \in X'$ be arbitrary and the associated optimal control be denoted U^*. At time $t+1$ consider $\tilde{U}(t+1) = [u^*_{t+1}, u^*_{t+2}, \ldots, u^*_{t+N-1}, -Kx^*_{t+N|t}]$, where $x^*_{t+k|t}$ is the state at time $t+k$ associated with U^*. Since U^* is N-step feasible, $x^*_{t+N|t} \in \Omega$. Hence, $\tilde{U}(t+1)$ is feasible and the tail of the trajectories remain feasible since Ω is positively invariant. Since $\hat{V}(x)$ is an upper bound on $V^*(x)$, standard arguments, [6], give:

$$\begin{aligned} V^*(x(t+1)) &\leq \hat{V}(x(t+1)) \\ &= \hat{V}(x(t)) - \|x(t)\|^2_{Qx} - \|u(t)\|^2_R - \|x^*_{t+N|t}\|^2_{Px} \\ &\quad + \|f(x^*_{t+N|t}, -Kx^*_{t+N|t}\|^2_{Px} + \|x^*_{t+N|t}\|^2_{Qx} + \|Kx^*_{t+N|t}\|^2_R \\ &\leq \hat{V}(x(t)) - \|x(t)\|^2_{Qx} - \|u(t)\|^2_R \end{aligned} \tag{4.29}$$

The first inequality is due to Theorem 1.3 in Chapter 1. The second inequality is due to Lemma 4.2 since the following holds:

$$\|f(x^*_{t+N|t}, -Kx^*_{t+N|t}\|^2_{Px} - \|x^*_{t+N|t}\|^2_{Px} \leq -\|x^*_{t+N|t}\|^2_{Qx} - \|Kx^*_{t+N|t}\|^2_R \tag{4.30}$$

Define $\Omega_{\alpha_2} = \{x \in \Omega |\, x^T P_x x \leq \alpha_2\}$ such that it is contained in the set of x where the explicit approximate NMPC coincides with $u = -Kx$. Such a set with non-empty interior exists due to Assumption 4.8. Then, for $x \notin \Omega_{\alpha_2}$ it follows from (4.29) and Assumption 4.9 that:

$$\begin{aligned} V^*(x(t+1)) - V^*(x(t)) &\leq \bar{\varepsilon} - \|x(t)\|^2_{Q_x} - \|u(t)\|^2_R \\ &\leq -(1-\gamma)\|x(t)\|^2_{Q_x} < 0 \end{aligned} \tag{4.31}$$

It follows that $x(t) \to \Omega_{\alpha_2}$ as $t \to \infty$. Asymptotic stability of the origin can be concluded due to Lemma 4.2 because $u = -Kx$ in the positively invariant set Ω_{α_2} such that the closed loop satisfies (4.26). □

As it is mentioned in [18], the tolerance $\bar{\varepsilon}$ can be chosen a priori for each hyper-rectangle X_0 to satisfy (4.28). Hence, one can guarantee a priori that the PWL feedback law generated by Algorithm 1.1, extended with Assumption 4.8, will be asymptotically stabilizing. The parameter γ in (4.28) determines the approximation accuracy and degree of sub-optimality. A γ close to one is sufficient for stability, but γ close to zero gives less approximation error and sub-optimality.

4.4 Application 1: Compressor Surge Regulation

Consider the following 2-nd order compressor model introduced in Chapter 3 ([16]) with x_1 being normalized mass flow, x_2 normalized pressure and u normalized mass flow through a close-coupled valve in series with the compressor:

$$\dot{x}_1 = B(\Psi_e(x_1) - x_2 - u) \tag{4.32}$$

$$\dot{x}_2 = \frac{1}{B}(x_1 - \Phi(x_2)) \tag{4.33}$$

The following compressor and valve characteristics are used:

$$\Psi_e(x_1) = \psi_{c0} + H\left(1 + 1.5\left(\frac{x_1}{W} - 1\right) - 0.5\left(\frac{x_1}{W} - 1\right)^3\right) \tag{4.34}$$

$$\Phi(x_2) = \gamma sign(x_2)\sqrt{|x_2|} \tag{4.35}$$

with $\gamma = 0.5$, $B = 1$, $H = 0.18$, $\psi_{c0} = 0.3$ and $W = 0.25$. Like in [17], the control objective is to avoid surge. This is formulated as [17]:

$$\begin{aligned} J(U, x(t)) = &\sum_{k=0}^{N-1} \left[\alpha(x_{t+k|t} - x^*)^T(x_{t+k|t} - x^*) + ku^2_{t+k}\right] + Rv^2 \\ &+ \beta(x_{t+N|t} - x^*)^T(x_{t+N|t} - x^*) \end{aligned} \tag{4.36}$$

with $\alpha, \beta, k, R \geq 0$ and the set-point $x_1^* = 0.4$, $x_2^* = 0.6$ corresponds to an unstable equilibrium point. We have chosen $\alpha = 1$, $\beta = 0$ and $k = 0.08$. The horizon is chosen as $T = 12$, which is split into $N = 15$ equal-sized intervals, leading to a piecewise constant control input parameterization. Valve capacity requires the following constraint to hold:

$$0 \leq u(t) \leq 0.3 \tag{4.37}$$

The pressure constraint:

$$x_2(t) \geq 0.4 - v \tag{4.38}$$

avoids operation too far left of the operating point. The variable $v \geq 0$ is a slack variable introduced in order to avoid infeasibility and $R = 8$ is a large weight. Numerical analysis of the cost function shows that it is non-convex [17]. It can be seen that this NMPC problem formulation differs from Problem 4.1 in Section 4.2 in the absence of a terminal constraint and in the use of a slack variable.

A version of Algorithm 1.2 in Section 1.1.5.2, based on parallel computations, is applied to obtain an explicit approximate solution to the NMPC problem formulated above. The resulting mp-NLP problem (4.20) has 16 free variables, 46 constraints, and 2 parameters, while the NLP problem in Procedure 1.3 has 46 free variables and 811 constraints. One internal region $X_0^1 \subset X_0$ is used in Procedure 1.1. In Procedure 1.3, it is chosen $\mu = 10$ and the control input only at the first sample is considered. The approximation tolerance is chosen to depend on X_0 such that:

$$\bar{\varepsilon}(X_0) = \max(\bar{\varepsilon}_a, \bar{\varepsilon}_r V^*_{\min}) \tag{4.39}$$

where $\bar{\varepsilon}_a = 0.0001$ and $\bar{\varepsilon}_r = 0.02$ can be interpreted as absolute and relative tolerances, respectively, and $V^*_{\min} = \min_{x \in X_0} V^*(x)$. Here, $V^*(x)$ denotes a close-to-global solution.

The partition of the approximate explicit NMPC controller is shown in Fig. 4.1 [13]. It has 595 regions and 12 levels of search. With one scalar comparison required at each level of the $k-d$ tree, 12 arithmetic operations are required in the worst case

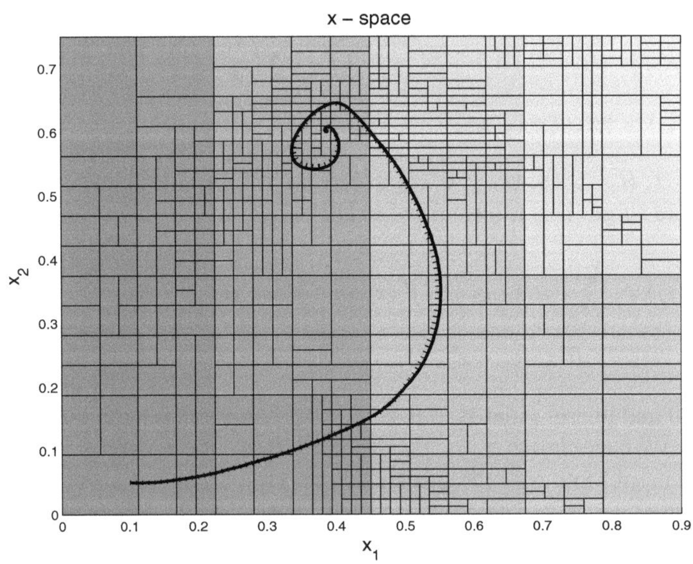

Fig. 4.1 State space partition of the approximate explicit NMPC. The solid curves are with the approximate explicit NMPC and the dotted curves are with the exact NMPC.

to determine which region the state belongs to. Totally, 16 arithmetic operations are needed in real-time to compute the control input and 1368 numbers needs to be stored in real-time computer memory.

The off-line computation of the partition is performed on a 0.8 GHz Intel(R) Core(TM) i7 CPU x8 cores. The CPU time corresponding to 11 consecutive iterations (steps 3 to 31) of Algorithm 1.2 in Section 1.1.5.2, performed by using single-core (non-parallel), dual-core, and quad-core computations, is shown in Fig. 4.2. The reason why the CPU time per iteration increases along the iterations is that the state space partition is refined with each iteration and the number of new regions, which have to be processed, increases. The average CPU time necessary to compute a single region of the partition is 74 sec with single-core computations, 39 sec with dual-core computations, and 22 sec with quad-core computations. As it should be expected, the partitioning algorithm based on parallel computations performs faster than the non-parallel algorithm. However, it should be noted that with this computer architecture, using more cores (for example 6 or 8) does not improve the computational efficiency of the partitioning algorithm. The reason for this may be the fact that the computational efficiency is influenced not only by the number of cores, but also by the amount of computer memory.

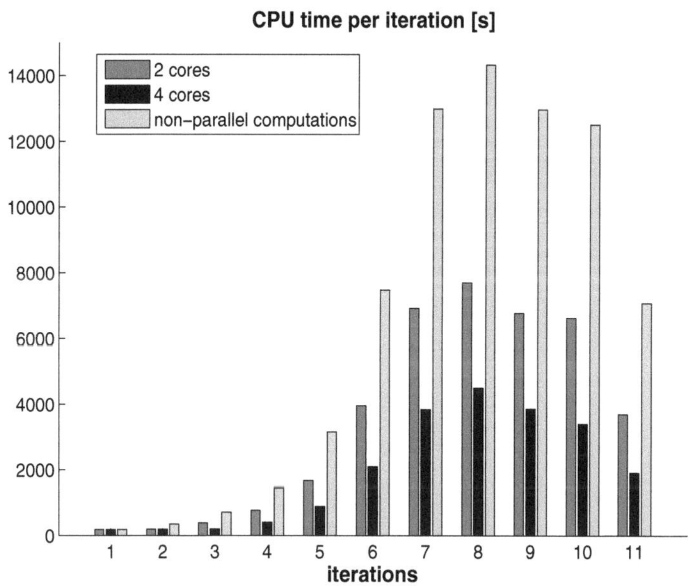

Fig. 4.2 CPU time corresponding to 11 consecutive iterations of Algorithm 1.2.

The performance of the closed-loop system is simulated for initial condition $x(0) = [0.1\ 0.05]^T$ and with sampling time $T_s = 0.02$. Euler integration with step size T_s is applied to solve the ordinary differential equations (4.32)–(4.33). The resulting closed-loop response is depicted in the state space (Fig. 4.1), as well as

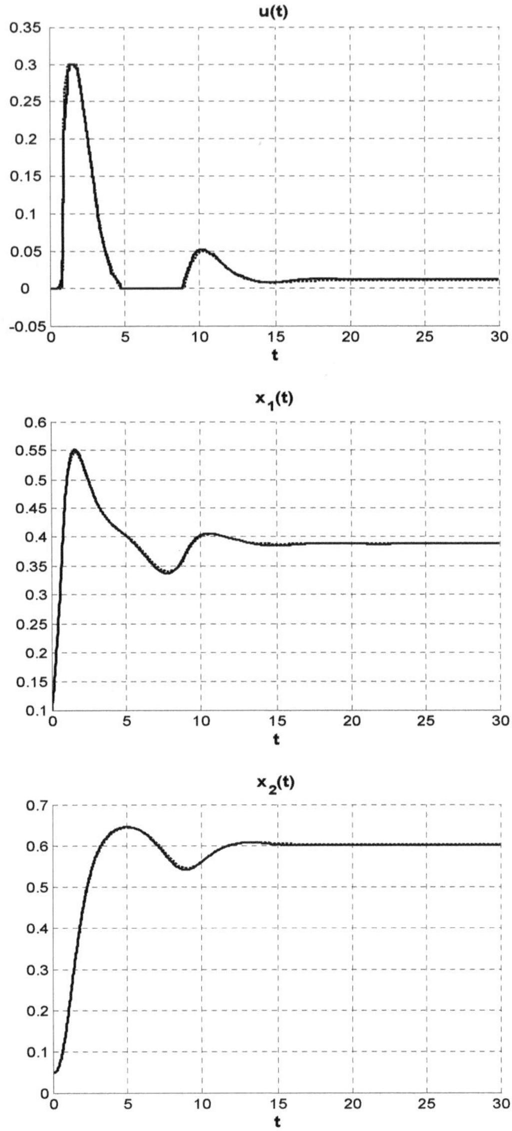

Fig. 4.3 From top to bottom: The control input, the state variable x_1, and the state variable x_2. The solid curves are with the explicit approximate NMPC and the dotted curves are with the exact NMPC.

trajectories in time (Fig. 4.3) [13]. In Fig. 4.1 and Fig. 4.3, the exact NMPC solution is also shown, which at each time step is determined by comparing the local minima of the cost function (4.36) corresponding to several initial guesses for the optimization variables.

The approximate and the optimal PWL feedback laws, as well as the associated cost functions are shown in Fig. 4.4. It can be seen from Fig. 4.4 that the cost function is non-convex.

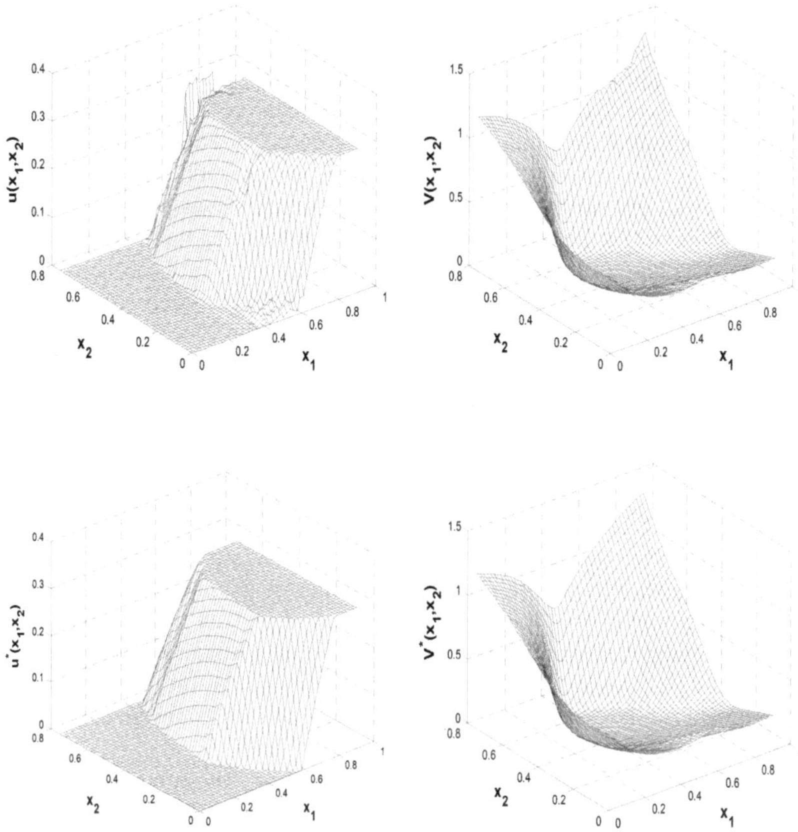

Fig. 4.4 The approximate PWL feedback law (top left), the associated sub-optimal cost function (top right), the optimal PWL feedback law (bottom left), and the optimal cost function (bottom right).

4.5 Application 2: Reference Tracking Control of an Electropneumatic Clutch Actuator Using On/Off Valves and Pulse-Width Modulation

In [14, 15], the design and performance of explicit NMPC for the position control of an electropneumatic clutch actuator for heavy duty trucks is considered. This clutch system is well suited for Automated Manual Transmissions (AMT) systems, which consist of an automated actuated clutch during gear shifts and a direct transmission through the clutch disc. Some of the AMT's largest advantages are low cost, high efficiency, reduced clutch wear and improved fuel consumption.

It is known that pneumatic actuators can be controlled both with proportional valves [23, 30, 19] and with on/off valves [27, 31, 28, 1, 26, 29, 21, 22]. The advantage of the on/off valves in comparison to proportional valves is that they are smaller and cheaper. For this reason, the electropneumatic clutch actuator, considered in this paper, is controlled by using on/off valves. The on/off valves can be controlled by pulse-width modulation (PWM) as it was described in [31, 28, 1]. This allows controllers designed for servo valves to be applied to control the on/off valves. Recently, sliding mode techniques are applied to control the on/off valves (see for example [26]). Alternatively, the case when the valves have a pure on/off behavior, i.e. fully opened and fully closed are the only possible states of the valves, is considered in [29, 21, 22]. Then in [29, 21], two controllers are designed to govern switches between these states based on backstepping and Lyapunov theory. Both controllers show promising results, but also have some weaknesses. Therefore in [22], a dual-mode controller is derived, which is a combination of the two switched controllers. It should be noted however, that the mentioned methods can not explicitly handle state constraints (constraints imposed on the actuator position). On the other hand, Nonlinear Model Predictive Control (NMPC) is an optimization based method for control which can handle both state and input constraints [24, 4]. This makes the NMPC methodology suitable to the optimal control of clutch actuators. Another advantage of the NMPC approach is that the effect of the tuning parameters is well known, while with the switched controllers in [29, 21, 22] not all tuning parameters are straightforward to choose. The fast dynamics of the considered electropneumatic clutch actuator would require the design of an explicit NMPC controller, where the only computation performed on-line would be a simple function evaluation. It should be mentioned that MPC approaches (based on on-line optimization or on pre-computed explicit solution) have been already applied to design controllers for hydraulic clutch actuators in cars (see for example [2, 5]).

Explicit NMPC is a highly promising control method. The motivation of the work in [15] is to illustrate how it allows improvements in control performance, and demonstrate how it can be implemented with low computational complexity in a fast mechatronic system. In [15], two different types of explicit NMPC controllers for reference tracking control of an electropneumatic clutch actuator using on/off valves are designed and compared. The first explicit controller has a *continuous* control input since it applies a PWM scheme. The other controller represents an explicit *quantized* NMPC, where the valves are allowed to be only fully opened or fully closed. The closed-loop performance of the two controllers is compared based on the experimentally validated 5-th order model in [19]. The real-time computational complexity and storage requirements of the two controllers are studied. Also, a performance comparison with a sliding mode controller and a PID controller is considered.

4.5.1 Mathematical Model of the Clutch Actuator Dynamics

A pneumatic actuator of an electropneumatic clutch system is considered. The pneumatic actuator acts on the clutch plates through the clutch spring, and the state of the clutch directly depends on the actuator position. The clutch actuator system is shown in Fig. 4.5. To control both supply to and exhaust from the clutch actuator

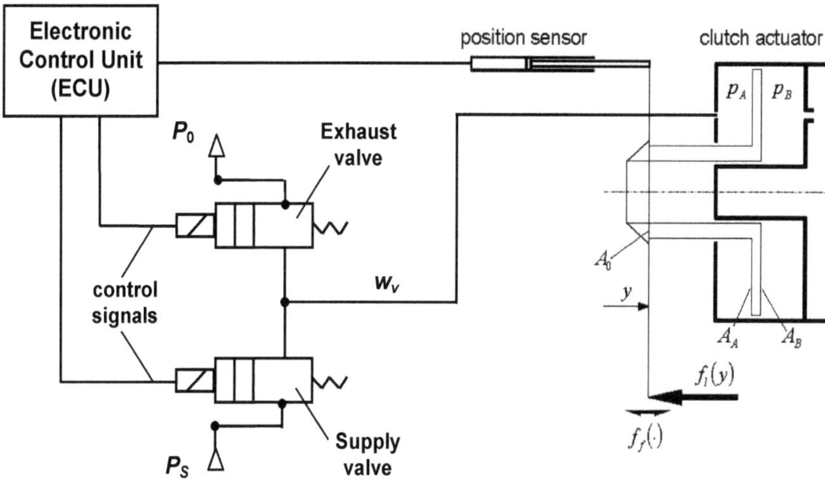

Fig. 4.5 Electropneumatic clutch actuator ([19], [29]).

chamber one pair of on/off valves is used. The electronic control unit (ECU) calculates and sets voltage signals to control the on/off valves. These signals control whether the valve should open or close, and thus also the flow into the actuator. A position sensor measures position and feeds it back to the ECU. To calculate the control signals, knowledge of other states of the system (actuator velocity, cylinder chamber pressures, friction state) are also needed, and these can be obtained either by sensors or by estimation [19]. The *full 5-th order* model of the clutch actuator dynamics, which has been experimentally validated in [19], is the following:

$$\dot{y} = v \tag{4.40}$$

$$\dot{v} = \frac{1}{M}(A_0 P_0 + A_A p_A - A_B p_B - f_f(v,z) - f_l(y)) \tag{4.41}$$

$$\dot{p}_A = -\frac{A_A}{V_A(y)} v p_A + \frac{RT_0}{V_A(y)} w_v(p_A, u) \tag{4.42}$$

$$\dot{p}_B = \frac{A_B}{V_B(y)} v p_B + \frac{RT_0}{V_B(y)} w_r(p_B) \tag{4.43}$$

$$\dot{z} = v - \frac{K_z}{F_C} |v|_q z \tag{4.44}$$

where y is the position, v is the velocity, p_A is the pressure in chamber A, p_B is the pressure in chamber B, z is the friction state, $w_v(p_A, u)$ is the mass flow rate to/from chamber A, $w_r(p_B)$ is the mass flow rate to/from chamber B, u is the control input. In (4.42), (4.43), $V_A(y) = V_{A0} + A_A y$ and $V_B(y) = V_{B0} - A_B y$ are the volumes of chambers A and B, and V_{A0}, V_{B0} are the dead volumes of these chambers. The meaning of the parameters is the following: A_A and A_B are the cross-section areas of chambers A and B, $A_0 = A_B - A_A$ is the cross-section piston area, M is piston mass, P_0 is the ambient pressure, T_0 is the temperature, R is the gas constant of air, K_z is asperity stiffness, F_C is Coulomb friction. In (4.44), $|v|_q = \sqrt{v^2+\sigma^2}$, where $\sigma > 0$ is an arbitrary small design parameter. In (4.41), $f_l(y)$ and $f_f(v,z)$ are the clutch load and the friction force, described by [19], [29]:

$$f_l(y) = K_l(1-e^{-L_l y}) - M_l y \tag{4.45}$$
$$f_f(v,z) = D_v v + K_z z + D_{\dot{z}}\dot{z}(v,z) \tag{4.46}$$

Here, K_l, L_l, M_l are the load characteristic parameters, and D_v and $D_{\dot{z}}$ are the viscous and the micro viscous effects.

Depending on the control input u at a given time instant, there can be one of the following three situations for the mass flow rate $w_v(p_A,u)$:

1) The exhaust valve is on and the supply valve is off $\Rightarrow w_v(p_A,u) = -w_{v,out}$.
2) Both the exhaust and the supply valves are off $\Rightarrow w_v(p_A,u) = 0$.
3) The exhaust valve is off and the supply valve is on $\Rightarrow w_v(p_A,u) = w_{v,in}$.

Here, $w_{v,out}$ is the mass flow rate from chamber A and $w_{v,in}$ is the mass flow rate to chamber A, and they are determined by the following expressions [19]:

$$w_{v,out} = \rho_0 C_{v,out}\,\psi(r, B_{v,out})\,p_A\,,\quad r = \frac{P_0}{p_A} \tag{4.47}$$
$$w_{v,in} = \rho_0 C_{v,in}\,\psi(r, B_{v,in})\,P_S\,,\quad r = \frac{p_A}{P_S} \tag{4.48}$$

Here, ρ_0 is the flow density, $C_{v,out}$ and $C_{v,in}$ are flow conductances, $B_{v,out}$ and $B_{v,in}$ are critical flow parameters, and P_S is the supply pressure.

The valve flow function $\psi(r, B_{v,in/out})$ is given by [19]:

$$\psi(r, B_{v,in/out}) = \psi_0(r,0) + B_{v,in/out}\begin{cases} \psi_0(r,B_0) - \psi_0(r,0), & B_{v,in/out} \geq 0 \\ \psi_0(r,0) + r - 1, & B_{v,in/out} < 0 \end{cases} \tag{4.49}$$

where:

$$\psi_0(r,B_0) = \begin{cases} 1\,, & r < B_0 \\ \sqrt{1-\left(\frac{r-B_0}{1-B_0}\right)^2}\,, & r \geq B_0 \end{cases} \tag{4.50}$$

Here, B_0 is the critical flow parameter for air. It follows from (4.50) that $\psi_0(r,0) = \sqrt{1-r^2}$.

The mass flow rate $w_r(p_B)$ to/from chamber B is [19]:

$$w_r(p_B) = w_{r,in}(p_B) - w_{r,out}(p_B) \tag{4.51}$$

where:

$$w_{r,in}(p_B) = \rho_0 C_r \psi_{in}(r) P_0\,,\quad r = \frac{p_B}{P_0} \tag{4.52}$$
$$w_{r,out}(p_B) = \rho_0 C_r \psi_{out}(r) p_B\,,\quad r = \frac{P_0}{p_B} \tag{4.53}$$

Here, C_r is the flow conductance. The restriction flow function $\psi_{in/out}(r)$ is given by [19]:

$$\psi_{in/out}(r) = \Omega_0(r) + b\Omega_1(r, sign(b)),\; b \in [-1, 1] \tag{4.54}$$

In (4.54), $\Omega_0(r)$ is defined by:

$$\Omega_0(r) = \begin{cases} \sqrt{1-r^2}\,, & r \in [0, 1] \\ 0 & , r > 1 \end{cases} \tag{4.55}$$

We have $b = 0.5$ and $\Omega_1(r, +1)$ is given by:

$$\Omega_1(r, +1) = -\Omega_0(r) + \begin{cases} 1 & , r \in [0, B_0] \\ \sqrt{1 - \left(\frac{r - B_0}{1 - B_0}\right)^2} & , r \in (B_0, 1] \\ 0 & , r > 1 \end{cases} \tag{4.56}$$

Further, in order to reduce the computational burden, the design of the explicit NMPC controllers is based on the following *simplified 3-rd order* model [29]:

$$\dot{y}^s = v^s \tag{4.57}$$

$$\dot{v}^s = \frac{1}{M}(-A_A P_0 + A_A p_A^s - f_f^*(v^s) - f_l(y^s)) \tag{4.58}$$

$$\dot{p}_A^s = -\frac{A_A}{V_A(y^s)} v^s p_A^s + \frac{RT_0}{V_A(y^s)} w_{v^s}(p_A^s, u) \tag{4.59}$$

where the states are the actuator position y^s, the velocity v^s and the pressure p_A^s in chamber A. The term $f_f^*(v^s)$ represents the static sliding friction characteristic [19]:

$$f_f^*(v^s) = D_v v^s + F_C \frac{v^s}{\sqrt{v^{s2} + \sigma^2}} \tag{4.60}$$

The values of the clutch actuator parameters are given in Table 4.1.

The on/off valves are controlled by applying a pulse-width modulation (PWM) scheme and their duty cycle depends on the control input u. The use of the traditional linear PWM scheme would result in the valves failing to respond when the on-time is less than their response time. In order to avoid this, the approach in [31] is applied. Thus, the duty cycle of either the exhaust or the supply valve is not allowed to fall below $d_{\min}$, the minimum possible duty cycle where the valve will still respond, given by:

$$d_{\min} = \frac{T_{vr}}{T_{PWM}} \cdot 100\% \tag{4.61}$$

Here, T_{vr} is the valve response time and T_{PWM} is the PWM period. For the considered clutch actuator, $T_{vr} = 0.002\,[s]$ and the PWM period is equal to the sampling time, i.e. $T_{PWM} = T_s = 0.01\,[s]$. Therefore, the minimum duty cycle is $d_{\min} = 20\%$. The PWM scheme is illustrated in Fig. 4.6.

Table 4.1 Parameters of the electropneumatic clutch actuator

Parameter	Value	Unit	Description
A_A	$1.7 \cdot 10^{-2}$	m^2	Area of chamber A
A_B	$1.9 \cdot 10^{-2}$	m^2	Area of chamber B
A_0	$0.2 \cdot 10^{-2}$	m^2	Piston area
V_{A0}	$0.8 \cdot 10^{-3}$	m^3	Dead volume of chamber A
V_{B0}	$5.7 \cdot 10^{-4}$	m^3	Dead volume of chamber B
P_0	$1.095 \cdot 10^5$	Pa	Ambient pressure
P_S	$9.5 \cdot 10^5$	Pa	Supply pressure
T_0	293	K	Temperature
R	288	$J/(kg \cdot K)$	Gas constant of air
M	10	kg	Mass of piston
D_v	5000	$N \cdot s/m$	Viscous effect
K_z	$1 \cdot 10^6$	N/m	Asperity stiffness
$D_{\dot{z}}$	5000	$N \cdot s/m$	Micro viscous effect
F_C	200	N	Coulomb friction
ρ_0	1.185	kg/m^3	Flow density
$B_{v,in}$	0.5	-	Critical flow parameter
$B_{v,out}$	-0.5	-	Critical flow parameter
$C_{v,in}$	$1.5 \cdot 10^{-8}$	$m^3/(Pa \cdot s)$	Conductance
$C_{v,out}$	$2 \cdot 10^{-8}$	$m^3/(Pa \cdot s)$	Conductance
B_0	0.528	-	Critical flow parameter for air
C_r	$1 \cdot 10^{-8}$	$m^3/(Pa \cdot s)$	Conductance
K_l	5000	N/m	Load characteristic parameter
L_l	500	-	Load characteristic parameter
M_l	25000	N/m	Load characteristic parameter

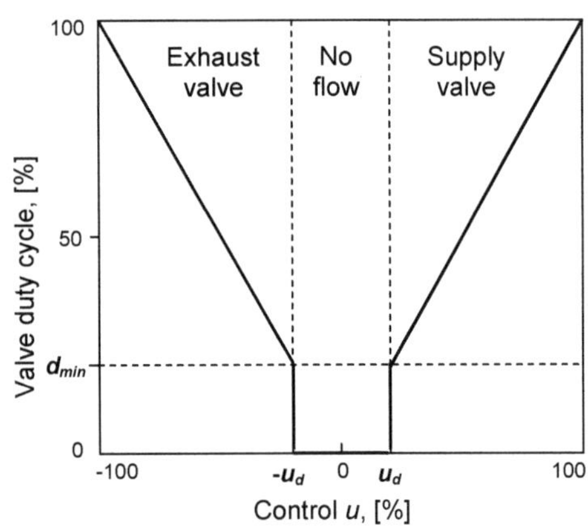

Fig. 4.6 Pulse-width modulation scheme.

It can be seen that the range of the control input u is divided into the following three parts:

1) $u \in [-100; -u_d)$.

In this case, the exhaust valve is on and its duty cycle d_e is computed from the linear characteristic $d_e = a_e u + b_e$. Let $t \in \mathbb{Z}_{\geq 0}$ be the discrete time and $\theta \in \mathbb{R}_{\geq 0}$ be the continuous time. Then, during the sampling time T_s the following applies:

$$w_v(p_A, u) = \begin{cases} -w_{v,out}, \text{ if } \theta \in [tT_s; (t + d_e/100)T_s] \\ 0, \text{ if } \theta \in ((t + d_e/100)T_s; (t+1)T_s] \end{cases} \tag{4.62}$$

where $w_{v,out}$ is determined from (4.47).

2) $u \in [-u_d; u_d]$.

In this range, both valves are off and there is no flow to/from chamber A, i.e. $w_v(p_A, u) = 0$ for $\theta \in [tT_s; (t+1)T_s]$.

3) $u \in (u_d; 100]$.

In this case, the supply valve is on and its duty cycle d_s is computed from the linear characteristic $d_s = a_s u + b_s$. Then, during the sampling time T_s the following applies:

$$w_v(p_A, u) = \begin{cases} w_{v,in}, \text{ if } \theta \in [tT_s; (t + d_s/100)T_s] \\ 0, \text{ if } \theta \in ((t + d_s/100)T_s; (t+1)T_s] \end{cases} \tag{4.63}$$

where $w_{v,in}$ is determined from (4.48).

In the PWM scheme, u_d is a design parameter which determines the slopes of the linear characteristics and the width of the "no flow" range. In our case, $u_d = 10\%$.

4.5.2 *Design and Performance of Explicit Reference Tracking Controller with Continuous Control Input*

In [15], an explicit NMPC controller for optimal reference tracking control of the electropneumatic clutch actuator is designed. The values of the prediction horizon and the weighting coefficients in the NMPC problem formulation are chosen among a set of values such that the corresponding optimal solution has the best reference tracking quality. The approximate mp-NLP approach described in Section 1.1.5.2 is applied to design an explicit NMPC with continuous control input for reference tracking control of the electropneumatic clutch actuator. The controller design is based on the *simplified 3-rd order* model (4.57)–(4.59) of the clutch actuator dynamics, introduced in Section 4.5.1. The sampling time is $T_s = 0.01$ $[s]$ and the PWM scheme described in Section 4.5.1 is applied to control the exhaust and the supply valves. The forward Euler method with stepsize $T_E = 0.0001$ $[s]$ is used to integrate the equations (4.57)–(4.59). The control objective is to have the actuator position y^s track a reference signal $r(t) > 0$, which is achieved by minimizing the following cost function:

$$J(U,y^s(t),r(t)) = \sum_{k=0}^{N-1}\left[Q_y\left(\frac{y^s_{t+k|t}-r(t)}{r(t)}\right)^2 + Rh(u_{t+k})^2\right] + P_y\left(\frac{y^s_{t+N|t}-r(t)}{r(t)}\right)^2 \tag{4.64}$$

where $h(u_{t+k}) = w_{v^s}(p^s_{A,t+k|t}, u_{t+k})/(w_{v^s,\max} - w_{v^s,\min})$. Here, $N=5$ is the horizon, $Q_y = P_y = 10$, $R=0.1$ are the weighting coefficients, and $w_{v^s,\max}$ and $w_{v^s,\min}$ are the maximal and the minimal mass flow rates to/from chamber A. The control input u determines the duty cycle of the on/off valves (see Section 4.5.1) and it should satisfy the constraint $-100\ \% \le u_{t+k} \le 100\ \%$, $k = 0,1,...,N-1$. The piston position y for this clutch actuator is allowed to be between 0 [m] and 0.025 [m], which leads to the constraint:

$$0\,[m] \le y^s_{t+k|t} \le 0.025\,[m]\,,\ k = 1,\ \dots\ , N \tag{4.65}$$

In order to further reduce the partition complexity of the explicit controller, a projection of the reference tracking error is used. Thus in Problem 4.3, the extended state vector is $\tilde{x}(t) = [e(t), v^s(t), p^s_A(t), r(t)] \in \mathbb{R}^4$, where $e(t) = r(t) - y^s(t)$ is the reference tracking error. Note, that the state vector $\tilde{x}(t)$ does not include $u(t-1)$ since it is not used in the expression (4.64) of the cost function. The state space to be partitioned is 4-dimensional and it is defined by $X = [-0.005; 0.005] \times [-0.05; 0.15] \times [P_0; P_S] \times [0.0001; 0.024]$ where P_0 and P_S are given in Table 4.1. The values of $e(t)$ outside this hyper-rectangle are projected onto its boundary. It has been shown in [18] that bounding the approximation error of the approximate explicit solution to regulation NMPC problems ensures stability of the sub-optimal closed-loop system. Here, a similar approach is applied and the cost function approximation tolerance is chosen as $\bar{\varepsilon}(X_0) = \max(\bar{\varepsilon}_a, \bar{\varepsilon}_r \min_{\tilde{x}\in X_0} V^*(\tilde{x}))$, where $\bar{\varepsilon}_a = 0.001$ and $\bar{\varepsilon}_r = 0.02$ are the absolute and the relative tolerances. The stability of the closed-loop system is verified by simulations. It should be noted that the possibility for implementation verification is a significant advantage of the explicit NMPC in comparison to NMPC based on real-time optimization. The partition of the explicit continuous NMPC controller has 9035 regions. Its performance was simulated for a typical clutch reference signal and compared with that of a sliding mode controller (SMC) and of a PID controller. The SMC controller is designed by applying the approach in [12]. The responses associated to the three controllers are depicted in Fig. 4.7 to Fig. 4.12. The simulations of the closed-loop system are based on the *full 5-th order* model (4.40)–(4.44). It is required for the tracking error to be less than 0.0002 [m] in the area where the clutch engages/disengages and it is relaxed outside this area. All three controllers meet this requirement. However, it can be observed from Fig. 4.7 that the control input of the explicit NMPC controller has less chattering in comparison to that of the SMC controller. Also, the explicit NMPC has a higher quality of the overall tracking performance compared to that of the PID controller (characterized with an overshoot).

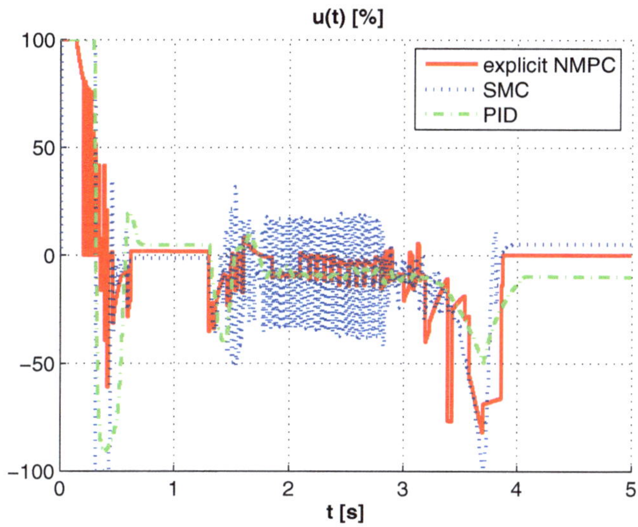

Fig. 4.7 The control input u.

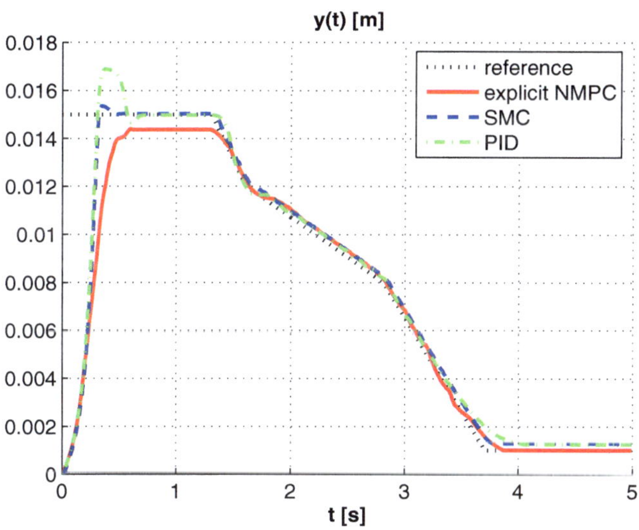

Fig. 4.8 The clutch actuator position y.

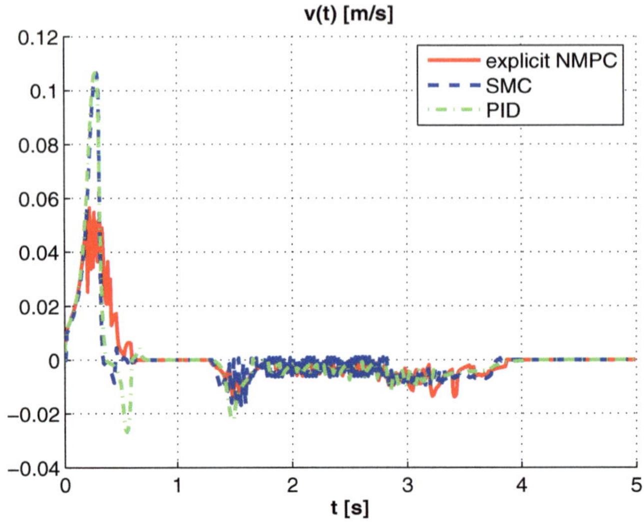

Fig. 4.9 The velocity v.

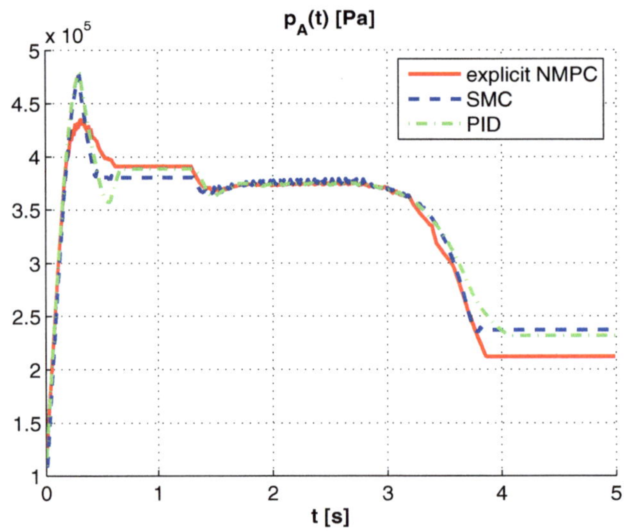

Fig. 4.10 The pressure p_A in chamber A.

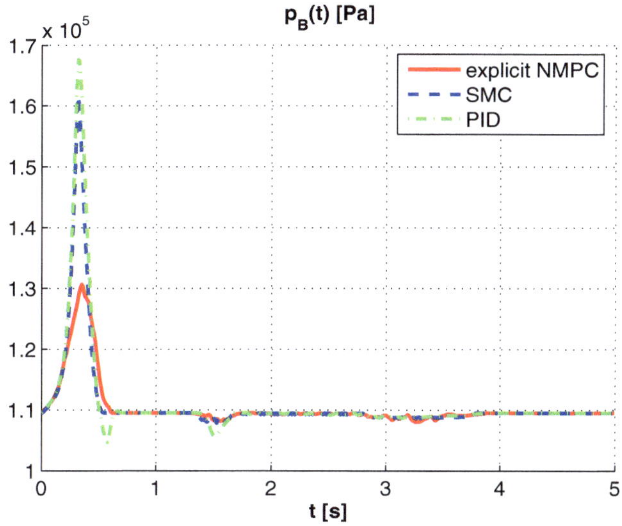

Fig. 4.11 The pressure p_B in chamber B.

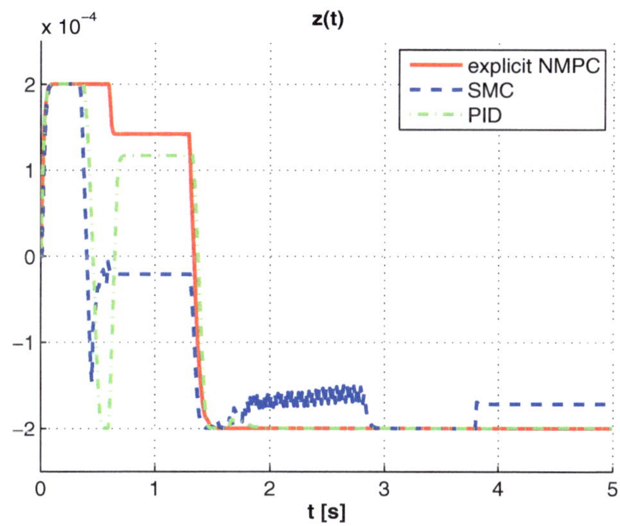

Fig. 4.12 The friction state z.

4.6 Time-Optimal Diver Decompression

Diving sickness (or decompression sickness) is known as an injury that affects ascending divers who have breathed gas which is at a higher pressure than the surface pressure due to the pressure of the surrounding water. To prevent decompression sickness one should set up a sufficiently large gradient for gas elimination. This leads to another question - how close to optimal in terms of total decompression time will a decompression procedure be, [11, 8]?

Using a physiological nonlinear dynamic model of the inert gas dynamics in the diver's tissue and blood, the minimum time decompression problem can be formulated as a time-optimal control problem subject to constraints that are set up to avoid inert gas bubbles of too large size or amount in the diver's blood stream, [11]. This optimal control problem can be discretized and solved to compute practical diving decompression procedures that will provide the diver with advice about stop depths and stop depths during ascent to the surface, [9, 11]. Since the model should be updated dynamically based on the water depth profile during the dive, the decompression profile should be updated at minute intervals in a receding horizon fashion in order to account for the actions of the diver, [11].

An obstacle for practical implementation of such procedures is the limited CPU capacity and power consumption requirements for a typical low-cost diving computer. The real-time nonlinear optimization and the receding horizon algorithm requires a large amount of floating point numerical computations per re-optimization, [9]. For safety reasons the convergence and correctness of the numerical optimization result needs to be verified. In [8], an mp-NLP problem is solved to pre-compute an explicit NMPC controller suitable for implementation in a diving computer. Rather than re-computing numerically the optimal solution at each sampling instant based on updated physiological state information, the solution is simply obtained by evaluating a piecewise approximation to the optimal solution function. Since such piecewise function representation can be precomputed off-line and stored in memory on the diving computer, the computational load on the diving computer will be dramatically relaxed. Moreover, since the solution function is pre-computed, it is easier to verify the correctness and will benefit in the safety-critical applications such as a diving computer, [8].

References

1. Ahn, K., Yokota, S.: Intelligent switching control of pneumatic actuator using on/off solenoid valves. Mechatronics 15, 683–702 (2005)
2. Amari, R., Tona, P., Alamir, M.: Experimental evaluation of a hybrid MPC strategy for vehicle start-up with an automated manual transmission. In: Proceedings of the European Control Conference, Budapest, Hungary, pp. 4271–4277 (2009)
3. Alessio, A., Bemporad, A.: A Survey on Explicit Model Predictive Control. In: Magni, L., Raimondo, D.M., Allgöwer, F. (eds.) Nonlinear Model Predictive Control: Towards New Challenging Applications. LNCIS, vol. 384, pp. 345–369. Springer, Heidelberg (2009)
4. Allgöwer, F., Zheng, A. (eds.): Nonlinear model predictive control. Progress in system and control theory, vol. 26. Birkhäuser, Basel (2000)

5. Bemporad, A., Borrelli, F., Glielmo, L., Vasca, F.: Hybrid control of dry clutch engagement. In: Proceedings of the European Control Conference, Porto, pp. 635–639 (2001)
6. Bemporad, A., Filippi, C.: Suboptimal explicit MPC via approximate quadratic programming. In: Proceedings of the IEEE Conference on Decision and Control, Orlando, vol. 5, pp. 4851–4856 (2001)
7. Chen, H., Allgöwer, F.: A quasi-infinite horizon nonlinear model predictive control scheme with guaranteed stability. Automatica 34, 1205–1217 (1998)
8. Feng, L., Gutvik, C.R., Johansen, T.A.: Optimal decompression through multi-parametric nonlinear programming. In: IFAC NOLCOS, Bologna, Italy (2010), `http://www.IFAC-PapersOnLine.net`
9. Feng, L., Gutvik, C.R., Johansen, T.A., Sui, D.: Barrier function nonlinear optimization for optimal decompression of divers. In: IEEE Conf. Decision and Control, Shanghai (2009)
10. Fiacco, A.V.: Introduction to sensitivity and stability analysis in nonlinear programming. Academic Press, Orlando (1983)
11. Gutvik, C.R., Johansen, T.A., Brubakk, A.O.: Optimal Decompression of Divers - Procedures for constraining predicted bubble growth. IEEE Control Systems Magazine 31(1), 19–28
12. Gjone, K.: Robustness tests and analysis of control strategies on an electropneumatic actuator. M.Sc. thesis, Norwegian University of Science and Technology, Trondheim, Norway (2007)
13. Grancharova, A., Johansen, T.A., Tøndel, P.: Computational aspects of approximate explicit nonlinear model predictive control. In: Findeisen, R., Allgöwer, F., Biegler, L. (eds.) Assessment and Future Directions of Nonlinear Model Predictive Control. LNCIS, vol. 358, pp. 181–192. Springer, Heidelberg (2007)
14. Grancharova, A., Johansen, T.A.: Explicit model predictive control of an electropneumatic clutch actuator using on/off valves and pulse-width modulation. In: Proceedings of the European Control Conference, Budapest, Hungary, pp. 4278–4283 (2009)
15. Grancharova, A., Johansen, T.A.: Design and comparison of explicit model predictive controllers for an electropneumatic clutch actuator using on/off valves. IEEE/ASME Transactions on Mechatronics 16, 665–673 (2011)
16. Gravdahl, J.T., Egeland, O.: Compressor surge control using a close-coupled valve and backstepping. In: Proceedings of the American Control Conference, Albuquerque, NM (1997)
17. Johansen, T.A.: On multi-parametric nonlinear programming and explicit nonlinear model predictive control. In: Proceedings of the IEEE Conference on Decision and Control, Las Vegas, NV, vol. 3, pp. 2768–2773 (2002)
18. Johansen, T.A.: Approximate explicit receding horizon control of constrained nonlinear systems. Automatica 40, 293–300 (2004)
19. Kaasa, G.O.: Nonlinear output-feedback control applied to electro-pneumatic clutch actuation in heavy-duty trucks. Ph.D. thesis, Norwegian University of Science and Technology, Trondheim, Norway (2006)
20. Kerrigan, E., Maciejowski, J.M.: Invariant sets for constrained nonlinear discrete-time systems with application to feasibility in model predictive control. In: Proceedings of the IEEE Conference on Decision and Control, Sydney (2000)
21. Langjord, H., Johansen, T.A., Hespanha, J.P.: Switched control of an electropneumatic clutch actuator using on/off valves. In: Proceedings of the American Control Conference, Seattle (2008)
22. Langjord, H., Johansen, T.A., Bratli, C.: Dual-mode switched control of an electropneumatic clutch actuator with input restrictions. In: Proceedings of the European Control Conference, Budapest, Hungary, pp. 2085–2090 (2009)

23. Marchant, J.A., Street, M.J., Gurney, P., Benson, J.A.: Design and testing of a servo controller for pneumatic cylinders. Proceedings of the Institution of Mechanical Engineers, Part E 203, 21–27 (1989)
24. Mayne, D.Q., Rawlings, J.B., Rao, C.V., Scokaert, P.O.M.: Constrained model predictive control: Stability and optimality. Automatica 36, 789–814 (2000)
25. Michalska, H., Mayne, D.Q.: Robust receding horizon control of constrained nonlinear systems. IEEE Transactions on Automatic Control 38, 1623–1633 (1993)
26. Nguyen, T., Leavitt, J., Jabbari, F., Bobrow, J.E.: Accurate sliding-mode control of pneumatic systems using low-cost solenoid valves. IEEE/ASME Transactions on Mechatronics 12, 216–219 (2007)
27. Paul, A.K., Mishra, J.K., Radke, M.G.: Reduced order sliding mode control for pneumatic actuator. IEEE Transactions on Control Systems Technology 2, 271–276 (1994)
28. Peterson, K.S., Stefanopoulou, A.G.: Extremum seeking control for soft landing of an electromechanical valve actuator. Automatica 40, 1063–1069 (2004)
29. Sande, H., Johansen, T.A., Kaasa, G.O., Snare, S.R., Bratli, C.: Switched backstepping control of an electropneumatic clutch actuator using on/off valves. In: Proceedings of the American Control Conference, New York, pp. 76–81 (2007)
30. Smaoui, M., Brun, X., Thomasset, D.: A study on tracking position control of an electropneumtaic system using backstepping design. Control Engineering Practice 14, 923–933 (2006)
31. van Varseveld, R.B., Bone, G.M.: Accurate position control of a pneumatic actuator using on/off solenoid valves. IEEE/ASME Transactions on Mechatronics 2, 195–204 (1997)

Chapter 5
Explicit MPC of Constrained Nonlinear Systems with Quantized Inputs

Abstract. This chapter presents an approximate multi-parametric Nonlinear Integer Programming (mp-NIP) approach to design explicit MPC controllers for constrained nonlinear systems with quantized control inputs. It is organized as follows. In Section 5.2, general regulation and reference tracking quantized NMPC problems are formulated and represented as an mp-NIP problem. Then, in Section 5.3, an approximate mp-NIP approach to explicit quantized NMPC is described. The idea of the approach is to construct a piecewise constant (PWC) approximation to the optimal solution of the mp-NIP problem on a hyper-rectangle of interest by imposing an orthogonal state space partition. In Section 5.4, an explicit quantized NMPC controller for the electropneumatic clutch actuator (described in Section 4.5) is designed and its performance is compared to that of the explicit NMPC with continuous control input. In Section 5.5, the approximate mp-NIP approach is applied to design an explicit quantized NMPC controller for optimal regulation of a continuous stirred tank reactor.

5.1 Introduction

In several control engineering problems, the system to be controlled is characterized by a finite set of possible control actions. Such systems are referred to as systems with *quantized* control input and the possible values of the input represent the levels of *quantization*. For example, hydraulic systems using on/off valves are systems with *quantized* input. In order to achieve a high quality of the control system performance it would be necessary to take into account the effect of the control input *quantization*. Thus, in [10] receding horizon optimal control ideas were proposed for synthesizing *quantized* control laws for *linear* systems with *quantized* inputs and quadratic optimality criteria. Further in [1], a method for *explicit* solution of optimal control problems with *quantized* control input was developed. It is based on solving multi-parametric Nonlinear Integer Programming (mp-NIP) problems, where the cost function and the constraints depend linearly on the vector of parameters. In [4, 3], an MPC problem for constrained *nonlinear* systems with *quantized*

A. Grancharova & T.A. Johansen: Explicit Nonlinear Model Predictive Ctrl., LNCIS 429, pp. 111–125.
springerlink.com

input is formulated and represented as an mp-NIP problem. Then, a computational method for *explicit* approximate solution of the resulting mp-NIP problem is suggested. The benefits of the *explicit* solution consist in efficient on-line computations using a binary search tree and verifiability of the design and implementation. The mp-NIP method proposed in [4, 3] is more general compared to the mp-NIP method in [1], since it allows the cost function and the constraints to depend nonlinearly on the vector of parameters.

Note that the term Nonlinear Integer Programming is used instead of the more general Mixed-Integer Nonlinear Programming (MI-NLP) since the problem formulation contains only integer free variables. This is possible since continuous variables are eliminated using a direct single shooting strategy, and all control inputs are assumed to be quantized. The general ideas can, however, be extended to MI-NLP to account for situations with mixed continuous and integer variables.

5.2 Formulation of the Quantized NMPC Problem as an mp-NIP Problem

Consider the discrete-time nonlinear system:

$$x(t+1) = f(x(t),u(t)) \tag{5.1}$$

$$y(t) = Cx(t), \tag{5.2}$$

where $x(t) \in \mathbb{R}^n$ is the state variable, $y(t) \in \mathbb{R}^p$ is the output variable, and $u(t) \in \mathbb{R}^m$ is the control input, which is constrained to belong to the finite set of values $U^A = \{\overline{u}_1, \overline{u}_2, \dots, \overline{u}_L\}$, $\overline{u}_i \in \mathbb{R}^m$, $\forall i = 1,2,\dots,L$, i.e. $u(t) \in U^A$. Here, $\overline{u}_1, \overline{u}_2, \dots, \overline{u}_L$ represent the levels of *quantization* of the control input u. In (5.1), $f : \mathbb{R}^n \times U^A \longmapsto \mathbb{R}^n$ is a nonlinear function. It is supposed that a full measurement of the state $x(t)$ is available at the current time t.

First, consider the optimal regulation problem where the goal is to steer the system state to the origin by minimizing a certain performance criterion. For the current $x(t)$, the *quantized* NMPC regulation solves the optimization problem:

Problem 5.1:

$$V^*(x(t)) = \min_{U \in U^B} J(U,x(t)) \tag{5.3}$$

subject to $x_{t|t} = x(t)$ and:

$$y_{\min} \le y_{t+k|t} \le y_{\max}, \quad k = 1, \dots, N \tag{5.4}$$

$$u_{t+k} \in U^A = \{\bar{u}_1, \bar{u}_2, \dots, \bar{u}_L\}, \quad k = 0, 1, \dots, N-1 \tag{5.5}$$

$$\|x_{t+N|t}\|^2 \le \delta_x \tag{5.6}$$

$$x_{t+k+1|t} = f(x_{t+k|t}, u_{t+k}), \quad k \ge 0 \tag{5.7}$$

$$y_{t+k|t} = Cx_{t+k|t}, \quad k \ge 0 \tag{5.8}$$

Here, $U = [u_t, u_{t+1}, \ldots, u_{t+N-1}] \in \mathbb{R}^{Nm}$ is the set of free control moves, $U^B \triangleq (U^A)^N = U^A \times \ldots \times U^A$. The set U^B is also represented as $U^B = \{\bar{U}_j \mid j = 1, 2, \ldots, M\}$, where $\bar{U}_j \in \mathbb{R}^{Nm}$ are the levels of quantization of the control vector U and $M = L^N$. The cost function is given by:

$$J(U, x(t)) = \sum_{k=0}^{N-1} \left[\|x_{t+k|t}\|_{Q_x}^2 + \|h(x_{t+k|t}, u_{t+k})\|_R^2 \right] + \|x_{t+N|t}\|_{P_x}^2 \tag{5.9}$$

Here, N is a finite horizon and $h : \mathbb{R}^n \times U^A \mapsto \mathbb{R}^s$ is a nonlinear function. It is assumed that $\delta_x > 0$ and $P_x, Q_x, R \succ 0$.

Now, consider the reference tracking problem where the goal is to have the output variable $y(t)$ track the reference signal $r(t) \in \mathbb{R}^p$. For the current $x(t)$, the reference tracking *quantized* NMPC solves the following optimization problem:

Problem 5.2:

$$V^*(x(t), r(t)) = \min_{U \in U^B} J(U, x(t), r(t)) \tag{5.10}$$

subject to $x_{t|t} = x(t)$ and:

$$y_{\min} \leq y_{t+k|t} \leq y_{\max}, \; k = 1, \ldots, N \tag{5.11}$$

$$u_{t+k} \in U^A = \{\overline{u}_1, \overline{u}_2, \ldots, \overline{u}_L\}, k = 0, 1, \ldots, N-1 \tag{5.12}$$

$$\left\| y_{t+N|t} - r(t) \right\| \leq \delta_y \tag{5.13}$$

$$x_{t+k+1|t} = f(x_{t+k|t}, u_{t+k}), \; k \geq 0 \tag{5.14}$$

$$y_{t+k|t} = C x_{t+k|t}, \; k \geq 0 \tag{5.15}$$

Here, $U = [u_t, u_{t+1}, \ldots, u_{t+N-1}] \in \mathbb{R}^{Nm}$ is the set of free control moves, $U^B = (U^A)^N = U^A \times \ldots \times U^A$ and the cost function is given by:

$$\begin{aligned} J(U, x(t), r(t)) = \sum_{k=0}^{N-1} \left[\|y_{t+k|t} - r(t)\|_{Q_y}^2 + \|h(x_{t+k|t}, u_{t+k})\|_R^2 \right] \\ + \|y_{t+N|t} - r(t)\|_{P_y}^2 \end{aligned} \tag{5.16}$$

Similar to above, N is a finite horizon and $h : \mathbb{R}^n \times U^A \longmapsto \mathbb{R}^s$ is a nonlinear function. It is assumed that $\delta_y > 0$ and $P_y, Q_y, R \succ 0$.

From a stability point of view it is desirable to choose δ_x in (5.6) or δ_y in (5.13) as small as possible [9]. However, in the case of quantized input, the equilibrium point of the closed-loop system may either have an offset from the reference, or there may be a limit cycle about the reference. Therefore, the feasibility of Problems 5.1 and 5.2 will rely on δ_x and δ_y being sufficiently large. A part of the NMPC design will be to address this tradeoff.

The optimization Problems 5.1 and 5.2 can be formulated in a compact form as follows:

Problem 5.3:

$$V^*(\tilde{x}(t)) = \min_{U \in U^B} J(U,\tilde{x}(t)) \text{ subject to } G(U,\tilde{x}(t)) \leq 0 \tag{5.17}$$

Here $\tilde{x}(t) \in \mathbb{R}^{\tilde{n}}$ and for the regulation Problem 5.1 it is:

$$\tilde{x}(t) = x(t),\ \tilde{n} = n \tag{5.18}$$

while for the reference tracking Problem 5.2 it is:

$$\tilde{x}(t) = [x(t), r(t)] \in \mathbb{R}^{\tilde{n}},\ \tilde{n} = n + p \tag{5.19}$$

Problem 5.3 defines a multi-parametric Nonlinear Integer Programming (mp-NIP) problem, since it is NIP in U parameterized by $\tilde{x}(t)$. An optimal solution to this problem is denoted $U^* = [u_t^*, u_{t+1}^*, \ldots, u_{t+N-1}^*]$ and the control input is chosen according to the receding horizon policy $u(t) = u_t^*$. Define the set of feasible parameters as follows:

$$X_f = \{\tilde{x} \in \mathbb{R}^{\tilde{n}} |\ G(U,\tilde{x}) \leq 0 \text{ for some } U \in U^B\} \tag{5.20}$$

For Problem 5.1, X_f is the set of N-step feasible initial states. If δ_x, δ_y and N are such that the Problem 5.1 or 5.2 is feasible, then X_f is a non-empty set. In parametric programming problems one seeks the solution $U^*(\tilde{x})$ as an explicit function of the parameters $\tilde{x}$ in a set $X \subseteq X_f \subseteq \mathbb{R}^{\tilde{n}}$ [2].

5.3 Approximate mp-NIP Approach to Explicit Quantized NMPC

5.3.1 Computation of Explicit Approximate Solution

We restrict our attention to a hyper-rectangle $X \subset \mathbb{R}^{\tilde{n}}$ where we seek to approximate the optimal solution $U^*(\tilde{x})$ to Problem 5.3. We require that the state space partition is orthogonal and can be represented as a $k-d$ tree. The main idea of the approximate mp-NIP approach in [4, 3] is to construct a *piecewise constant* (PWC) approximation $\widehat{U}(\tilde{x})$ to $U^*(\tilde{x})$ on X, where the constituent constant functions are defined on hyper-rectangles covering X. The solution of Problem 5.3 is computed at the $2^{\tilde{n}}$ vertices of a considered hyper-rectangle X_0, as well as at some interior points. These additional points represent the vertices and the facets centers of one or more hyper-rectangles contained in the interior of X_0. The Procedure 1.1 is used to generate a set of points $W_0 = \{w_0, w_1, w_2, \ldots, w_{N_1}\}$ associated to a hyper-rectangle X_0. Then, a close-to-global solution $U^*(w_i)$ of Problem 5.3 at a point $w_i \in W_0$ is computed by using the routine 'glcSolve' of the TOMLAB optimization environment in

Matlab [7]. The routine 'glcSolve' implements an extended version of the DIRECT algorithm [8], that handles problems with both nonlinear and integer constraints. The DIRECT algorithm (DIviding RECTangles) [8] is a deterministic sampling algorithm for searching for the global minimum of a multivariate function subject to constraints, using no derivative information. It is a modification of the standard Lipschitzian approach that eliminates the need to specify a Lipschitz constant.

Based on the close-to-global solutions $U^*(w_i)$ at all points $w_i \in W_0$, a local constant approximation $\widehat{U}_0(\tilde{x}) = K_0$ to the optimal solution $U^*(\tilde{x})$, valid in the whole hyper-rectangle X_0, is determined by applying the following procedure [4]:

Procedure 5.1 (computation of explicit approximate solution). *Consider any hyper-rectangle $X_0 \subseteq X$ with a set of points $W_0 = \{w_0, w_1, \dots, w_{N_1}\}$ determined by applying Procedure 1.1. Compute K_0 by solving the following NIP:*

$$\min_{K_0 \in U^B} \sum_{i=0}^{N_1} (J(K_0, w_i) - V^*(w_i)) \;\; \textit{subject to} \;\; G(K_0, w_i) \leq 0, \forall w_i \in W_0 \quad (5.21)$$

5.3.2 Estimation of Error Bounds

Suppose that a constant function $\widehat{U}_0(\tilde{x}) = K_0$, associated to the hyper-rectangle X_0 has been determined by applying Procedure 5.1. Then, for the cost function approximation error in X_0 we have:

$$\varepsilon(\tilde{x}) = \widehat{V}(\tilde{x}) - V^*(\tilde{x}) \leq \varepsilon_0 \;, \;\; \tilde{x} \in X_0 \quad (5.22)$$

where $\widehat{V}(\tilde{x}) = J(\widehat{U}_0(\tilde{x}), \tilde{x})$ is the sub-optimal cost and $V^*(\tilde{x})$ denotes the cost corresponding to the close-to-global solution $U^*(\tilde{x})$, i.e. $V^*(\tilde{x}) = J(U^*(\tilde{x}), \tilde{x})$. The following procedure can be used to obtain an estimate $\widehat{\varepsilon}_0$ of the maximal approximation error ε_0 in X_0.

Procedure 5.2 (computation of error bound approximation). *Consider any hyper-rectangle $X_0 \subseteq X$ with a set of points $W_0 = \{w_0, w_1, \dots, w_{N_1}\}$ determined by applying Procedure 1.1. Compute an estimate $\widehat{\varepsilon}_0$ of the error bound ε_0 through the following maximization:*

$$\widehat{\varepsilon}_0 = \max_{i \in \{0,1,2,\dots,N_1\}} (\widehat{V}(w_i) \quad V^*(w_i)) \quad (5.23)$$

5.3.3 Approximate mp-NIP Algorithm

Assume the tolerance $\bar{\varepsilon} > 0$ of the cost function approximation error is given. The following algorithm is proposed to design explicit reference tracking *quantized* NMPC [4]:

Algorithm 5.1. Explicit reference tracking quantized NMPC.

Step 1. Initialize the partition to the whole hyper-rectangle, i.e. $\Pi = \{X\}$. Mark the hyper-rectangle X as unexplored.

Step 2. Select any unexplored hyper-rectangle $X_0 \in \Pi$. If no such hyper-rectangle exists, terminate.

Step 3. Generate a set of points $W_0 = \{w_0, w_1, w_2, ..., w_{N_1}\}$ associated to X_0 by applying Procedure 1.1.

Step 4. Compute a solution to Problem 5.3 for $\tilde{x}$ fixed to each of the points w_i, $i = 0,1,2,...,N_1$ by using routine 'glcSolve' of TOMLAB optimization environment. If Problem 5.3 has a feasible solution at all these points, go to step 7. Otherwise, go to step 5.

Step 5. Compute the size of X_0 using some metric. If it is smaller than some given tolerance, mark X_0 infeasible and explored and go to step 2. Otherwise, go to step 6.

Step 6. If at least one of the points w_i, $i = 0,1,2,...,N_1$ is feasible, split X_0 into hyper-rectangles X_1, X_2, ... , X_{N_s} by applying the Heuristic splitting rule 1.1. Mark X_1, X_2, ... , X_{N_s} unexplored, remove X_0 from Π, add X_1, X_2, ... , X_{N_s} to Π, and go to step 2. If none of the points w_i, $i = 0,1,2,...,N_1$ are feasible, split X_0 into two hyper-rectangles X_1 and X_2 by a hyperplane through its center point and orthogonal to an arbitrary axis. Mark X_1 and X_2 unexplored, remove X_0 from Π, add X_1 and X_2 to Π, and go to step 2.

Step 7. Compute a constant function $\widehat{U}_0(\tilde{x})$ using Procedure 5.1, as an approximation to be used in X_0. If a feasible solution was found, go to step 8. Otherwise, split X_0 into two hyper-rectangles X_1 and X_2 by a hyperplane through its center point and orthogonal to an arbitrary axis. Mark X_1 and X_2 unexplored, remove X_0 from Π, add X_1 and X_2 to Π, and go to step 2.

Step 8. Compute an estimate $\widehat{\varepsilon}_0$ of the error bound ε_0 in X_0 by applying Procedure 5.2. If $\widehat{\varepsilon}_0 \leq \bar{\varepsilon}$, mark X_0 as explored and feasible and go to step 2. Otherwise, split X_0 into two hyper-rectangles X_1 and X_2 by applying a procedure that is similar to Procedure 1.5. Mark X_1 and X_2 unexplored, remove X_0 from Π, add X_1 and X_2 to Π, and go to step 2.

5.4 Application 1: Reference Tracking Quantized Control of an Electropneumatic Clutch Actuator Using On/Off Valves

Consider the electropneumatic clutch actuator, whose mathematical model is described in Section 4.5.1. With the *quantized* control input the two valves are only allowed to be fully open or fully closed (no pulse-width modulation is used). Thus, the control input is an integer variable which can take only three values, i.e. $u \in U^A = \{1,2,3\}$. This is related to the mass flow rate $w_v(p_A, u)$ to/from chamber A in the following way:

$$
\begin{aligned}
u = 1 &\Rightarrow w_v(p_A, 1) = -w_{v,out}, \text{ for } \theta \in [tT_s; (t+1)T_s] \\
u = 2 &\Rightarrow w_v(p_A, 2) = 0, \text{ for } \theta \in [tT_s; (t+1)T_s] \\
u = 3 &\Rightarrow w_v(p_A, 3) = w_{v,in}, \text{ for } \theta \in [tT_s; (t+1)T_s]
\end{aligned}
\tag{5.24}
$$

where $w_{v,out}$ and $w_{v,in}$ are determined from (5.7)-(5.8), and θ is the time variable. Therefore, $u = 1$ corresponds to maximal flow from chamber A, $u = 2$ means no flow, and $u = 3$ corresponds to maximal flow to chamber A during the *whole* sampling period T_s.

5.4.1 Design of Explicit NMPC with Quantized Control Input

In [4, 5], an explicit *quantized* NMPC controller for the clutch actuator is designed, which is based on the *simplified 3-rd order* model (4.57)–(4.59), introduced in Section 4.5.1. The *quantized* NMPC has sampling time $T_s = 0.01\,[s]$ and it minimizes the cost function (4.64) in Section 4.5.2 (with u_{t+k} and U being here the quantized control input and the quantized control input sequence, respectively), subject to the system equations (4.57)–(4.59) and the constraint (4.65). In (4.64), the horizon is $N = 10$ and the weighting coefficients are $Q_y = P_y = 1$, $R = 0.1$. The extended state vector $\tilde{x}(t)$ and the state space X to be partitioned are the same as for the NMPC controller with *continuous* control input, designed in Section 4.5.2. The cost function approximation tolerance is $\bar{\varepsilon}(X_0) = \max(\bar{\varepsilon}_a, \bar{\varepsilon}_r \min_{\tilde{x} \in X_0} V^*(\tilde{x}))$, where $\bar{\varepsilon}_a = 0.001$ and $\bar{\varepsilon}_r = 0.02$. The partition has 10871 regions. The performance of the explicit *quantized* NMPC was simulated for the typical clutch reference signal and the resulting response is depicted in Fig. 5.1 and Fig. 5.2. The simulations of the closed-loop system are based on the *full 5-th order* model (4.40)–(4.44) of the clutch actuator dynamics, described in Section 4.5.1.

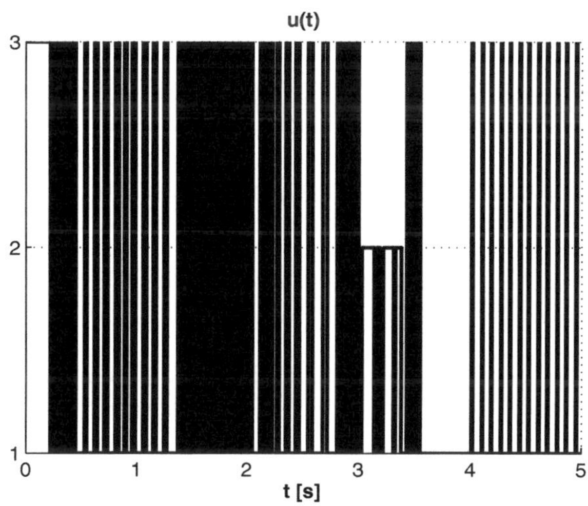

Fig. 5.1 The control input u.

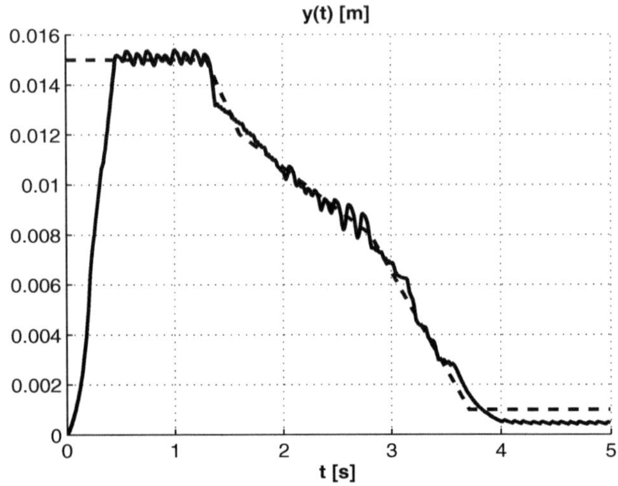

Fig. 5.2 The clutch actuator position y with the explicit *quantized* NMPC (the dashed curve is the reference signal).

5.4.2 Comparison between the Explicit NMPC with Quantized Control Input and the Explicit NMPC with Continuous Control Input

In [5], a comparative study of the explicit *quantized* NMPC controller and the explicit NMPC controller with continuous control input for reference tracking control of the electropneumatic clutch actuator is made.

5.4.2.1 Chattering

The chattering of the explicit *quantized* NMPC controller, designed in this section, and the explicit NMPC controller with continuous control input (using a PWM scheme), the SMC controller, and the PID controller, considered in Section 4.5.2, is studied. In Table 5.1, statistics about the chattering in the control input u (only for the controllers which generate a continuous control input) and in the actuator position y is given. The chattering is expressed as:

$$\Delta u(t) = |u(t) - u(t-1)|\,,\ \Delta y(t) = |y(t) - y(t-1)| \tag{5.25}$$

where $t = 2, 3, ..., 500$ is the discrete time instant. It can be seen that the control input chattering of the explicit NMPC with PWM is comparable to that of the PID controller and it is significantly smaller than that of the SMC controller. Also, the explicit NMPC with PWM leads to the smallest position chattering among the four studied controllers.

Table 5.1 Statistics about chattering

Controller	Average $\Delta u\ [\%]$	Maximal $\Delta u\ [\%]$	Average $\Delta y\ [m]$	Maximal $\Delta y\ [m]$
NMPC with PWM	4.65	81.99	$5.66 \cdot 10^{-5}$	$6.22 \cdot 10^{-4}$
SMC controller	12.16	168.96	$5.94 \cdot 10^{-5}$	$10.63 \cdot 10^{-4}$
PID controller	1.23	52.49	$6.63 \cdot 10^{-5}$	$10.63 \cdot 10^{-4}$
Quantized NMPC	–	–	$9.83 \cdot 10^{-5}$	$7.23 \cdot 10^{-4}$

5.4.2.2 Tracking Performance

In Table 5.2, statistics about the absolute reference tracking error $e_y(t) = |y(t) - r(t)|$ and the sum squared relative reference tracking error S_y for the four controllers are given. The error $e_y(t)$ is considered after the position settles near the first reference value $r = 0.015\ [m]$ (after 0.8 $[s]$ of time). The reason is that the trajectories from the initial state $y = 0\ [m]$ to a neighborhood of $r = 0.015\ [m]$ for the four controllers are characterized by the same maximal reference tracking error $e_y = 0.015\ [m]$. The error S_y is computed on the entire transients as:

$$S_y = \frac{1}{500} \sum_{t=1}^{500} \left(\frac{y(t) - r(t)}{r(t)} \right)^2 \tag{5.26}$$

It can be seen that the explicit NMPC with PWM provides the highest quality of tracking performance.

Table 5.2 Statistics about reference tracking error

Controller	Average $e_y(t)\ [m]$	Maximal $e_y(t)\ [m]$	Sum squared error S_y
NMPC with PWM	$2.48 \cdot 10^{-4}$	$6.89 \cdot 10^{-4}$	$0.431 \cdot 10^{-1}$
SMC controller	$3.05 \cdot 10^{-4}$	$6.30 \cdot 10^{-4}$	$0.679 \cdot 10^{-1}$
PID controller	$2.77 \cdot 10^{-4}$	$10.36 \cdot 10^{-4}$	$0.839 \cdot 10^{-1}$
Quantized NMPC	$3.27 \cdot 10^{-4}$	$10.39 \cdot 10^{-4}$	$1.063 \cdot 10^{-1}$

5.4.2.3 Real-Time Computational Complexity and Storage Requirements

The explicit approximate solutions of the two explicit NMPC controllers are implemented as binary search trees by applying the method in [11]. In Table 5.3, the real-time computational complexity (the worst-case number of arithmetic operations needed to compute the control input) and the storage requirements (in terms of numbers that have to be stored), associated to the binary search trees of the two controllers, are given. It can be observed that the number of on-line arithmetic operations is negligibly small with both controllers. The explicit NMPC with PWM requires significantly more storage in comparison to the explicit *quantized* NMPC controller. It can be explained with the fact that for each region of the partition of

Table 5.3 On-line computational complexity and storage requirements

Controller	**arithmetic ops. per sample**	**stored reals**	**stored integers**
Explicit NMPC with PWM	152	39960	36771
Explicit *quantized* NMPC	143	295	7831

this controller, an affine control law needs to be stored (while only one constant needs to be stored with the *quantized* controller). Further, since the total number of solutions for the *quantized* controller is only 3, merging of regions with the same solutions into one convex region leads to a significant decrease of the complexity of the search tree.

5.5 Application 2: Regulation of a Continuous Stirred Tank Reactor with Quantized Control Input

In [3], the approximate mp-NIP approach (described in Section 5.3) is applied to design an explicit *quantized* NMPC controller for optimal regulation of a continuous stirred tank reactor (CSTR). In the CSTR, a first-order irreversible reaction $A \rightarrow B$ takes place (Fig. 5.3). The mathematical model of CSTR and the values of the parameters are taken from [6]. The mass and heat balance of CSTR expressed through dimensionless concentration $\tilde{c}$ and temperature $\tilde{T}$ are [6]:

$$\frac{d\tilde{c}}{dt} = \frac{(1-\tilde{c})}{q} - k_0 e^{-\frac{E}{\tilde{T}}} \tilde{c} \tag{5.27}$$

$$\frac{d\tilde{T}}{dt} = \frac{(\tilde{T}_f - \tilde{T})}{q} + k_0 e^{-\frac{E}{\tilde{T}}} \tilde{c} - \alpha u(\tilde{T} - \tilde{T}_c) \tag{5.28}$$

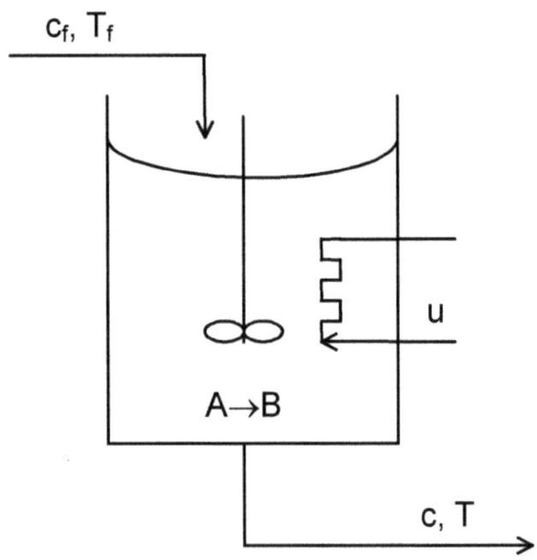

Fig. 5.3 Continuous stirred tank reactor.

where the dimensionless quantities $\tilde{c}$, $\tilde{T}$, $\tilde{T}_c$ and $\tilde{T}_f$ are defined as follows:

$$\tilde{c} = \frac{c}{c_f},\ \tilde{T} = \frac{T}{Jc_f},\ \tilde{T}_c = \frac{T_c}{Jc_f},\ \tilde{T}_f = \frac{T_f}{Jc_f} \tag{5.29}$$

The coolant flowrate u is a *quantized* control variable. The values of the parameters are taken from [6] and are $q = 10$, $c_f = 1$, $T_c = 290$, $T_f = 300$, $J = 100$, $E = 25.2$, $k_0 = 300$, $\alpha = 1.95 \cdot 10^{-4}$.

We consider the set point $\tilde{c}^* = 0.41$, $\tilde{T}^* = 3.3$. Then, the model of the reactor can be written in the form:

$$\frac{dx_1}{dt} = \frac{(1-\tilde{c}^* - x_1)}{q} - k_0 e^{-\frac{E}{(\tilde{T}^*+x_2)}}(\tilde{c}^* + x_1) \tag{5.30}$$

$$\frac{dx_2}{dt} = \frac{(\tilde{T}_f - \tilde{T}^* - x_2)}{q} + k_0 e^{-\frac{E}{(\tilde{T}^*+x_2)}}(\tilde{c}^* + x_1) - \alpha u(\tilde{T}^* + x_2 - \tilde{T}_c) \tag{5.31}$$

where x_1 and x_2 denote the deviations of the concentration and temperature from the set point values ($x_1 = \tilde{c} - \tilde{c}^*$, $x_2 = \tilde{T} - \tilde{T}^*$). The forward Euler method with step size $T_E = 0.01$ is used to integrate the equations (5.30)–(5.31).

The coolant flowrate u is *quantized* with the following levels of *quantization*:

$$u \in U^A = \{u_{\min},\ u_{st},\ u_{\max}\} \tag{5.32}$$

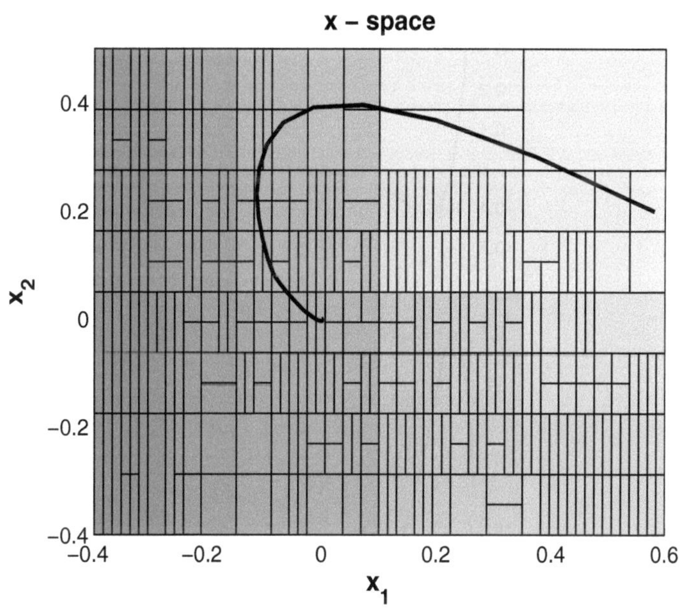

Fig. 5.4 State space partition of the explicit approximate quantized NMPC.

where $u_{\min} = 250$, $u_{\max} = 500$, and $u_{st} = 370$ is the steady state value corresponding to the set point $\tilde{c}^* = 0.41$, $\tilde{T}^* = 3.3$.

The suggested approximate mp-NIP approach is applied to design an explicit *quantized* NMPC controller for this reactor. The NMPC minimizes the cost function (5.9) subject to the system equations (5.30)–(5.31) and the input constraint

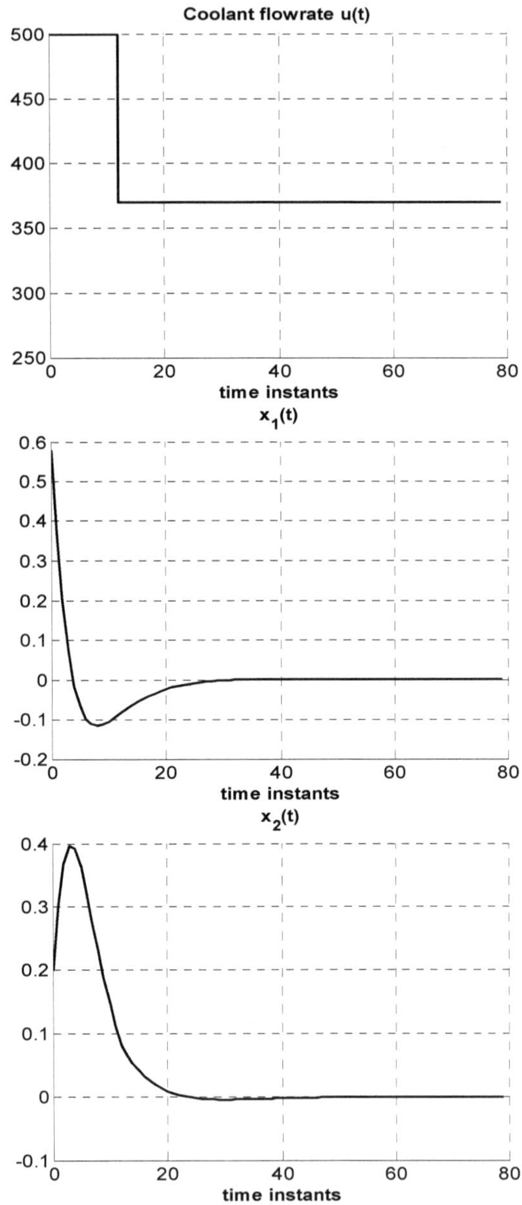

Fig. 5.5 From top to bottom: The control input, the state variable x_1, and the state variable x_2 for $\alpha = 1.95 \cdot 10^{-4}$.

(5.32). In (5.9), $h(x_{t+k|t}, u_{t+k}) \equiv u_{t+k} - u_{st}$ and the cost matrices are $Q_x = P_x = \mathrm{diag}\{100, 300\}$, $R = 1 \cdot 10^{-4}$. The horizon is $N = 30$ with a sampling time for the control input $T_s = 1$. In (5.6), it is chosen $\delta_x = 0.002$. The state space to be partitioned is defined by $X = [-0.4, 0.6] \times [-0.4, 0.5]$. The state space partition of the explicit *quantized* NMPC controller is shown in Fig. 5.4. It has 341 regions and 14 levels of search. Thus, 14 arithmetic operations are needed in real-time to compute the control input (14 comparisons). Due to *quantization*, it would be straightforward to join neighboring regions with the same solution at the first sample of the control trajectory in a postprocessing step. This would lead to a significant reduction of the complexity of the partition.

The performance of the closed-loop system was simulated for initial condition $x(0) = [0.58, 0.2]^T$. The resulting closed-loop response corresponding to the explicit approximate *quantized* NMPC (the solid curves) and to the exact *quantized* NMPC (the dotted curves) is depicted in the state space (Fig. 5.4), as well as trajectories in time (Fig. 5.5). The results show that the exact and the approximate solutions are indistinguishable.

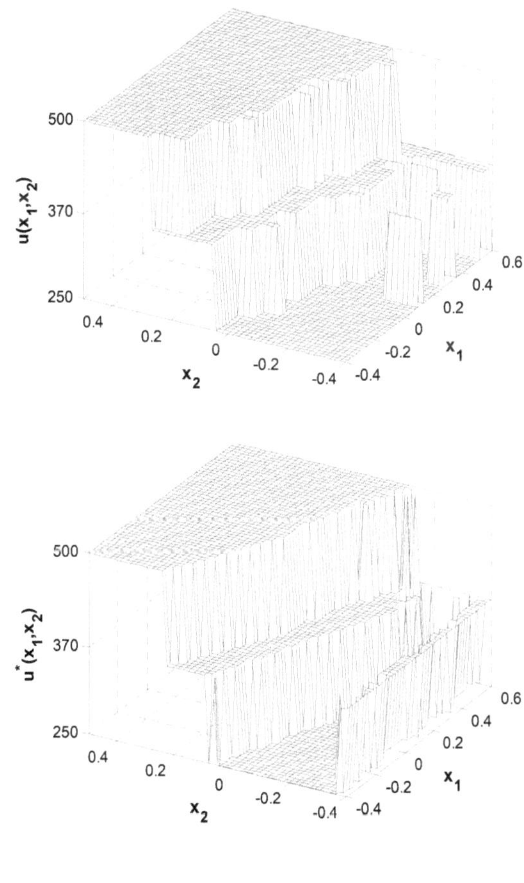

Fig. 5.6 Top: The suboptimal state feedback law $\hat{u}(x_1, x_2)$. Bottom: The optimal state feedback law $u^*(x_1, x_2)$.

The suboptimal and the optimal state feedback laws are shown in Fig. 5.6.

In order to study the robustness of the explicit controller, we assume that the real value of the heat transfer coefficient is $\alpha = 1.85 \cdot 10^{-4}$ (instead of $\alpha = 1.95 \cdot 10^{-4}$ used to design the controller). The closed-loop response corresponding to $\alpha = 1.85 \cdot 10^{-4}$ and initial condition $x(0) = [0.58,\ 0.2]^T$ is depicted in Fig. 5.7. It can be seen

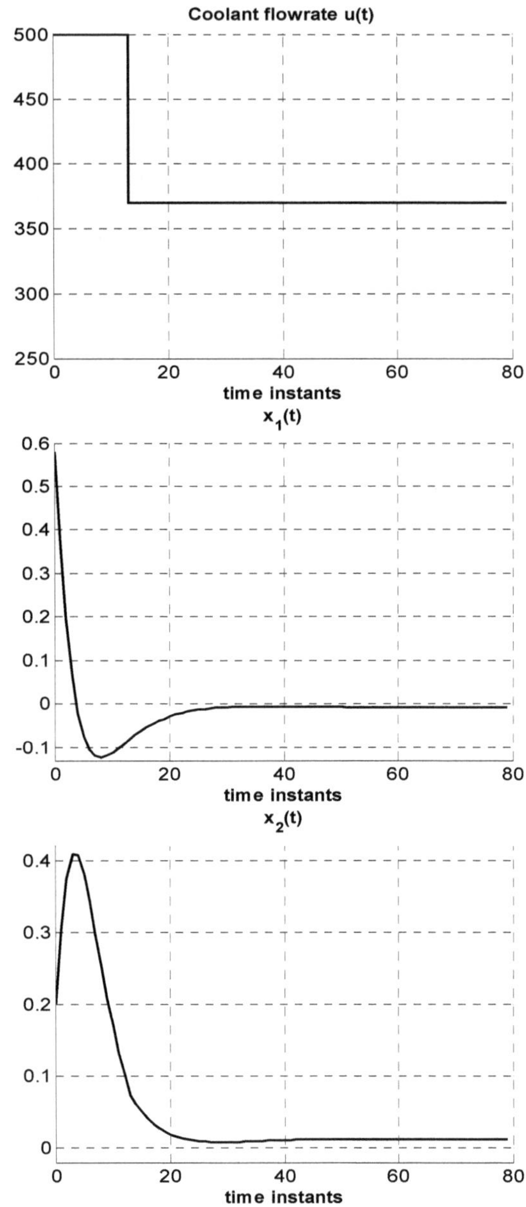

Fig. 5.7 From top to bottom: The control input, the state variable x_1, and the state variable x_2 for $\alpha = 1.85 \cdot 10^{-4}$.

that the closed-loop trajectory has an offset due to the fact that the steady state value u_{st} (cf. equation (5.32)) corresponding to the set point $\tilde{c}^* = 0.41$, $\tilde{T}^* = 3.3$ is a function of the coefficient α.

References

1. Bemporad, A.: Multiparametric nonlinear integer programming and explicit quantized optimal control. In: Proceedings of the IEEE Conference an Decision and Control, Maui, Hawaii, USA (2003)
2. Fiacco, A.V.: Introduction to sensitivity and stability analysis in nonlinear programming. Academic Press, Orlando (1983)
3. Grancharova, A., Johansen, T.A.: Explicit solution of regulation control problems for nonlinear systems with quantized inputs. In: Proceedings of the International Conference on Automatics and Informatics, Sofia, Bulgaria, pp. I-15–I-18 (2008)
4. Grancharova, A., Johansen, T.A.: Explicit Approximate Model Predictive Control of Constrained Nonlinear Systems with Quantized Input. In: Magni, L., Raimondo, D.M., Allgöwer, F. (eds.) Nonlinear Model Predictive Control: Towards New Challenging Applications. LNCIS, vol. 384, pp. 371–380. Springer, Heidelberg (2009)
5. Grancharova, A., Johansen, T.A.: Design and comparison of explicit model predictive controllers for an electropneumatic clutch actuator using on/off valves. IEEE/ASME Transactions on Mechatronics 16, 665–673 (2011)
6. Hicks, G., Ray, W.: Approximation methods for optimal control synthesis. The Canadian Journal of Chemical Engineering 49, 522–528 (1971)
7. Holmström, K., Göran, A.O., Edvall, M.M.: User's Guide for TOMLAB (2007)
8. Jones, D.R.: The DIRECT global optimization algorithm. In: Floudas, C.A., Pardalos, P.M. (eds.) Encyclopedia of Optimization, vol. 1, pp. 431–440. Kluwer, Dordrecht (2001)
9. Mayne, D.Q., Rawlings, J.B., Rao, C.V., Scokaert, P.O.M.: Constrained model predictive control: Stability and optimality. Automatica 36, 789–814 (2000)
10. Picasso, B., Pancanti, S., Bemporad, A., Bicchi, A.: Receding-horizon control of LTI systems with quantized inputs. In: Proceedings of the IFAC Conference on Analysis and Design of Hybrid Systems, Saint Malo, France (2002)
11. Tøndel, P., Johansen, T.A., Bemporad, A.: Evaluation of piecewise affine control via binary search tree. Automatica 39, 743–749 (2003)

Chapter 6
Explicit Min-Max MPC of Constrained Nonlinear Systems with Bounded Uncertainties

Abstract. This chapter considers two approaches to explicit min-max NMPC of general constrained nonlinear discrete-time systems in the presence of bounded disturbances and/or parameter uncertainties. The approach in Section 6.2 is based on an *open-loop* min-max NMPC formulation and constructs a piecewise linear (PWL) approximation of the optimal solution. An explicit *open-loop* min-max NMPC controller is designed for a continuous stirred tank reactor, whose heat transfer coefficient is an uncertain parameter. The approach in Section 6.3 adopts a *closed-loop* (also referred to as *feedback*) min-max NMPC formulation and builds a piecewise nonlinear (PWNL) approximation of the optimal sequence of feedback control policies. The approach is applied to design an explicit *feedback* min-max NMPC controller for a cart and spring system in the presence of bounded disturbances.

6.1 Introduction

Models are only an approximation of the real process, and therefore it is important for NMPC to be robust with respect to model uncertainties and disturbances. One approach to robust NMPC design is to optimize the *nominal* performance while guaranteeing robust feasibility and robust stability of the closed-loop system. Thus in [25], a Lyapunov based robust NMPC design for input-affine nonlinear systems subject to uncertainty and input constraints is developed, which allows for an explicit characterization of the closed-loop stability region. Another robust NMPC strategy consists of solving a min-max problem to optimize the *robust* performance while enforcing the state and input constraints for all possible uncertainties. The min-max robust MPC was first proposed in [5]. There are two formulations of min-max NMPC: the *open-loop* and the *closed-loop* (also referred to as *feedback*) formulation (see [22] for review of the min-max NMPC approaches). The *open-loop* min-max NMPC [26, 19, 22] guarantees the robust stability and the robust feasibility of the system, but it may be very conservative since the control sequence has to ensure constraints fulfillment for all possible uncertainty scenarios without considering the fact that future measurements of the state contain information about past

A. Grancharova & T.A. Johansen: Explicit Nonlinear Model Predictive Ctrl., LNCIS 429, pp. 127–156.
springerlink.com

uncertainty values. As a result, the *open-loop* min-max NMPC controllers may have a small feasible set and sub-optimal performance. An approximate multi-parametric Nonlinear Programming (mp-NLP) approach to explicit solution of *open-loop* min-max NMPC problems has been suggested in [8]. This approach is considered in Section 6.2.

The conservativeness of the *open-loop* approaches is overcome by the *closed-loop* min-max NMPC [21, 22, 20], where the optimization is performed over a sequence of feedback control policies. With the *closed-loop* approach, the min-max NMPC problem represents a differential game where the controller is the minimizing player and the disturbance is the input of the maximizing player ('the nature') [21]. The controller chooses the control input as a function of the current state so as to ensure that the effect of the disturbance on the system output is sufficiently small for any choice made by 'the nature'. In this way, the *closed-loop* min-max NMPC would guarantee a larger feasible set and a higher level of performance compared to the *open-loop* min-max NMPC [21]. Recently, several approaches have been developed for explicit solution of min-max MPC problems for special classes of uncertain nonlinear systems. Thus, for constrained linear systems with polytopic uncertainty, approaches for explicit solution of the *open-loop* and the *closed-loop* min-max MPC problems have been developed, respectively in [6] and in [31, 4, 29]. The method in [2] applies to linear systems with polyhedral parametric uncertainty and additive bounded disturbances and both the *open-loop* and the *closed-loop* min-max control problems are solved explicitly. Approaches for explicit solution of *robust* finite horizon optimal control problems for constrained piecewise affine systems with bounded disturbances have been proposed, based on an *open-loop* formulation in [27], and on a *closed-loop* formulation in [16, 30]. Methods for explicit solution of min-max MPC or H_∞ problems for constrained linear systems with additive bounded uncertainties are suggested in [28] for the *open-loop* formulation, and in [15, 24] for the *closed-loop* formulation. In [11], an approximate mp-NLP approach to explicit solution of *closed-loop* min-max NMPC problems for general nonlinear systems with state and input constraints has been developed. This approach is considered in Section 6.3.

6.2 Explicit Open-Loop Min-Max MPC of Constrained Nonlinear Systems with Bounded Uncertainties

This section considers the approximate mp-NLP approach [8] to explicit solution of *open-loop* min-max NMPC problems for constrained nonlinear systems in the presence of model uncertainty. It is based on an orthogonal search tree structure of the state space partition and thus represents an extension of the approach in [14]. The explicit NMPC controller is designed by formulating a min-max optimization problem, i.e. by minimizing the worst-case with respect to the uncertain parameters cost function value. The controller formulation is *robust* in the sense that all constraints are attempted satisfied for all possible values of the uncertain parameters.

6.2.1 Formulation of the Open-Loop Min-Max NMPC Problem as an mp-NLP Problem

Consider the discrete-time nonlinear system:

$$x(t+1) = f(x(t), u(t), \theta) \tag{6.1}$$

$$y(t) = Cx(t) \tag{6.2}$$

where $x(t) \in \mathbb{R}^n$, $u(t) \in \mathbb{R}^m$, and $y(t) \in \mathbb{R}^p$ are the state, input and output variable, θ is the vector of time-invariant uncertain parameters that is assumed to belong to a bounded polyhedral set $\theta \in \Theta^A \subset \mathbb{R}^s$. It is assumed that the function f is sufficiently smooth. It is also supposed that a full measurement of the state $x(t)$ is available at the current time t. We consider the following *open-loop* robust NMPC problem: For the current $x(t)$, NMPC minimizes the worst-case cost function through the following optimization:

Problem 6.1:

$$V^*_{\max}(x(t)) = \min_{U} \max_{\theta \in \Theta^A} J(U, x(t), \theta) \tag{6.3}$$

subject to $x_{t|t} = x(t)$ and:

$$y_{\min} \le y_{t+k|t} \le y_{\max}\,,\ \forall \theta \in \Theta^A\,,\ k = 1, \ldots, N \tag{6.4}$$

$$u_{\min} \le u_{t+k} \le u_{\max}\,,\ k = 0, 1, \ldots, N-1 \tag{6.5}$$

$$x^T_{t+N|t} x_{t+N|t} \le \delta\,,\ \forall \theta \in \Theta^A \tag{6.6}$$

$$x_{t+k+1|t} = f(x_{t+k|t}, u_{t+k}, \theta)\,,\ \theta \in \Theta^A\,,\ k \ge 0 \tag{6.7}$$

$$y_{t+k|t} = Cx_{t+k|t}\,,\ k \ge 0 \tag{6.8}$$

with $U = [u_t, u_{t+1}, \ldots, u_{t+N-1}]$ and the cost function given by:

$$J(U, x(t), \theta) = \sum_{k=0}^{N-1} \left[x^T_{t+k|t} Q x_{t+k|t} + u^T_{t+k} R u_{t+k} \right] + x^T_{t+N|t} P x_{t+N|t} \tag{6.9}$$

Here, N is a finite horizon. The formulation implies that a direct single shooting strategy is employed, see Section 2.2.2.1, i.e. the equality constraints (6.7)–(6.8) are substituted and eliminated in the cost and constraint functions. In (6.3), the existence of the minimum and maximum are implicitly assumed. From a stability point of view it is desirable to choose δ in (6.6) as small as possible [23]. However, due to the fact that $x_{t+N|t}$ depends on the unknown θ, the feasibility of Problem 6.1 will rely on δ being sufficiently large. A part of the NMPC design will be to address this trade-off. If the system is asymptotically stable (or pre-stabilized), N is large, and possibly an integral action is introduced to account for the steady-state effect of the uncertainty, then it is more likely that the choice of a small δ will be possible.

The following assumptions are made:

Assumption 6.1. $P, Q, R \succ 0$.

Assumption 6.2. $y_{\min} < 0 < y_{\max}$.

Assumption 6.3. *θ is time-invariant uncertainty that belongs to a bounded polyhedral set, i.e. $\theta = \text{const} \in \Theta^A$. The polyhedral set Θ^A is defined by $\Theta^A = \{\theta \in \mathbb{R}^s \,|\, \theta^L \leq \theta \leq \theta^U\}$, where θ^L and θ^U represent given lower and upper bounds on θ.*

Assumption 6.4. *For each $\theta \in \Theta^A$ there exists $u_{st} \in \mathbb{R}^m$ satisfying $u_{\min} \leq u_{st} \leq u_{\max}$, and such that $f(0, u_{st}, \theta) = 0$.*

Assumption 6.4 means that the point $x = 0$, $u = u_{st}$ is a feasible steady state point for system (6.1)–(6.2). It also implies that the steady state value of the control input may be different for the different values of the uncertain parameters.

The worst-case value of cost function (6.9) with respect to the uncertain parameters is denoted by:

$$V_{\max}(U, x(t)) = \max_{\theta \in \Theta^A} J(U, x(t), \theta) \tag{6.10}$$

An optimal solution to the min-max NMPC Problem 6.1 is denoted $U^* = [u_t^*, u_{t+1}^*, ..., u_{t+N-1}^*]$ and the control input is chosen according to the receding horizon policy $u(t) = u_t^*$. The optimization problem can be formulated in a compact form as follows:

Problem 6.2:

$$V_{\max}^*(x(t)) = \min_U \max_{\theta \in \Theta^A} J(U, x(t), \theta) \text{ subject to } G(U, x(t), \theta) \leq 0, \forall \theta \in \Theta^A \tag{6.11}$$

This min-max NMPC problem defines an mp-NLP, since it is NLP in U parameterized by x. Since the equality constraints are eliminated by the direct single shooting strategy, (6.11) contains only inequality constraints. Define the set of N-step robustly feasible initial states as follows:

$$X_f = \{x \in \mathbb{R}^n \,|\, G(U, x, \theta) \leq 0, \forall \theta \in \Theta^A \text{ for some } U \in \mathbb{R}^{Nm}\} \tag{6.12}$$

If Assumption 6.4 is satisfied and δ is chosen such that the Problem 6.1 is feasible, then X_f is a non-empty set. Then, due to Assumption 6.2, the origin is an interior point in X_f.

6.2.2 *Approximate mp-NLP Approach to Explicit Open-Loop Min-Max NMPC*

The numerical computations involved in constructing the approximate explicit state feedback are simplified under the following convexity assumption:

Assumption 6.5. *$G(U,x,\theta)$ is jointly convex for all $(U, x, \theta) \in U^A \times X \times \Theta^A$, where $U^A = [u_{\min}, u_{\max}]^N$ is the set of admissible inputs and $X \subseteq X_f \subseteq \mathbb{R}^n$ is a polytopic set.*

We exploit the result in [12], where it has been shown that if the constraint function $G(U,x,\theta)$ is jointly convex in U and θ, and there is U that is feasible at the vertices of Θ^A, then U is feasible for all $\theta \in \Theta^A$. This is formulated in the following lemma:

Lemma 6.1. *Suppose Assumptions 6.3 and 6.5 hold and denote the vertices of the polyhedron $\Theta^A \subset \mathbb{R}^s$ with $\{\theta_1, \theta_2, ..., \theta_L\}$. Denote also $\widetilde{G}^i(U,x) = G(U,x,\theta_i)$. If there exist U that satisfies the following constraints:*

$$\widetilde{G}^i(U,x) \leq 0,\ i \in \{1, 2, ..., L\} \tag{6.13}$$

then U satisfies the constraints in (6.11).

Thus, we can replace the infinite number of constraints in (6.11) with the following finite set of jointly convex constraints which are function only of U and x:

$$\widetilde{G}(U,x) \leq 0,\ \widetilde{G}(U,x) = \{\widetilde{G}^i(U,x),\ i = 1, 2, ..., L\} \tag{6.14}$$

Then, the Problem 6.2 can be reformulated as:

Problem 6.3:

$$V^*_{\max}(x) = \min_U V_{\max}(U,x) \text{ subject to } \widetilde{G}(U,x) \leq 0 \tag{6.15}$$

where $V_{\max}(U,x)$ is defined by (6.10).

Problem 6.3 defines a mp-NLP problem, since it is an NLP in U parameterized by x. In case the Problem 6.3 is convex, its approximate solution can be found by applying the approximate mp-NLP approach, described in Section 1.1.5.1. Otherwise, the approximate mp-NLP approach from Section 1.1.5.2 should be used, where in addition to the set of vertices of a given hyper-rectangle in the parameter space, the optimal solution is also searched for at several interior points and global optimization methods are applied. Further, if Assumption 6.5 does not hold, then it would not be sufficient to consider the constraints $G(U,x,\theta)$ only at the vertices of the set Θ^A, i.e. it would not be possible to apply Lemma 6.1, but it would be advisable to impose these constraints also at a finite set of interior points of the set Θ^A.

6.2.3 Application 1: Min-Max MPC of a Continuous Stirred Tank Reactor

The considered approximate mp-NLP approach is applied to design an explicit min-max NMPC controller for the continuous stirred tank reactor (CSTR), described in Section 5.5. We consider the set point $\tilde{c}^* = 0.41$, $\tilde{T}^* = 3.3$. Then, the model of the reactor can be written in the form [13]:

$$\frac{dx_1}{dt} = \frac{(1-\tilde{c}^* - x_1)}{q} - k_0 e^{-\frac{E}{(\tilde{T}^*+x_2)}} (\tilde{c}^* + x_1) \tag{6.16}$$

$$\frac{dx_2}{dt} = \frac{(\tilde{T}_f - \tilde{T}^* - x_2)}{q} + k_0 e^{-\frac{E}{(\tilde{T}^*+x_2)}} (\tilde{c}^* + x_1) - \alpha u(\tilde{T}^* + x_2 - \tilde{T}_c) \tag{6.17}$$

where x_1 and x_2 denote the deviations of the dimensionless concentration and temperature from the set point values ($x_1 = \tilde{c} - \tilde{c}^*$, $x_2 = \tilde{T} - \tilde{T}^*$). The coolant flow-rate u is a real-valued control variable. The heat transfer coefficient α is an uncertain parameter that belongs to the interval:

$$1.9 \cdot 10^{-4} \leq \alpha \leq 2.5 \cdot 10^{-4} \tag{6.18}$$

The values of the other parameters are given in Section 5.5. The coolant flow-rate is constrained to be:

$$0 \leq u \leq 600 \tag{6.19}$$

We discretize the model (6.16)–(6.17) using a sampling time $T_s = 1$. The forward Euler method with step size $T_E = 0.01$ is used to integrate the equations (6.16)–(6.17).

The mp-NLP formulation described in Section 6.2.2 is applied to design an explicit *open-loop* min-max NMPC controller for this reactor. The NMPC minimizes the worst-case (maximal) value with respect to the uncertain parameter $\theta = \alpha$ of the cost function (6.9) subject to the system equations (6.16)–(6.17) and the input constraint (6.19). In (6.9), the cost matrices are $Q = P = \mathrm{diag}\{100, 300\}$, $R = 1 \cdot 10^{-6}$. The horizon is $N = 30$. In (6.6), it is chosen $\delta = 0.002$. The state space to be partitioned is defined by $X = [-0.4, 0.6] \times [-0.4, 0.5]$.

The state space partition of the approximate min-max NMPC controller resulting from the algorithms and procedures in Section 1.1.5.2 is shown in Fig. 6.1. It has 94 regions and 10 levels of search. With one scalar comparison required at each level of the $k-d$ tree, 10 arithmetic operations are required in the worst case to determine which region the state belongs to. Totally, 14 arithmetic operations are needed in real-time to compute the control input (10 comparisons, 2 multiplications and 2 additions).

The performance of the closed-loop system was simulated for initial condition $x(0) = [0.58,\ 0.3]^T$ and for three values of the uncertain parameter ($\alpha = 1.9 \cdot 10^{-4}$, $\alpha = 2.2 \cdot 10^{-4}$, $\alpha = 2.5 \cdot 10^{-4}$). The resulting closed-loop response is depicted in the state space (Fig. 6.1), as well as trajectories in time (Fig. 6.2 and Fig. 6.3). It can be seen that the explicit approximate min-max NMPC controller brings the reactor to the desired set point despite of the model uncertainty, and the constraints imposed on the system are satisfied. In order to avoid a possible offset, the dual-mode control strategy of [26] was applied and a locally stabilizing control law was used in a neighborhood of the origin.

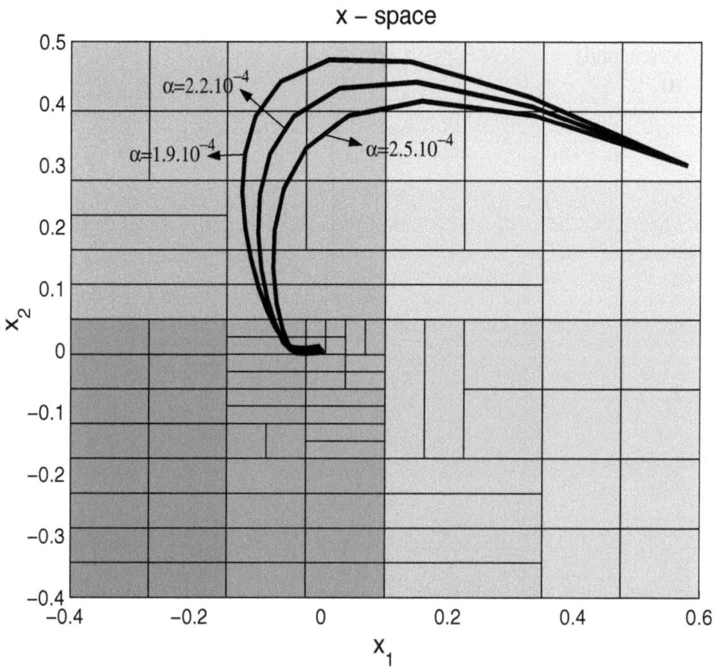

Fig. 6.1 State space partition of the approximate explicit open-loop min-max NMPC and the state trajectories corresponding to $\alpha = 1.9 \cdot 10^{-4}$, $\alpha = 2.2 \cdot 10^{-4}$, $\alpha = 2.5 \cdot 10^{-4}$.

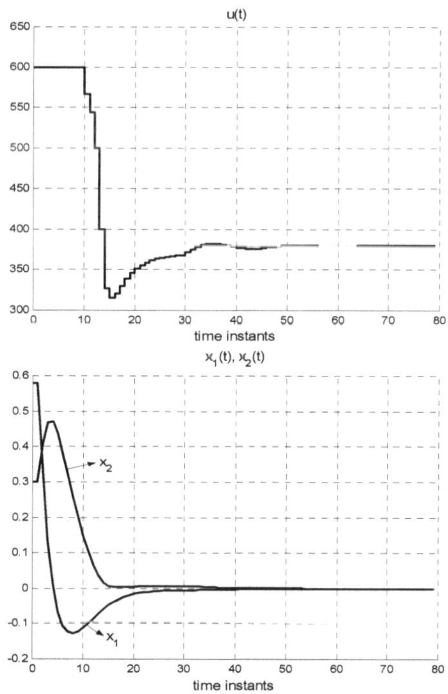

Fig. 6.2 Control input and state trajectory corresponding to $\alpha = 1.9 \cdot 10^{-4}$.

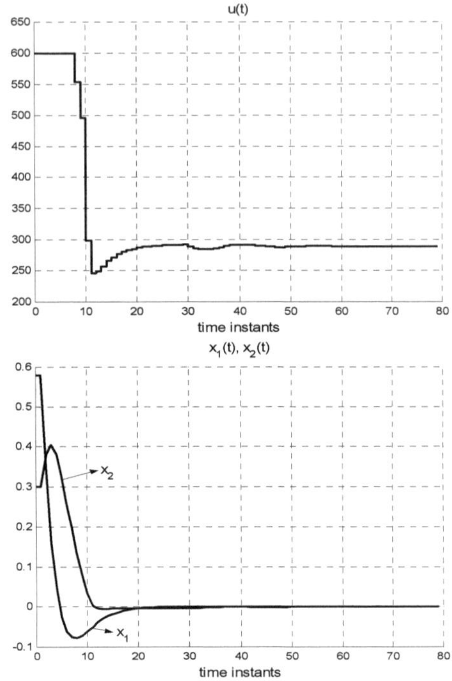

Fig. 6.3 Control input and state trajectory corresponding to $\alpha = 2.5 \cdot 10^{-4}$.

6.3 Explicit Closed-Loop Min-Max MPC of Constrained Nonlinear Systems with Bounded Uncertainties

This section considers the approximate mp-NLP approach [11] to explicit solution of *closed-loop* (*feedback*) min-max NMPC problems for general constrained nonlinear discrete-time systems in the presence of bounded disturbances and/or parameter uncertainties. The approach consists in constructing a *piecewise nonlinear* (PWNL) approximation to the optimal sequence of feedback control policies, defined on an orthogonal state space partition. Conditions guaranteeing the l_2-stability of the closed-loop system are derived.

6.3.1 Formulation of the Closed-Loop Min-Max NMPC Problem as an mp-NLP Problem

Consider the discrete-time nonlinear system:

$$\begin{aligned} x(t+1) &= f(x(t), u(t), w(t)) \\ y(t) &= h(x(t), u(t), w(t)), \end{aligned} \tag{6.20}$$

where $x(t) \in \mathbb{R}^n$, $u(t) \in \mathbb{R}^m$, $y(t) \in \mathbb{R}^r$ and $w(t) \in \mathbb{R}^q$ are the state, input, output and disturbance variable. The following constraints are imposed:

$$u_{\min} \le u(t) \le u_{\max},\ y_{\min} \le y(t) \le y_{\max} \tag{6.21}$$

Following [21],

Assumption 6.6. *f and h are C^2 functions with $f(0,0,0) = 0$, $h(0,0,0) = 0$.*

Assumption 6.7. *$y_{\min} < 0 < y_{\max}$ and $u_{\min} < 0 < u_{\max}$.*

Assumption 6.8. *Let $\widetilde{X}$ be a non-empty set containing the origin as an interior point, and let t_0 be a positive integer. The system (6.20) is zero-state detectable in $\widetilde{X}$, i.e. $\forall x(0) \in \widetilde{X}$ and $\forall u(\cdot)$ such that constraints (6.21) are satisfied $\forall t \ge 0$ and $x(t) \in \widetilde{X}$, $\forall t \ge t_0$, we have $y(t)|_{w=0} = 0$, $\forall t \ge t_0 \Rightarrow \lim_{t\to\infty} x(t) = 0$.*

Assumption 6.9. *There exists a positive constant γ_Δ, such that the disturbance w satisfies:*

$$\|w(t)\|^2 \le \gamma_\Delta^2 \|y(t)\|^2,\ t \ge t_0. \tag{6.22}$$

Let $x(t) = x$ and $u(t) = u$. Then, the space of the admissible disturbances is denoted by $W^A(u,x) \subset \mathbb{R}^q$. As mentioned in [21], inequality (6.22) can also represent a wide class of modeling errors. As in [21], first a H_∞ control problem is defined:

Definition 6.1 (H_∞ control problem). Design a state-feedback control law:

$$u = k(x) \tag{6.23}$$

guaranteeing that the closed-loop system (6.20)–(6.23) with input $w \in W^A(u,x)$ and output y has a finite l_2-gain $\le \gamma$ in a bounded positively invariant set Ω, that is, $\forall x(t) \in \Omega$:

i. $x(t+i) \in \Omega,\ \forall i > 0$.
ii. $u_{\min} \le k(x(t+i)) \le u_{\max}$ and $y_{\min} \le h(x(t+i), k(x(t+i)), w(t+i)) \le y_{\max}$, $\forall i \ge 0$.
iii. There exists a positive definite function $\beta(x(t))$, such that $\forall T \ge 0$:

$$\sum_{i=0}^{T} \|y(t+i)\|^2 \le \gamma^2 \sum_{i=0}^{T} \|w(t+i)\|^2 + \beta(x(t)) \tag{6.24}$$

for any non-zero $w \in W^A(u,x)$.

The following assumption is also made [21]:

Assumption 6.10. *Suppose that there exists an auxiliary control law $u = k_a(x)$ that solves the H_∞ control problem, with a domain of attraction Ω_a, whose boundary is assumed to be a level curve of a positive function $V_{k_a}(x)$ such that:*

$$\begin{aligned} &V_{k_a}(f(x,k_a(x),w)) - V_{k_a}(x) \le -\frac{1}{2}(\|y\|^2 - \gamma^2\|w\|^2),\\ &\forall x \in \Omega_a,\ \forall w \in W^A(u,x) \end{aligned} \tag{6.25}$$

and $V_{k_a}(0) = 0$.

Definition 6.2 (Admissible disturbance realization). Let $K = \{k_0, k_1, \dots, k_{N-1}\} \triangleq \{k_0(x_{t|t}), k_1(x_{t+1|t}), \dots, k_{N-1}(x_{t+N-1|t})\}$ be a vector of feedback control policies and N be a finite horizon. Consider the closed-loop system for $i = 0, 1, 2, \dots, N-1$:

$$\begin{aligned} x_{t+i+1|t} &= f(x_{t+i|t}, k_i(x_{t+i|t}), w_{t+i}) \\ y_{t+i|t} &= h(x_{t+i|t}, k_i(x_{t+i|t}), w_{t+i}) \end{aligned} \tag{6.26}$$

with initial state $x_{t|t} = x$. Then, the disturbance realization $W = \{w_t, \dots, w_{t+N-1}\} \in \mathbb{R}^{qN}$ is admissible for the given K and x if the following holds:

$$\|w_{t+i}\|^2 \leq \gamma_\Delta^2 \|y_{t+i|t}\|^2, \; i = 0, 1, 2, \dots, N-1. \tag{6.27}$$

The space of the admissible disturbance realizations over horizon N and corresponding to the given K and x is denoted by $W^B(K,x) \subset \mathbb{R}^{qN}$.

It is supposed that a full measurement x of the state is available at the current time t. We consider the *feedback* min-max NMPC problem [22]:

Definition 6.3 (Constrained feedback min-max NMPC problem). Suppose that Assumptions 6.6–6.10 hold. For the current x, the *feedback* min-max NMPC solves the following optimization problem:

$$V^o_{\max}(x) = \min_K \max_{W \in W^B(K,x)} J(K, x, W) \tag{6.28}$$

subject to $x_{t|t} = x$ and:

$$y_{\min} \leq y_{t+i|t} \leq y_{\max}, \; i = 1, \dots, N-1 \tag{6.29}$$

$$u_{\min} \leq u_{t+i} \leq u_{\max}, \; i = 0, 1, \dots, N-1 \tag{6.30}$$

$$x_{t+N|t} \in \Omega_a \tag{6.31}$$

$$u_{t+i} = k_i(x_{t+i|t}), \; i = 0, 1, \dots, N-1 \tag{6.32}$$

$$x_{t+i+1|t} = f(x_{t+i|t}, u_{t+i}, w_{t+i}), \; w_{t+i} \in W^A(u_{t+i}, x_{t+i|t}), \; 0 \leq i \leq N-1 \tag{6.33}$$

$$y_{t+i|t} = h(x_{t+i|t}, u_{t+i}, w_{t+i}), \; w_{t+i} \in W^A(u_{t+i}, x_{t+i|t}), \; 0 \leq i \leq N-1 \tag{6.34}$$

and the cost function given by:

$$J(K, x, W) = \frac{1}{2} \sum_{i=0}^{N-1} \left[\|y_{t+i|t}\|^2 - \gamma^2 \|w_{t+i}\|^2 \right] + V_{k_a}(x_{t+N|t}) \tag{6.35}$$

Here, N is the finite horizon and γ is the l_2-gain which is interpreted as the disturbance attenuation level. Note that in (6.28)–(6.35) w_{t+i} denotes a single disturbance at time instant $t+i$, while W is an admissible disturbance realization as specified in Definition 6.2. An auxiliary control law $k_a(x)$ is typically obtained by solving the H_∞ control problem for the linearized system [26]. Thus, a practical way to compute a nonlinear control $k_a(x)$ satisfying Assumption 6.10 is suggested in [21].

An optimal solution to the *feedback* min-max NMPC problem (6.28)–(6.35) is denoted $K^o = \{k_0^o, k_1^o, \dots, k_{N-1}^o\} \triangleq \{k_0^o(x_{t|t}), k_1^o(x_{t+1|t}), \dots, k_{N-1}^o(x_{t+N-1|t})\}$ and the control input is chosen according to the receding horizon policy $u(x_{t|t}) = k_0^o(x_{t|t})$. It is assumed that:

Assumption 6.11. *Each feedback control policy* $k_i(x_{t+i|t})$, $i = 0, 1, \dots, N-1$, *has the form:*

$$k_i(x_{t+i|t}) = \alpha_i k_a(x_{t+i|t}) + r_i(\xi_i, x_{t+i|t}) = g_i(p_i, x_{t+i|t}), \tag{6.36}$$

where $p_i = [\alpha_i^T \ \xi_i^T]^T \in \mathbb{R}^{n_i}$ *are the parameters that need to be optimized,* $k_a(x_{t+i|t})$ *is an auxiliary control law that satisfies Assumption 6.10, and* $r_i(\xi_i, x_{t+i|t})$ *is a continuous function with* $r_i(\xi_i, 0) = 0$.

In general, the parameterization of the form (6.36) would lead to an approximate solution to the *feedback* min-max NMPC problem (6.28)–(6.35). Denote with P the whole set of parameters that need to be determined, i.e. $P = [p_0^T \ p_1^T \ \dots \ p_{N-1}^T]^T \in \mathbb{R}^{n_p}$, where $n_p = \sum_{i=0}^{N-1} n_i$. Then, the worst-case value of cost function (6.35) with respect to the disturbances is denoted by:

$$V_{\max}(P, x) = \max_{W \in W^B(P,x)} J(P, x, W) \tag{6.37}$$

Note that the argument K is now substituted with the argument P. Using a direct single shooting strategy to eliminate all the equality constraints (6.32)–(6.34), the optimization problem (6.28)–(6.35) can be formulated in a compact form as follows [11]:

Problem 6.4:

$$V_{\max}^o(x) = \min_P \max_{W \in W^B(P,x)} J(P, x, W) \tag{6.38}$$

$$\text{subject to } G(P, x, W) \le 0, \ \forall W \in W^B(P, x) \tag{6.39}$$

Problem 6.4 defines an mp-NLP, since it is NLP in P parameterized by x. We remark that the constraints function $G(P, x, W)$ in (6.39) is implicitly defined by (6.29)–(6.34). Define the set of N-step robustly feasible initial states:

$$X_f = \{x \in \mathbb{R}^n \,|\, G(P, x, W) \le 0, \forall W \in W^B(P, x) \text{ for some } P \in \mathbb{R}^{n_p}\} \tag{6.40}$$

If the problem (6.28)–(6.35) is feasible, then X_f is a non-empty set. Then, due to Assumption 6.7, the origin is an interior point in X_f.

As mentioned in Chapter 1, in parametric programming problems one seeks the solution $P^o(x)$ as an explicit function of the parameters x in some set $\underline{X} \subseteq X_f \subseteq \mathbb{R}^n$ [7]. However, in the general case, an *exact* explicit solution of Problem 6.4 with the associated shape of the state space partition can not be found. Therefore, it would be necessary to use methods for *approximate* explicit solution by preliminary specifying the structure of the partition. In [9, 11], practical computational methods for constructing an explicit *approximate* solution of *feedback* min-max NMPC problems for general constrained nonlinear systems are suggested, which are based on an orthogonal structure of the state space partition. Since the regions in the partition do not overlap (except at the boundary), the approximation corresponds to orthogonal basis-functions that form a complete basis on the space of continuous functions. This ensures an arbitrarily good approximation if the optimal solution is a continuous function. Note that this type of partition does not impose any restrictions on the class of problems that can be solved.

6.3.2 Approximate mp-NLP Approach to Explicit Closed-Loop Min-Max NMPC

In [9, 11], an approximate mp-NLP approach to explicit solution of the *feedback* (*closed-loop*) min-max NMPC problem (Definition 6.3) is proposed. In contrast to the method in [8] (considered in Section 6.2) where a sequence of control actions is optimized, here the optimization is performed over a sequence of feedback control policies. Another difference from most approximate mp-NLP approaches, where a *piecewise linear* solution is obtained, is that the presented method constructs an explicit approximate solution, which represents a *piecewise nonlinear* function.

6.3.2.1 Non-convexity and Close-to-Global Solutions

From a physical insight on the considered system (6.20), it is supposed that the disturbance w can vary in the range:

$$w_{\min} \leq w(t) \leq w_{\max}, \tag{6.41}$$

with known $w_{\min}$, $w_{\max}$. The procedure used to generate a discrete set of admissible disturbance realizations is the following [11]:

Procedure 6.1 (Generation of discrete set of admissible disturbance realizations). *Consider system (6.20), where* $w(t) \in [w_{\min}; w_{\max}]$. *Let* N *be a finite horizon and* $K = \{k_0, k_1, ..., k_{N-1}\}$ *be a vector of feedback control policies where each feedback function* $k_i(x)$, $i = 0, ..., N-1$, *has the form (6.36). Suppose that the initial state of the system (6.20) is* $x_{t|t} = x$ *and let* $j_{\max}$ *be a positive integer. Then, for a given vector* $P = [p_0^T \ p_1^T \ ... \ p_{N-1}^T]^T$ *of parameters of* K, *a finite set* $W^0(P,x) = \{W_1, W_2, ..., W_{N_W}\}$ *of admissible disturbance realizations is generated where each realization* $W_s = \{w_t^s, w_{t+1}^s, ..., w_{t+i}^s, ..., w_{t+N-1}^s\}$, $s = 1, 2, ..., N_W$ *is determined by applying Algorithm 6.1.*

Algorithm 6.1. Generation of an admissible disturbance realization.

Input: N, $P = [p_0^T \; p_1^T \; \dots \; p_{N-1}^T]^T$, x, $j_{\max}$.
Output: $W_s = \{w_t^s, w_{t+1}^s, \dots, w_{t+i}^s, \dots, w_{t+N-1}^s\}$.
1. Let $i = 0$.
2. **while** $i \leq N-1$ **do**
3. Let $flag = 0$, $j = 0$.
4. **while** $flag = 0$ **do**
5. Generate value $w_{t+i}^s \in [w_{\min}; w_{\max}]$ by using random generator with uniform distribution.
6. $j = j+1$.
7. **if** $\|w_{t+i}^s\|^2 \leq \gamma_\Delta^2 \|h(x_{t+i|t}, k_i(x_{t+i|t}), w_{t+i}^s)\|^2$ **then**
8. Compute $x_{t+i+1|t} = f(x_{t+i|t}, k_i(x_{t+i|t}), w_{t+i}^s)$.
9. $flag = 1$.
10. **else**
11. **if** $j > j_{\max}$, **terminate** (an admissible disturbance realization is not found).
12. **end if**
13. **end while**
14. $i = i+1$.
15. **end while**

In Algorithm 6.1, the parameter $j_{\max}$ denotes the maximal allowed number of unsuccessful iterations and it is typically chosen to be $j_{\max} = 100q$, where q is the dimension of w. A special case is the case when the disturbance is of the form $w(t) = d^T y(t)$, where $d \in \mathbb{R}^r$ is a vector of uncertain parameters with $d_{\min} \leq d \leq d_{\max}$. Then, the set of the admissible disturbance realizations can be generated by simulating the closed-loop system response for different values $d^s \in [d_{\min}; d_{\max}]$, $s = 1, 2, \dots, N_W$ of d.

The procedure used to approximate Problem 6.4 is [11]:

Procedure 6.2 (Approximation of Problem 6.4). *Suppose that Assumptions 6.6–6.11 hold. Let P be a given vector of parameters of the sequence K of feedback control policies. Suppose that a finite set $W^0(P,x) = \{W_1, W_2, \dots, W_{N_W}\}$ of admissible disturbance realizations has been determined by applying Procedure 6.1. An estimate $\widetilde{V}_{\max}(P,x)$ of $V_{\max}(P,x)$ is computed as follows:*

$$\widetilde{V}_{\max}(P,x) = \max_{W_i \in W^0(P,x)} J(P,x,W_i) \tag{6.42}$$

Denote with $\widetilde{G}(P,x)$ the set of constraints functions:

$$\widetilde{G}(P,x) = \{G(P,x,W_i),\, W_i \in W^0(P,x)\} \tag{6.43}$$

Then Problem 6.4 is approximated with the following mp-NLP problem:

Problem 6.5:

$$\widetilde{V}^{o}_{\max}(x) = \min_{P} \widetilde{V}_{\max}(P,x) \ \ \textit{subject to} \ \ \widetilde{G}(P,x) \leq 0. \tag{6.44}$$

Thus, we can approximate the infinite number of constraints (6.39) with a finite amount of constraints which are functions only of P and x. For a given min-max NMPC problem it would be necessary to analyze how the size of the set of admissible disturbance realizations generated with Procedure 6.1 would effect the worst-case cost function value and the satisfaction of constraints in Problem 6.4. It should be expected that with the increase of the number of the generated disturbance sequences, the probability of satisfaction of the constraints in Problem 6.4 would be higher. On the other hand, this will lead to an increase of the computational efforts related to the design of the explicit NMPC controller. Therefore, for every specific min-max NMPC problem, a tradeoff should be made and a reasonable number of admissible disturbance realizations should be determined. Hereafter, let $X \subset \mathbb{R}^n$ be a hyper-rectangle where we seek an explicit approximate solution of Problem 6.5.

Problem 6.5 can be non-convex with multiple local minima. Therefore, it would be necessary to apply an efficient initialization of Problem 6.5 so to find a close-to-global solution. One possible way to obtain this is to find a close-to-global solution at a point $v_0 \in X_0$ (where X_0 is a hyper-rectangle in the state space) by comparing the local minima corresponding to several initial guesses and then to use this solution as an initial guess at the neighboring points $v_i \in X_0$, $i = 1,2,\ldots,N_1$, i.e. to propagate the solution. For this purpose, Procedures 1.1 and 1.2 from Chapter 1 can be used to generate a set of points $V_0 = \{v_0, v_1, v_2, \ldots, v_{N_1}\}$, associated to X_0, and to find a close-to-global solution at these points, respectively. It should be noted that the notation used here is different from the one in Chapter 1. Thus here, the points and the set of points are denoted with v_i and V_0 (instead of w_i and W_0), the vector of optimization variables is P (instead of z), the objective function and the constraints function in the mp-NLP problem are $\widetilde{V}_{\max}(\cdot,\cdot)$ and $\widetilde{G}(\cdot,\cdot)$ (instead of $f(\cdot,\cdot)$ and $g(\cdot,\cdot)$).

6.3.2.2 Computation of Explicit Approximate Solution

We restrict our attention to a hyper-rectangle $X \subset \mathbb{R}^n$ where we seek to approximate the close-to-global sequence of control policies $K^* = \{k_0^*, k_1^*, \ldots, k_{N-1}^*\} \triangleq \{k_0^*(x_{t|t}), k_1^*(x_{t+1|t}), \ldots, k_{N-1}^*(x_{t+N-1|t})\}$. We require that the state space partition is orthogonal and can be represented as a $k-d$ tree [3]. The main idea of the approximate mp-NLP approach is to construct a *piecewise nonlinear* (PWNL) approximation $\widehat{K} = \{\widehat{k}_0, \widehat{k}_1, \ldots, \widehat{k}_{N-1}\} \triangleq \{\widehat{k}_0(x_{t|t}), \widehat{k}_1(x_{t+1|t}), \ldots, \widehat{k}_{N-1}(x_{t+N-1|t})\}$ to the close-to-global feedback $K^* = \{k_0^*, k_1^*, \ldots, k_{N-1}^*\}$ on X. The constituent sequences of *nonlinear* control policies are denoted with $\widehat{K}_{X_i} = \{\widehat{k}_{0,X_i}, \ldots, \widehat{k}_{N-1,X_i}\} \triangleq \{\widehat{k}_{0,X_i}(x_{t|t}), \ldots, \widehat{k}_{N-1,X_i}(x_{t+N-1|t})\}$ and are defined on hyper-rectangles X_i covering

X. This means that a sequence $\widehat{K}_{X_i}$ is applied for $\forall x_{t|t} \in X_i$. Let $\widehat{K}_{X_0} = \{\widehat{k}_{0,X_0}, ..., \widehat{k}_{N-1,X_0}\}$ be an approximation to the close-to-global solution $K^* = \{k_0^*, ..., k_{N-1}^*\}$, valid in X_0. Denote with $P_{X_0} = [p_{0,X_0}^T \cdots p_{N-1,X_0}^T]^T$ the parameters of $\widehat{K}_{X_0}$. According to Assumption 6.11, $\widehat{k}_{i,X_0}(x_{t+i|t}) = g_i(p_{i,X_0}, x_{t+i|t})$, $i = 0, 1, ..., N-1$. Let $\widehat{\widetilde{V}}_{\max}(P_{X_0}, x)$ be the cost function value due to initial state $x = x_{t|t}$ and sequence $\widehat{K}_{X_0}$ of control policies, i.e.

$$\widehat{\widetilde{V}}_{\max}(P_{X_0}, x) = \max_{W_i \in W^0(P_{X_0}, x)} J(P_{X_0}, x, W_i). \tag{6.45}$$

Then, the approximate sequence

$$\widehat{K}_{X_0} = \{\widehat{k}_{0,X_0}, ..., \widehat{k}_{N-1,X_0}\} \triangleq \{g_0(p_{0,X_0}, x_{t|t}), ..., g_{N-1}(p_{N-1,X_0}, x_{t+N-1|t})\}, \tag{6.46}$$

valid for $\forall x_{t|t} \in X_0$, is computed with the following procedure [11]:

Procedure 6.3 (Computation of explicit approximate solution). *Suppose that Assumptions 6.6–6.11 hold. Consider any hyper-rectangle $X_0 \subseteq X$ with a set of points $V_0 = \{v_0, v_1, v_2, ..., v_{N_1}\}$ determined with Procedure 1.1. Suppose that a close-to-global solution of Problem 6.5 at the points $v_i \in V_0$, $i = 0, 1, 2, ..., N_1$ has been obtained by applying Procedure 1.2 and let $\widetilde{V}_{\max}^*(v_i)$, $i = 0, 1, 2, ..., N_1$ be the close-to-global cost function values. Compute the parameters $P_{X_0} = [p_{0,X_0}^T \cdots p_{N-1,X_0}^T]^T$ of the sequence $\widehat{K}_{X_0} = \{\widehat{k}_{0,X_0}, ..., \widehat{k}_{N-1,X_0}\}$ by solving the NLP:*

$$\min_{P_{X_0}} \sum_{i=0}^{N_1} \left(\widehat{\widetilde{V}}_{\max}(P_{X_0}, v_i) - \widetilde{V}_{\max}^*(v_i) + \mu \| g_0(p_{0,X_0}, v_i) - k_0^*(v_i) \|_2^2 \right) \tag{6.47}$$

$$\text{subject to} \quad \widetilde{G}(P_{X_0}, v_i) \leq 0, \ \forall v_i \in V_0. \tag{6.48}$$

In (6.47), the parameter $\mu > 0$ is a weighting coefficient. Note that the sequence $\widehat{K}_{X_0} - \{\widehat{k}_{0,X_0}, ..., \widehat{k}_{N-1,X_0}\}$, computed with Procedure 6.3, satisfies the constraints in Problem 6.5 only for the discrete set of points $V_0 \subset X_0$.

6.3.2.3 Estimation of Error Bounds

Suppose that the parameters P_{X_0} of the sequence $\widehat{K}_{X_0}$, valid in X_0, has been computed with Procedure 6.3. Then, for the cost function approximation error in X_0 we have:

$$\widetilde{\varepsilon}(x) = \widehat{\widetilde{V}}_{\max}(P_{X_0}, x) - \widetilde{V}_{\max}^*(x) \leq \varepsilon_0, \ x \in X_0. \tag{6.49}$$

The following procedure can be used to obtain an estimate $\widehat{\varepsilon}_0$ of the maximal approximation error ε_0 in X_0 [11]:

Procedure 6.4 (Computation of error bound approximation). *Consider a hyper-rectangle $X_0 \subseteq X$ with a set of points $V_0 = \{v_0, v_1, v_2, \dots, v_{N_1}\}$ determined by applying Procedure 1.1. Compute an estimate $\widehat{\varepsilon}_0$ of the error bound ε_0 through the following maximization:*

$$\widehat{\varepsilon}_0 = \max_{i \in \{0,1,2,\dots,N_1\}} (\widehat{\widetilde{V}}_{\max}(P_{X_0}, v_i) - \widetilde{V}^*_{\max}(v_i)). \tag{6.50}$$

The estimate $\widehat{\varepsilon}_0$ represents an approximate degree of sub-optimality, since it depends on the finite set of admissible disturbance realizations generated with Procedure 6.1.

6.3.2.4 Approximate mp-NLP Algorithm

Assume the tolerance $\bar{\varepsilon} > 0$ of the cost function approximation error is given. Denote with S_{X_0} the volume of a given hyper-rectangular region $X_0 \subset X \subset \mathbb{R}^n$, i.e. $S_{X_0} = \prod_{i=1}^{n} \Delta x_i$, where Δx_i is the size of X_0 along the state variable x_i. Let $S_{\min}$ be the minimal allowed volume of the regions in the partition of X. The following algorithm is proposed to compute the explicit approximate *feedback* min-max NMPC controller on X [11]:

Algorithm 6.2. Explicit feedback min-max NMPC.

Input: Data to Problem 6.5, the number N_0 of internal regions (used in Procedure 1.1), the parameter μ (used in Procedure 6.3), the approximation tolerance $\bar{\varepsilon}$.

Output: Partition $\Pi = \{X_1, X_2, \dots, X_{N_X}\}$ and associated PWNL control function $\widehat{K}_\Pi = \{\widehat{K}_{X_1}, \widehat{K}_{X_2}, \dots, \widehat{K}_{X_{N_X}}\}$.

1. Initialize the partition to the whole hyper-rectangle, i.e. $\Pi = \{X\}$. Mark the hyper-rectangle X as unexplored, $flag := 1$.
2. **while** $flag = 1$ **do**
3.   **while** $\exists$ unexplored hyper-rectangles in Π **do**
4.   Select any unexplored hyper-rectangle $X_0 \in \Pi$.
5.   Compute a solution to Problem 6.5 at the center point v_0 of X_0 by applying Procedure 1.2a.
6.   **if** Problem 6.5 has a feasible solution at v_0 **then**
7.   Define a set of points $V_0 = \{v_0, v_1, v_2, \dots, v_{N_1}\}$ by applying Procedure 1.1.
8.   Compute a solution to Problem 6.5 for x fixed to each of the points v_i, $i = 1, 2, \dots, N_1$ by applying Procedure 1.2b.
9.   **if** Problem 6.5 has a feasible solution at all points v_i, $i = 1, 2, \dots, N_1$ **then**
10.   **if** $0 \in X_0$ **then**
11.   Let $\widehat{K}_{X_0} = k_a(x)$.
12.   **If** $X_0 \subseteq \Omega_a$, mark X_0 as explored and feasible. **Otherwise**, mark X_0 to be split.
13.   **else**

14. Compute a sequence $\widehat{K}_{X_0} = \{\widehat{k}_{0,X_0}, \ldots, \widehat{k}_{N-1,X_0}\}$ of control policies using Procedure 6.3, as an approximation to be used in X_0.
15. **if** a sequence of control policies was found **then**
16. Compute an estimate $\widehat{\varepsilon}_0$ of the error bound ε_0 in X_0 by applying Procedure 6.4.
17. **If** $\widehat{\varepsilon}_0 > \bar{\varepsilon}$, mark the hyper-rectangle X_0 to be split. **Otherwise**, mark X_0 as explored and feasible.
18. **else**
19. Compute the volume S_{X_0} of the hyper-rectangle X_0. **If** $S_{X_0} < S_{\min}$, mark X_0 infeasible and explored. **Otherwise**, mark X_0 to be split.
20. **end if**
21. **end if**
22. **else**
23. Compute the volume S_{X_0} of the hyper-rectangle X_0. **If** $S_{X_0} < S_{\min}$, mark X_0 infeasible and explored. **Otherwise**, mark X_0 to be split.
24. **end if**
25. **else**
26. Compute the volume S_{X_0} of the hyper-rectangle X_0. **If** $S_{X_0} < S_{\min}$, mark X_0 infeasible and explored. **Otherwise**, mark X_0 to be split.
27. **end if**
28. **end while**
29. $flag := 0$
30. **if** $\exists$ hyper-rectangles in Π that are marked to be split **then**
31. $flag := 1$
32. **while** $\exists$ hyper-rectangles in Π that are marked to be split **do**
33. Select any hyper-rectangle $X_0 \in \Pi$ marked to be split.
34. Split X_0 into hyper-rectangles $X_1, \ldots, X_{N_s}$ by applying heuristic splitting rules. Mark $X_1, \ldots, X_{N_s}$ unexplored, remove X_0 from Π, and add $X_1, \ldots, X_{N_s}$ to Π.
35. **end while**
36. **end if**
37. **end while**

In step 34, the heuristic splitting rules from [10] (described in details in Section 1.1.5.2) are applied to partition a given hyper-rectangle X_0. Thus, if a sequence of control policies valid in X_0 is computed, but the required accuracy is not achieved, then X_0 is split by a hyperplane through its center and orthogonal to that axis where a maximal reduction of the approximation error can be achieved. If there is no feasible solution of Problem 6.5 at the center point v_0 of X_0, or the NLP problem (6.47)–(6.48) is infeasible, then X_0 is split by a hyperplane through its center and orthogonal to an arbitrary axis. If some of the points associated to X_0 are feasible and others are not, then X_0 is split into hyper-rectangles such that some of them will include only feasible points.

6.3.3 Stability

6.3.3.1 Computation of Approximate Region of Attraction for the Sub-optimal Closed-Loop System

Let $X_\Pi = \bigcup_{i=1}^{N_X} X_i$, $X_i \in \Pi$ be the set associated to the partition Π obtained with Algorithm 6.2. Consider the suboptimal closed-loop system:

$$x(t+1) = f(x(t), \widehat{k}_0(x(t)), w(t)) \tag{6.51}$$

$$y(t) = h(x(t), \widehat{k}_0(x(t)), w(t)), \tag{6.52}$$

where $\widehat{k}_0(x(t))$ is the approximate PWNL feedback law determined with Algorithm 6.2 and is defined on the set X_Π. The fact that the explicit NMPC controller is specified for an initial condition $x(t) \in X_\Pi$ does not imply that $x(t)$ is within the region of attraction for the system (6.51)–(6.52). Therefore, the set X_Π may not be a domain of attraction for this system. In fact, although a feasible control law exists at state $x(t) \in X_\Pi$, the successor state $x(t+1)$ may go out of the set X_Π. Moreover, the set X_Π may not be convex (see the simulation example in Section 6.3.4). Therefore, first it would be useful to find a set $\Omega_1 \subseteq X_\Pi$, which is an inner convex approximation of the set X_Π. Then, a convex set $\Omega_2 \subseteq \Omega_1$ should be determined such that $\Omega_2 \supset \Omega_a$ and for every initial state that belongs to the set Ω_2, the state trajectory of the system (6.51)–(6.52) will lie in the set Ω_1. This is specified in the following definition [11].

Definition 6.4 (Approximate region of attraction for the suboptimal closed-loop system). Let $\Pi = \{X_1, X_2, ..., X_{N_X}\}$, $X_\Pi = \bigcup_{i=1}^{N_X} X_i$, $X_i \in \Pi$ and $\widehat{K} = \{\widehat{k}_0, \widehat{k}_1, ..., \widehat{k}_{N-1}\}$ be respectively the state space partition, the associated set in the state space and the approximate PWNL sequence of feedback control policies, determined with Algorithm 6.2. Let $\widehat{P} = [\widehat{p}_0^T \; \widehat{p}_1^T \; ... \; \widehat{p}_{N-1}^T]^T$ be the parameters of $\widehat{K}$. Assume that X_Π is a non-empty set. Suppose that there exist polyhedral sets Ω_1 and Ω_2, such that $\Omega_a \subset \Omega_2 \subseteq \Omega_1 \subseteq X_\Pi$. Let $E_{\Omega_2} = \{x^j \,|\, x^j \in \Omega_2, \; j = 1, 2 ..., N_{p2}\}$ denote a finite set of randomly generated points. Let the state of the system (6.51)–(6.52) at time t be $x_{t|t} = x^j \in E_{\Omega_2}$. Consider a finite set $W^0(\widehat{P}, x^j) = \{W_1, W_2 ..., W_{N_W}\}$ of admissible disturbance realizations $W_s = \{w_t^s, w_{t+1}^s, ..., w_{t+N-1}^s\}$, $s = 1, 2 ..., N_W$, generated by applying Procedure 6.1. Let $X^{s,j} = \{x_{t+1|t}^{s,j}, x_{t+2|t}^{s,j}, ..., x_{t+N|t}^{s,j}\}$ denote the state trajectory of the system (6.51)–(6.52) obtained with $\widehat{K}$ and corresponding to initial state $x^j \in E_{\Omega_2}$ and disturbance realization $W_s \in W^0(\widehat{P}, x^j)$, i.e.:

$$x_{t+i+1|t}^{s,j} = f(x_{t+i|t}^{s,j}, \widehat{k}_0(x_{t+i|t}^{s,j}), w_{t+i}^s), \; i = 0, 1, 2, ..., N-1, \tag{6.53}$$

Then, if:

$$X^{s,j} \in \Omega_1, \; \forall x^j \in E_{\Omega_2} \text{ and } \forall W_s \in W^0(\widehat{P}, x^j), \tag{6.54}$$

the set Ω_2 is referred to as an approximate region of attraction for the suboptimal closed-loop system (6.51)–(6.52).

Let S_{Ω_1} and S_{Ω_2} denote the volumes of the polyhedral sets Ω_1 and Ω_2 defined as their Lebesgue measures, i.e. $S_{\Omega_1} = \int_{\Omega_1} dx$ and $S_{\Omega_2} = \int_{\Omega_2} dx$. The volume of the set X_Π is $S_{X_\Pi} = \sum_{i=1}^{N_X} \int_{X_i} dx$, i.e. it represents the sum of the Lebesgue measures of all regions $X_i \in \Pi$. Then, the following procedure is applied to compute an approximate region of attraction for the closed-loop system (6.51)–(6.52) [11]:

Procedure 6.5 (Computation of approximate region of attraction for the sub-optimal closed-loop system). *Let* $\Pi = \{X_1, X_2, \ldots, X_{N_X}\}$, $X_\Pi = \bigcup_{i=1}^{N_X} X_i$, $X_i \in \Pi$ *and* $\widehat{K} = \{\widehat{k}_0, \widehat{k}_1, \ldots, \widehat{k}_{N-1}\}$ *be respectively the state space partition, the associated set in the state space and the approximate PWNL sequence of feedback control policies, determined with Algorithm 6.2. Assume the set* X_Π *is non-empty. Suppose that there exist polyhedral sets* $\Omega_1 = \{x \in X_\Pi \,|\, a_i^1 \leq h_i^1 x \leq b_i^1,\, i = 1,2,\ldots,N_{\Omega_1}\}$ *and* $\Omega_2 = \{x \in X_\Pi \,|\, a_i^2 \leq h_i^2 x \leq b_i^2,\, i = 1,2,\ldots,N_{\Omega_2}\}$, *such that* $\Omega_a \subset \Omega_2 \subseteq \Omega_1 \subseteq X_\Pi$. *Let* $E_{\Omega_1} = \{x^k \,|\, x^k \in \Omega_1,\, k = 1,2,\ldots,N_{p1}\}$ *and* $E_{\Omega_2} = \{x^j \,|\, x^j \in \Omega_2,\, j = 1,2,\ldots,N_{p2}\}$ *denote finite sets of randomly generated points. Then, for specified* N_{Ω_1}, N_{Ω_2}, N_{p1} *and* N_{p2}, *the approximate region of attraction for the closed-loop system (6.51)–(6.52) is computed by implementing the following steps:*

1. *Determine the polyhedron* $\Omega_1^* = \{x \in X_\Pi \,|\, a_i^{1^*} \leq h_i^{1^*} x \leq b_i^{1^*},\, i = 1,2,\ldots,N_{\Omega_1}\}$, *where* $a_i^{1^*}$, $h_i^{1^*}$, $b_i^{1^*}$, $i = 1,2,\ldots,N_{\Omega_1}$ *are computed by solving the optimization problem:*

$$\{a_i^{1^*}, h_i^{1^*}, b_i^{1^*}, i = 1,2,\ldots,N_{\Omega_1}\} = \arg \min_{a_i^1, h_i^1, b_i^1, i=1,\ldots,N_{\Omega_1}} \left| S_{\Omega_1} - S_{X_\Pi} \right|$$
$$\textit{subject to } E_{\Omega_1} \subseteq X_\Pi. \tag{6.55}$$

2. *Determine the approximate region of attraction as the following polyhedron* $\Omega_2^* = \{x \in X_\Pi \,|\, a_i^{2^*} \leq h_i^{2^*} x \leq b_i^{2^*},\, i = 1,2,\ldots,N_{\Omega_2}\}$, *where* $a_i^{2^*}$, $h_i^{2^*}$, $b_i^{2^*}$, $i = 1,2,\ldots,N_{\Omega_2}$ *are computed by solving the optimization problem:*

$$\{a_i^{2^*}, h_i^{2^*}, b_i^{2^*}, i = 1,2,\ldots,N_{\Omega_2}\} = \arg \min_{a_i^2, h_i^2, b_i^2, i=1,\ldots,N_{\Omega_2}} \left| S_{\Omega_2} - S_{\Omega_1} \right|$$
$$\textit{subject to } E_{\Omega_2} \subset \Omega_1,\, \Omega_a \subset \Omega_2, \textit{ and condition (6.54)}. \tag{6.56}$$

Problems (6.55) and (6.56) are nonlinear programming problems and nonlinear programming techniques [1] can be used to solve them. Further in the paper, the sets Ω_2^* and Ω_1^* determined with Procedure 6.5 will be denoted as Ω_2 and Ω_1.

After Procedure 6.5 is implemented, a partition $\Pi^{RH} = \{R_1, R_2, \ldots, R_{N_R}\}$ is built such that $\Omega_1 = \bigcup_{i=1}^{N_R} R_i$. Each region $R_i \in \Pi^{RH}$ represents either a hyper-rectangular region, i.e. $R_i \equiv X_j$ or a polyhedral region, i.e. $R_i = X_j \cap \Omega_1$, where $X_j \in \Pi$. The PWNL function associated to the partition Π^{RH} is defined as $\widehat{K}_{\Pi^{RH}} = \{\widehat{K}_{R_1}, \widehat{K}_{R_2}, \ldots, \widehat{K}_{R_{N_R}}\}$, where $\widehat{K}_{R_i} \equiv \widehat{K}_{X_j}$, $\widehat{K}_{X_j} \in \widehat{K}_\Pi$, given that $R_i \equiv X_j$ or $R_i = X_j \cap \Omega_1$. As result, we obtain a partition Π^{RH} and an approximate PWNL sequence of feedback control policies $\widehat{K}^{RH} = \{\widehat{k}_0^{RH}, \widehat{k}_1^{RH}, \ldots, \widehat{k}_{N-1}^{RH}\}$ defined on the set Ω_1.

6.3.3.2 Stability Result

This section considers the stability of the closed-loop system:

$$x(t+1) = f(x(t), \widehat{k}_0^{RH}(x(t)), w(t)) \tag{6.57}$$

$$y(t) = h(x(t), \widehat{k}_0^{RH}(x(t)), w(t)), \tag{6.58}$$

where $\widehat{k}_0^{RH}(x(t))$ is the approximate PWNL feedback law determined with Algorithm 6.2 and Procedure 6.5 and is defined on the approximate region of attraction Ω_2 computed with Procedure 6.5.

The following notation is introduced. Let N be the prediction horizon and $x_{t|t} = x$ is the initial state of the system (6.57)–(6.58). For any $x \in \Omega_1$, let $\widehat{K}_N \equiv \widehat{K}^{RH} = \{\widehat{k}_0^{RH}(x_{t|t}), \ldots, \widehat{k}_{N-1}^{RH}(x_{t+N-1|t})\}$ denote the approximate solution to the optimization Problem 6.5. Let $X_N = \{x_{t+1|t}, \ldots, x_{t+N|t}\}$ and $Y_N = \{y_{t|t}, \ldots, y_{t+N-1|t}\}$ denote the state and output trajectories of system (6.57)–(6.58) obtained with $\widehat{K}_N$ and corresponding to a disturbance realization $W_N = \{w_t, \ldots, w_{t+N-1}\} \in W^B(\widehat{K}_N, x)$ ($W^B(\widehat{K}_N, x) \subset \mathbb{R}^{qN}$ is the set of the admissible disturbance realizations over horizon N). Let $\widehat{V}_{\max}(x, N)$ be the worst-case cost function value due to initial state $x_{t|t} = x$ and sequence $\widehat{K}_N$, i.e.:

$$\widehat{V}_{\max}(x, N) = \max_{W_N \in W^B(\widehat{K}_N, x)} J(x, \widehat{K}_N, W_N, N), \tag{6.59}$$

where

$$J(x, \widehat{K}_N, W_N, N) = \frac{1}{2} \sum_{i=0}^{N-1} [\|y_{t+i|t}\|^2 - \gamma^2 \|w_{t+i}\|^2] + V_{k_a}(x_{t+N|t}). \tag{6.60}$$

Consider the sequence

$$\widehat{K}_{N+1} = \{\widehat{k}_0^{RH}(x_{t|t}), \ldots, \widehat{k}_{N-1}^{RH}(x_{t+N-1|t}), k_a(x_{t+N|t})\} \tag{6.61}$$

for the Problem 6.5 with horizon $N+1$. Then, $X_{N+1} = \{x_{t+1|t}, \ldots, x_{t+N|t}, x_{t+N+1|t}\}$ and $Y_{N+1} = \{y_{t|t}, \ldots, y_{t+N-1|t}, y_{t+N|t}\}$ are the associated state and output trajectories of the system (6.57)–(6.58) corresponding to initial state $x_{t|t} = x$ and a disturbance realization $W_{N+1} = \{w_t, \ldots, w_{t+N-1}, w_{t+N}\} \in W^C(\widehat{K}_{N+1}, x)$ ($W^C(\widehat{K}_{N+1}, x) \subset \mathbb{R}^{q(N+1)}$ is the set of the admissible disturbance realizations over horizon $N+1$). Let $\widehat{V}_{\max}(x, N+1)$ be the worst-case cost function value due to initial state $x_{t|t} = x$ and sequence $\widehat{K}_{N+1}$, i.e.:

$$\widehat{V}_{\max}(x, N+1) = \max_{W_{N+1} \in W^C(\widehat{K}_{N+1}, x)} J(x, \widehat{K}_{N+1}, W_{N+1}, N+1), \tag{6.62}$$

where $J(x,\widehat{K}_{N+1},W_{N+1},N+1) = \frac{1}{2}\sum_{i=0}^{N}[\|y_{t+i|t}\|^2 - \gamma^2\|w_{t+i}\|^2] + V_{k_a}(x_{t+N+1|t})$.

Let $\widehat{P}^{RH}$ be the parameters of $\widehat{K}^{RH}$. The following assumption is made on the solution $\widehat{K}^{RH}$ and the sets Ω_1 and Ω_2 resulting from Algorithm 6.2 and Procedure 6.5 [11]:

Assumption 6.12 (Constraints satisfaction). *The constraints $G(\widehat{P}^{RH},x,W) \leq 0$ are satisfied for all $x \in \Omega_1$ and all $W \in W^B(\widehat{P}^{RH},x)$. The sets Ω_1 and Ω_2 are such that $\Omega_a \subset \Omega_2 \subseteq \Omega_1 \subseteq X_\Pi$ and $x_{t+i+1|t} = f(x_{t+i|t},\widehat{k}_i^{RH}(x_{t+i|t}),w_{t+i}) \in \Omega_1$, $\forall x_{t|t} \in \Omega_2$, $\forall w_{t+i} \in W^A(\widehat{k}_i^{RH}(x_{t+i|t}),x_{t+i|t})$, $i = 0,1,2,\ldots,N-1$.*

Here, the stability result is formulated [11]:

Theorem 6.1. *Given an auxiliary control law $k_a(x)$ and an associated invariant set Ω_a, consider two positive constants γ and γ_Δ with $\gamma_\Delta\gamma < 1$. Suppose that a non-empty region of attraction Ω_2 and associated set Ω_1 have been determined by applying Procedure 6.5. Let $\widehat{K}^{RH}$ with parameters $\widehat{P}^{RH}$ be the approximate PWNL feedback law determined with Algorithm 6.2 and Procedure 6.5. Consider the closed-loop system (6.57)–(6.58), where $\widehat{k}_0^{RH}(x(t)) = [I\, 0 \ldots 0]\,\widehat{K}^{RH}$. Then, under Assumptions 6.6–6.12, the following holds for the closed-loop system (6.57)–(6.58):*

i). In the absence of disturbance the origin is asymptotically stable for all $x \in \Omega_2$.
ii). In the presence of disturbance it has l_2-gain less than or equal to γ for all $x \in \Omega_2$.

Proof ([11]).

i). From Assumption 6.9 it follows that $\|y_{t+i|t}\|^2 \geq \frac{\|w_{t+i}\|^2}{\gamma_\Delta^2}$, $i \geq 0$. Then, the condition $\gamma_\Delta\gamma < 1$ leads to $\|y_{t+i|t}\|^2 > \gamma^2\|w_{t+i}\|^2$, $i \geq 0$. Therefore, the stage cost $L(y_{t+i|t},w_{t+i}) = \frac{1}{2}(\|y_{t+i|t}\|^2 - \gamma^2\|w_{t+i}\|^2)$ is a positive definite function. Then, by taking into account that $V_{k_a}(x)$ is a positive definite function too (cf. Assumption 6.10), it follows:

$$\widehat{V}_{\max}(x,N) \geq 0, \forall x \in \Omega_2. \tag{6.63}$$

In the absence of disturbance, the stage cost is $L(y,0) = L(h(x,\widehat{k}_0^{RH}(x),0),0)$ and it is a positive definite function defined on the set Ω_2 which contains the origin in its interior (according to Assumption 6.7). Then, it follows from Lemma 4.3 from [17] that there exist a $\mathcal{K}$-function $\alpha_1(\|x\|)$ such that $L(h(x,\widehat{k}_0^{RH}(x),0),0) \geq \alpha_1(\|x\|)$, $\forall x \in \Omega_2$. Similarly, there exists a $\mathcal{K}$-function $\alpha_2(\|x\|)$ such that $V_{k_a}(x) \leq \alpha_2(\|x\|)$, $\forall x \in \Omega_a$ (the reader is referred to [17] for the definition of $\mathcal{K}$-functions). Assumption 6.10 holds also in the case of absence of disturbance and therefore the set Ω_a is a positively invariant set for the nominal system (system (6.57)–(6.58) with $w(t) = 0$) in closed-loop with the auxiliary control law $k_a(x)$ and the inequality (6.25) takes the form:

$$V_{k_a}(f(x,k_a(x),0)) - V_{k_a}(x) + L(h(x,k_a(x),0),0) \leq 0, \; \forall x \in \Omega_a. \tag{6.64}$$

Therefore, according to Theorem 1 with Assumption 1 in [18] $x = 0$ is asymptotically stable for all $x \in \Omega_2$ when $w(t) = 0$.

ii). In a way similar to that in [21], it can be proved that for the worst-case cost function values defined by (6.59) and (6.62) the following holds:

$$\widehat{V}_{\max}(x,N+1) \leq \widehat{V}_{\max}(x,N), \forall x \in \Omega_2. \tag{6.65}$$

Following similar arguments as in [21] and by taking into account (6.65), for $\forall x \in \Omega_2$ and for $w_t \in W^A(\widehat{k}_0^{RH}(x),x)$, we have:

$$\begin{aligned}\widehat{V}_{\max}(x,N) &\geq \widehat{V}_{\max}(f(x,\widehat{k}_0^{RH}(x),w_t),N-1) + \frac{1}{2}\{\|h(x,\widehat{k}_0^{RH}(x),w_t)\|^2 - \gamma^2\|w_t\|^2\} \\ &\geq \widehat{V}_{\max}(f(x,\widehat{k}_0^{RH}(x),w_t),N) + \frac{1}{2}\{\|h(x,\widehat{k}_0^{RH}(x),w_t)\|^2 - \gamma^2\|w_t\|^2\}.\end{aligned} \tag{6.66}$$

Inequality (6.66) can be represented:

$$\begin{aligned}&\widehat{V}_{\max}(f(x,\widehat{k}_0^{RH}(x),w_t),N) - \widehat{V}_{\max}(x,N) \leq \\ &-\frac{1}{2}\{\|h(x,\widehat{k}_0^{RH}(x),w_t)\|^2 - \gamma^2\|w_t\|^2\}.\end{aligned} \tag{6.67}$$

Further, by considering that $x_{t+1|t} = f(x,\widehat{k}_0^{RH}(x),w_t)$ and $y_{t|t} = h(x,\widehat{k}_0^{RH}(x),w_t)$, the inequality (6.67) is written in the form:

$$\widehat{V}_{\max}(x_{t+1|t},N) - \widehat{V}_{\max}(x,N) \leq -\frac{1}{2}\{\|y_{t|t}\|^2 - \gamma^2\|w_t\|^2\}. \tag{6.68}$$

In a similar way, it can be shown that:

$$\begin{aligned}&\widehat{V}_{\max}(x_{t+i+1|t},N) - \widehat{V}_{\max}(x_{t+i|t},N) \leq -\frac{1}{2}\{\|y_{t+i|t}\|^2 - \gamma^2\|w_{t+i}\|^2\} \\ &i = 0,1,\ldots,T.\end{aligned} \tag{6.69}$$

After summing the inequalities (6.69) and by taking into account (6.63), we obtain:

$$\sum_{i=0}^{T}\frac{1}{2}\|y_{t+i|t}\|^2 \leq \gamma^2\sum_{i=0}^{T}\frac{1}{2}\|w_{t+i}\|^2 + \widehat{V}_{\max}(x,N) \tag{6.70}$$

$\forall x \in \Omega_2$, $\forall T \geq 0$, $\forall W_N \in W^B(\widehat{K}_N,x)$. Therefore, the closed-loop system (6.57)–(6.58) has l_2-gain less than or equal to γ in Ω_2. □

In the case when Assumption 6.12 does not hold, no guarantee on the l_2-gain can be given and only an estimate of its upper bound can be computed.

6.3.4 Application 2: Min-Max MPC of Cart and Spring System

Consider a cart with a mass M moving on a plane [21], shown in Fig. 6.4.

The carriage is attached to the wall via a spring with elasticity $\rho = \rho_0 e^{-x_1}$, where x_1 is the displacement of the carriage from the equilibrium position associated with

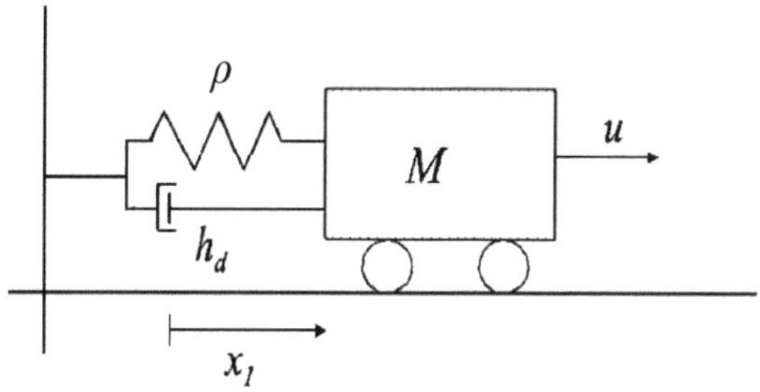

Fig. 6.4 Cart and spring system [21].

the external force $u = 0$. A damper with damping factor h_d affects the system in a resistive way. The damping factor h_d is an uncertain parameter and it is only known that $h_d = \overline{h}_d + \Delta h_d$, where $\overline{h}_d = 1.1$ and $-0.5 \leq \Delta h_d \leq 0.5$. The system is described by the nonlinear discrete-time model [21]:

$$x_1(t+1) = x_1(t) + T_s x_2(t) \tag{6.71}$$

$$x_2(t+1) = x_2(t) - T_s \frac{\rho_0}{M} e^{-x_1(t)} x_1(t) - T_s \frac{\overline{h}_d}{M} x_2(t) + T_s \frac{u(t)}{M} + T_s w(t), \tag{6.72}$$

where x_2 is the carriage velocity, $w(t) = -\frac{\Delta h_d}{M} x_2(t)$, $T_s = 0.4$ is the sampling time, $M = 1$ and $\rho_0 = 0.33$. Like in [21], we choose $y = [x_1 \; x_2 \; u]^T$ and it follows that $w(t) = [0 \; -\frac{\Delta h_d}{M} \; 0] y(t)$. Therefore, $\|w(t)\|^2 \leq \gamma_\Delta^2 \|y(t)\|^2$ with $\gamma_\Delta = 0.5$, according to Assumption 6.9. The following input and state constraints are imposed on the system:

$$-4 \leq u \leq 4, \; -1.3 \leq x_2 \leq 1.3. \tag{6.73}$$

Therefore, the disturbances vary in the range $-1.3\gamma_\Delta \leq w \leq 1.3\gamma_\Delta$. The horizon is $N = 15$ and the terminal constraint is:

$$x_{t+N|t} \in \Omega_a, \; \Omega_a = \{x \in \mathbb{R}^n \,|\, x^T \Sigma x \leq \delta\}, \tag{6.74}$$

where $\delta = 0.001$ [21] and $\Sigma = \begin{bmatrix} 1.3 & 1.9 \\ 1.9 & 3.0 \end{bmatrix}$.

In [11], the approximate mp-NLP approach (described in Section 6.3.2) is applied to design an explicit *feedback* min-max NMPC controller for the cart. The NMPC minimizes the worst-case of the cost function (6.35) subject to the system equations (6.71)–(6.72) and the constraints (6.73)–(6.74). In (6.35), it is chosen $\gamma = 1$ and the terminal penalty is $V_{k_a} = x^T \Sigma x$ [21]. Like in [21], the feedback functions $k_i(x)$, $i = 0, 1, \ldots, N-1$ have the form:

$$k_i(p_i, x) = \alpha_i k_a(x) + \xi_{i,1} x_1^2 + \xi_{i,2} x_2^2, \tag{6.75}$$

where $p_i = [\alpha_i \; \xi_{i,1} \; \xi_{i,2}]^T$ are the parameters that need to be optimized and $k_a(x)$ is the auxiliary control law. The expression (6.75) implies that for relatively small absolute deviations from the equilibrium (small $x^2_{1,t+i|t}$ and $x^2_{2,t+i|t}$) the control input value will be generated mainly by the auxiliary control law $k_a(x_{t+i|t})$. The control law $k_a(x_{t+i|t})$ is determined by applying the method in [21]:

$$k_a(x_{t+i|t}) = -\begin{bmatrix}1 & 0\end{bmatrix} R^{-1} \begin{bmatrix} F_2^T \\ F_3^T \end{bmatrix} \Sigma f_1(x_{t+i|t}), \tag{6.76}$$

where:

$$f_1(x_{t+i|t}) = \begin{bmatrix} x_{1,t+i|t} + T_s x_{2,t+i|t} \\ x_{2,t+i|t} - T_s \frac{\rho_0}{M} e^{-x_{1,t+i|t}} x_{1,t+i|t} - T_s \frac{\bar{h}_d}{M} x_{2,t+i|t} \end{bmatrix} \tag{6.77}$$

$$F_2 = \begin{bmatrix} 0 \\ \frac{T_s}{M} \end{bmatrix},\ F_3 = \begin{bmatrix} 0 \\ T_s \end{bmatrix},\ R = \begin{bmatrix} F_2^T \Sigma F_2 + I & F_2^T \Sigma F_3 \\ (F_2^T \Sigma F_3)^T & F_3^T \Sigma F_3 - \alpha^2 I \end{bmatrix} \tag{6.78}$$

A set of three admissible disturbance realizations is generated which correspond to three values for the uncertain parameter Δh_d ($\Delta h_d = -0.5$, $\Delta h_d = 0$, $\Delta h_d = 0.5$). One internal region $X_0^1 \subset X_0$ is used in Procedure 1.1. In (6.47), it is chosen $\mu = 10$. The approximation tolerance is chosen to be $\bar{\varepsilon}(X_0) = \max(\bar{\varepsilon}_a, \bar{\varepsilon}_r \min_{x \in X_0} \widetilde{V}^*_{\max}(x))$, where $\bar{\varepsilon}_a = 0.003$ and $\bar{\varepsilon}_r = 0.01$ are the absolute and the relative tolerances.

The state space partition of the *feedback* min-max NMPC controller (the set X_Π) and the associated sets Ω_1 and Ω_2 are shown in Fig. 6.5. It is noticed that in some part of the set $X = [-3, 5] \times [-2, 2]$ a feasible solution does not exist. The number of

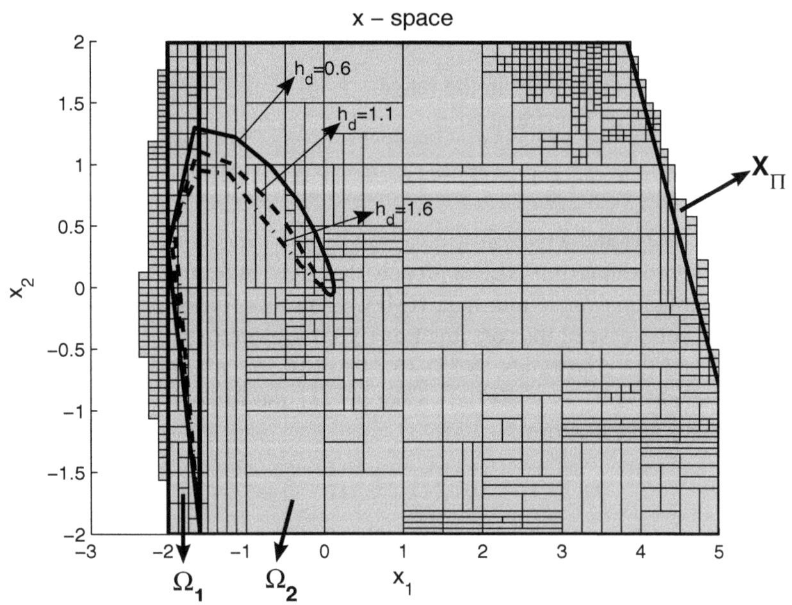

Fig. 6.5 State space partition of the explicit feedback min-max NMPC (the set X_Π), the associated sets Ω_1 and Ω_2, and the state trajectories for $h_d = 0.6$, $h_d = 1.1$, $h_d = 1.6$.

the inequalities describing the sets Ω_1 and Ω_2 is specified to be 5 and Procedure 6.5 is applied to determine them. The set Ω_1 is obtained graphically by minimizing the difference between its area and the area of the set X_Π. The computations of the state trajectories of the suboptimal closed-loop system, performed according to equation (6.53), have shown that the set Ω_2 can be determined simply by increasing the bound in one of the inequalities describing the set Ω_1. The partition (the set Ω_1 in Fig. 6.5) has 537 regions and 14 levels of search. Totally, 32 arithmetic operations are needed in real-time to compute the control input (14 comparisons, 11 multiplications, 6 additions and 1 exponential).

In Fig. 6.6, the optimal and the suboptimal feedback functions, respectively $u^*(x_1,x_2) = k_0^*(x_1,x_2)$ and $\widehat{u}(x_1,x_2) = \widehat{k}_0(x_1,x_2)$, are shown. In Fig. 6.7, the optimal cost function and the cost function approximation error, associated with the explicit approximate *feedback* min-max NMPC, are shown.

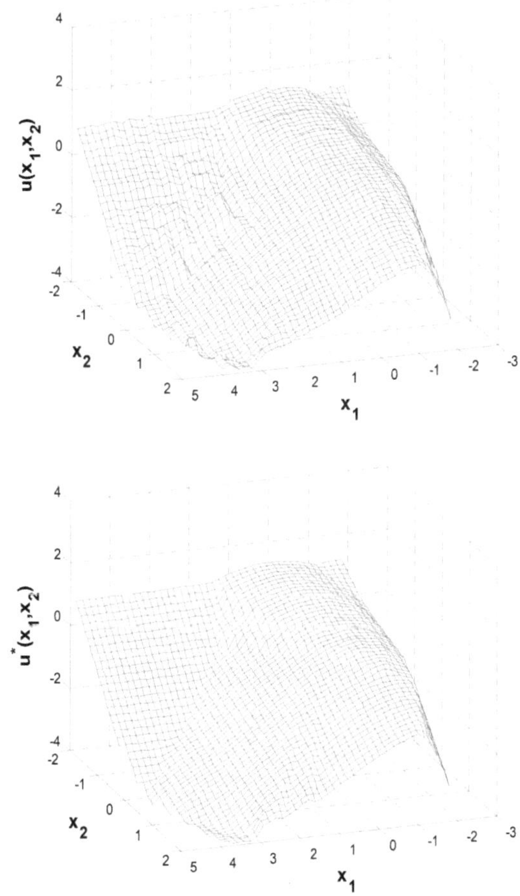

Fig. 6.6 The suboptimal (top) and the optimal (bottom) feedback functions (views rotated on 140°).

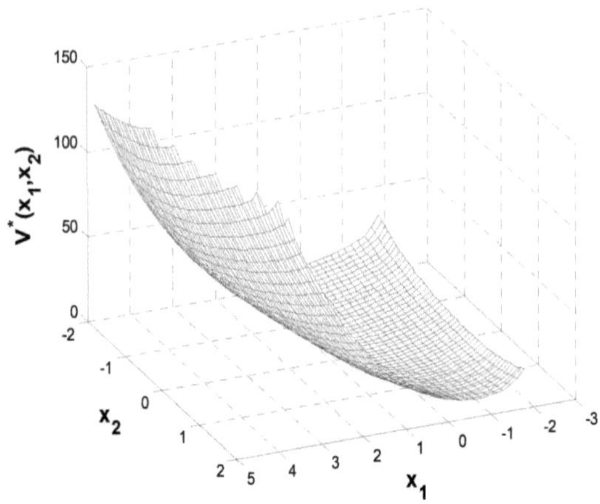

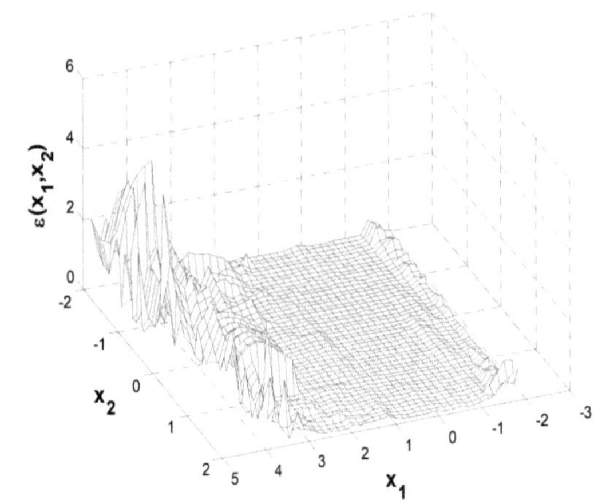

Fig. 6.7 The optimal cost function (top) and the cost function approximation error (bottom) (views rotated on 140°).

The performance of the suboptimal closed-loop system was simulated for initial state $x(0) = [-1.6\ \ -2]^T$ and for three values of h_d. The response is depicted in the state space (Fig. 6.5) and as trajectories in time (Fig. 6.8 and Fig. 6.9). It can be seen that the explicit *feedback* min-max NMPC controller brings the cart to the equilibrium despite of the presence of disturbance, and the constraints imposed on the system are satisfied. It can also be observed that the state trajectory does not leave the set Ω_1.

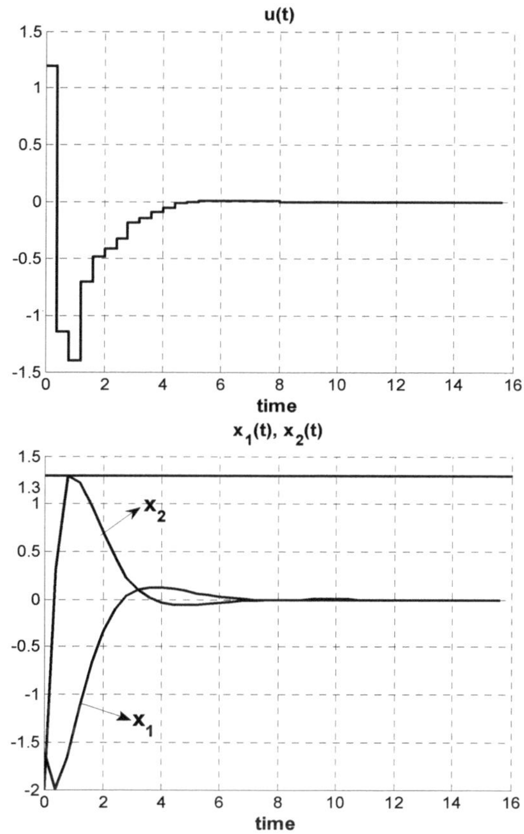

Fig. 6.8 Control input and state trajectory for $h_d = 0.6$.

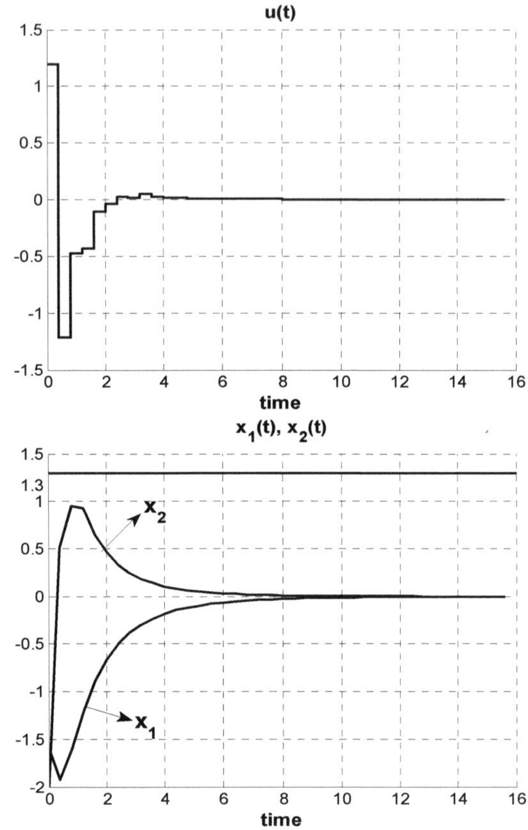

Fig. 6.9 Control input and state trajectory for $h_d = 1.6$.

References

1. Bazaraa, M.S., Sherali, H.D., Shetty, C.M.: Non-linear programming: theory and algorithms. Wiley, New York (1993)
2. Bemporad, A., Borrelli, F., Morari, M.: Min-max control of constrained uncertain discrete-time linear systems. IEEE Transactions on Automatic Control 48, 1600–1606 (2003)
3. Bentley, J.L.: Multidimensional binary search trees used for associative searching. Communications of the ACM 18, 509–517 (1975)
4. Björnberg, J., Diehl, M.: Approximate robust dynamic programming and robustly stable MPC. Automatica 42, 777–782 (2006)
5. Campo, P.J., Morari, M.: Robust model predictive control. In: Proceedings of the American Control Conference, Minneapolis, Minn., vol. 2, pp. 1021–1026 (1987)
6. Cychowski, M., O'Mahony, T.: Efficient off-line solutions to robust model predictive control using orthogonal partitioning. In: Proceedings of the 16th IFAC World Congress, Prague, Czech Republic (2005), www.IFAC-PapersOnLine.net
7. Fiacco, A.V.: Introduction to sensitivity and stability analysis in nonlinear programming. Academic Press, Orlando (1983)

8. Grancharova, A., Johansen, T.A.: Explicit min-max model predictive control of constrained nonlinear systems with model uncertainty. In: Proceedings of the 16th IFAC World Congress, Prague, Czech Republic (2005), www.IFAC-PapersOnLine.net
9. Grancharova, A., Johansen, T.A.: Explicit approximate approach to feedback min-max model predictive control of constrained nonlinear systems. In: Proceedings of the IEEE Conference on Decision and Control, San Diego, USA, pp. 4848–4853 (2006)
10. Grancharova, A., Johansen, T.A., Tøndel, P.: Computational aspects of approximate explicit nonlinear model predictive control. In: Findeisen, R., Allgöwer, F., Biegler, L. (eds.) Assessment and Future Directions of Nonlinear Model Predictive Control. LNCIS, vol. 358, pp. 181–192. Springer, Heidelberg (2007)
11. Grancharova, A., Johansen, T.A.: Computation, approximation and stability of explicit feedback min-max nonlinear model predictive control. Automatica 45, 1134–1143 (2009)
12. Grossmann, I.E., Halemane, K.P., Swaney, R.E.: Optimization strategies for flexible chemical processes. Computers and Chemical Engineering 7, 439–462 (1983)
13. Hicks, G., Ray, W.: Approximation methods for optimal control synthesis. The Canadian Journal of Chemical Engineering 49, 522–528 (1971)
14. Johansen, T.A.: Approximate explicit receding horizon control of constrained nonlinear systems. Automatica 40, 293–300 (2004)
15. Kerrigan, E.C., Maciejowski, J.M.: Feedback min-max model predictive control using a single linear program: Robust stability and the explicit solution. International Journal of Robust and Nonlinear Control 14, 395–413 (2004)
16. Kerrigan, E.C., Mayne, D.Q.: Optimal control of constrained piecewise affine systems with bounded disturbances. In: Proceedings of the IEEE Conference on Decision and Control, Las Vegas, NV, pp. 1552–1557 (2002)
17. Khalil, H.K.: Nonlinear systems, 3rd edn. Prentice Hall, USA (2002)
18. Lazar, M., Heemels, W.P.M.H., Bemporad, A., Weiland, S.: Discrete-time non-smooth nonlinear MPC: Stability and robustness. In: Findeisen, R., Allgöwer, F., Biegler, L. (eds.) Assessment and Future Directions of Nonlinear Model Predictive Control: Towards New Challenging Applications. LNCIS, vol. 358, pp. 93–103. Springer, Heidelberg (2007)
19. Limon, D., Alamo, T., Camacho, E.F.: Input-to-state stable MPC for constrained discrete-time nonlinear systems with bounded additive uncertainties. In: Proceedings of the IEEE Conference on Decision and Control, Las Vegas, NV, pp. 4619–4624 (2002)
20. Limon, D., Alamo, T., Salas, F., Camacho, E.F.: Input-to-state stability of min max MPC controllers for nonlinear systems with bounded uncertainties. Automatica 42, 797–803 (2006)
21. Magni, L., de Nicolao, G., Scattolini, R., Allgöwer, F.: Robust model predictive control for nonlinear discrete-time systems. International Journal of Robust and Nonlinear Control 13, 229–246 (2003)
22. Magni, L., Scattolini, R.: Robustness and robust design of MPC for nonlinear discrete-time systems. In: Findeisen, R., Allgöwer, F., Biegler, L. (eds.) Assessment and Future Directions of Nonlinear Model Predictive Control. LNCIS, vol. 358, pp. 239–254. Springer, Heidelberg (2007)
23. Mayne, D.Q., Rawlings, J.B., Rao, C.V., Scokaert, P.O.M.: Constrained model predictive control: Stability and optimality. Automatica 36, 789–814 (2000)
24. Mayne, D.Q., Raković, S.V., Vinter, R.B., Kerrigan, E.C.: Characterization of the solution to a constrained H_∞ optimal control problem. Automatica 42, 371–382 (2006)
25. Mhaskar, P.: Robust model predictive control design for fault-tolerant control of process systems. Industrial Engineering & Chemistry Research 45, 8565–8574 (2006)

26. Michalska, H., Mayne, D.Q.: Robust receding horizon control of constrained nonlinear systems. IEEE Transactions on Automatic Control 38, 1623–1633 (1993)
27. Mukai, M., Azuma, T., Kojima, A., Fujita, M.: Approximate robust receding horizon control for piecewise linear systems via orthogonal partitioning. In: Proceedings of the European Control Conference, Cambridge, U.K. (2003)
28. Muñoz de la Peña, D., Alamo, T., Ramirez, D.R., Camacho, E.F.: Min-max model predictive control as a quadratic program. IET Control Theory and Applications 1, 328–333 (2007)
29. Muñoz de la Peña, D., Bemporad, A., Filippi, C.: Robust explicit MPC based on approximate multi-parametric convex programming. IEEE Transactions on Automatic Control 51, 1399–1403 (2006)
30. Raković, S.V., Kerrigan, E.C., Mayne, D.Q.: Optimal control of constrained piecewise affine systems with state- and input-dependent disturbances. In: Proceedings of the 16th International Symposium on Mathematical Theory of Networks and Systems, Leuven, Belgium (2004)
31. Wan, Z., Kothare, M.: An efficient off-line formulation of robust model predictive control using linear matrix inequalities. Automatica 39, 837–846 (2003)

Chapter 7
Explicit Stochastic NMPC

Abstract. This chapter considers two approaches to explicit stochastic NMPC of general constrained nonlinear discrete-time systems in the presence of disturbances and/or parameter uncertainties with known probability distributions. In Section 7.2, an approach to explicit solution of *closed-loop* (*feedback*) stochastic NMPC problems for constrained nonlinear systems, described by stochastic *parametric* models, is considered. The approach constructs a piecewise nonlinear (PWNL) approximation to the optimal sequence of feedback control policies. It is applied to design an explicit *feedback* stochastic NMPC controller for the cart and spring system. In Section 7.3, an explicit approximate approach to *open-loop* stochastic NMPC based on Gaussian process models is presented. The Gaussian process models are *non-parametric* probabilistic black-box models, whose advantage in comparison to the stochastic *parametric* models is that they provide information about the prediction uncertainty. The approach in Section 7.3 constructs a piecewise linear (PWL) approximation to the optimal control sequence and it is applied to design an explicit stochastic NMPC reference tracking controller for a combustion plant.

7.1 Introduction

Mathematical models of engineering systems usually contain some amount of uncertainty (typically unknown additive disturbances and/or uncertain model parameters). In the robust MPC problem formulation, the model uncertainty is taken into account. In some applications, the system to be controlled is described by a stochastic model where the probabilistic distribution of the uncertainty is assumed to be known. Several approaches for constrained *open-loop* MPC based on stochastic *parametric* models are proposed in [32, 44, 46, 31, 7, 8, 5, 6, 24]. The approaches [32, 44, 46] are based on linear state space models with stochastic parameters and/or additive noise and they optimize the expected value of the cost function subject to hard input constraints [32] or probabilistic constraints [44, 46]. In [31, 7, 8, 5], stochastic linear MPC approaches incorporating a probabilistic cost and probabilistic constraints are developed. The method suggested in [31] is based on a moving

A. Grancharova & T.A. Johansen: Explicit Nonlinear Model Predictive Ctrl., LNCIS 429, pp. 157–186.
springerlink.com

average (MA) model with random coefficients. It was further extended to linear time-varying MA models [8] and to state space models with stochastic uncertainty in the output or the input map [7, 5]. Methods for *open-loop* stochastic MPC for nonlinear systems have been proposed in [6, 24].

The stochastic MPC methods mentioned above employ an *open-loop* formulation, which guarantees the robust stability and the robust feasibility of the system, but it may be conservative. This is related to the fact that the control sequence has to ensure constraints fulfillment for all possible uncertainty scenarios without considering the fact that future measurements of the state contain information about past uncertain values. Similar to the *closed-loop* min-max NMPC approaches [35, 34, 36], the conservativeness of the *open-loop* stochastic NMPC can be overcome by a *closed-loop* stochastic NMPC formulation, where the optimization is performed over a sequence of feedback control policies. In [18, 1], methods for *closed-loop* stochastic NMPC based on on-line optimization have been proposed and an approximate mp-NLP approach to explicit *closed-loop* min-max NMPC has been suggested in [21]. Based on the approach in [21], in [22] the explicit solution of *closed-loop* (*feedback*) stochastic NMPC problems for constrained nonlinear systems in the presence of uncertainty is considered by employing stochastic *parametric* models. The approach [22] constructs a piecewise nonlinear (PWNL) approximation to the optimal sequence of feedback control policies for efficient online implementation. This approach is considered in Section 7.2.

The stochastic MPC approaches [32, 44, 46, 31, 7, 8, 5, 6, 24, 18, 1, 22] are based on *parametric* probabilistic models. Alternatively, the stochastic systems can be modeled with *non-parametric* models which can offer a significant advantage compared to the *parametric* models. This is related to the fact that the *non-parametric* probabilistic models provide information about prediction uncertainties which are difficult to evaluate appropriately with the *parametric* models. The Gaussian process model is an example of a *non-parametric* probabilistic black-box model and up to now it has been applied to model mainly static nonlinearities. Its use and properties for modeling are reviewed in [41]. The use of Gaussian processes in the modeling of dynamic systems is a relatively recent development e.g. [17, 43, 26, 2, 29, 28, 39]. An on-line optimization approach and an approximate explicit approach to *open-loop* stochastic NMPC based on Gaussian process models have been proposed in [38, 30, 33] and in [19, 20], respectively. The approach [19, 20] constructs a piecewise linear (PWL) approximation to the optimal control sequence and it is considered in Section 7.3. A recent state-of-the-art survey of control algorithms based on Gaussian process models is provided in [27].

7.2 Explicit Stochastic NMPC Based on Parametric Probabilistic Models

This section considers the approximate mp-NLP approach [22] to explicit solution of *closed-loop* (*feedback*) stochastic NMPC problems for constrained nonlinear systems, described by stochastic *parametric* models. It is assumed that the discrete probability distribution of the uncertainty is known. The approach constructs a piecewise nonlinear (PWNL) approximation to the optimal sequence of feedback control policies, defined on an orthogonal partition of the state space.

7.2.1 Formulation of the Feedback Stochastic NMPC Problem as an mp-NLP Problem

Consider the discrete-time nonlinear system:

$$\begin{aligned} x(t+1) &= f(x(t), u(t), w(t)) \\ y(t) &= h(x(t), u(t), w(t)), \end{aligned} \tag{7.1}$$

where $x(t) \in \mathbb{R}^n$, $u(t) \in \mathbb{R}^m$, $y(t) \in \mathbb{R}^s$ and $w(t) \in \mathbb{R}^n$ are the state, input, output and uncertainty variables, and $t \in \mathbb{Z}_{\geq 0}$ is the discrete time. The input and the output variables are required to fulfill the following constraints:

$$u_{\min} \leq u \leq u_{\max}, \; y_{\min} \leq y \leq y_{\max}. \tag{7.2}$$

The following assumptions are made [22]:

Assumption 7.1. *f and h are C^2 functions with $f(0,0,0) = 0$, $h(0,0,0) = 0$.*

Assumption 7.2. *The uncertainty $w(t) = [w_1(t), w_2(t), \ldots, w_n(t)]$ includes both internal (state-dependent) and external uncertainty, i.e. it has the form $w_i(t) = \lambda_i x_i(t) + \gamma_i(t)$, where $x_i(t)$ is the i-th element of the state vector $x(t)$. Here λ_i is a discrete random parameter, associated to the internal (model) uncertainty. It takes values in the set $\Lambda_i = \{\lambda_i^1, \lambda_i^2, \ldots, \lambda_i^{n_{\lambda_i}}\}$ and is characterized with the probability mass function $\varphi_i : \Lambda_i \to [0, 1]$:*

$$\varphi_i(\lambda_i^j) = \Pr(\lambda_i = \lambda_i^j), \, j = 1, 2, \ldots, n_{\lambda_i} \; with \; \sum_{j=1}^{n_{\lambda_i}} \varphi_i(\lambda_i^j) = 1. \tag{7.3}$$

The external uncertainty $\gamma_i(t)$ is stochastic and piecewise constant with infrequent changes in the sense that $\gamma_i(t) = const$ for periods of time, which are not less than N_γ ($N_\gamma \in \mathbb{N}$ is supposed to be sufficiently large). It is assumed that γ_i takes values in the set $\Gamma_i = \{\gamma_i^1, \gamma_i^2, \ldots, \gamma_i^{n_{\gamma_i}}\}$, which contains the nominal value $\gamma_i = 0$, and it is characterized with the probability mass function $\psi_i : \Gamma_i \to [0, 1]$:

$$\psi_i(\gamma_i^j) = \Pr(\gamma_i = \gamma_i^j), \, j = 1, 2, \ldots, n_{\gamma_i} \; with \; \sum_{j=1}^{n_{\gamma_i}} \psi_i(\gamma_i^j) = 1. \tag{7.4}$$

The overall vector of uncertain model parameters is denoted $\lambda = [\lambda_1, \lambda_2, \ldots, \lambda_n] \in \Lambda$ with $\Lambda = \Lambda_1 \times \Lambda_2 \times \ldots \times \Lambda_n$. Given $\lambda^j = [\lambda_1^{i_1}, \lambda_2^{i_2}, \ldots, \lambda_n^{i_n}]$, $i_1 \in \{1, 2, \ldots, n_{\lambda_1}\}$, ... , $i_n \in \{1, 2, \ldots, n_{\lambda_n}\}$, the probability mass function $\varphi : \Lambda \to [0, 1]$ is:

$$\varphi(\lambda^j) = \Pr(\lambda = \lambda^j) = \varphi_1(\lambda_1^{i_1})\varphi_2(\lambda_2^{i_2})...\varphi_n(\lambda_n^{i_n}) \tag{7.5}$$

with $j = 1, ..., n_\lambda$ and $n_\lambda = n_{\lambda_1} n_{\lambda_2} ... n_{\lambda_n}$. Similarly, the overall vector of external uncertainty is denoted $\gamma = [\gamma_1, \gamma_2, ..., \gamma_n] \in \Gamma$ with $\Gamma = \Gamma_1 \times \Gamma_2 \times ... \times \Gamma_n$. Given $\gamma^j = [\gamma_1^{j_1}, \gamma_2^{j_2}, ..., \gamma_n^{j_n}]$, $i_1 \in \{1, 2, ..., n_{\gamma_1}\}$, ... , $i_n \in \{1, 2, ..., n_{\gamma_n}\}$, the probability mass function $\psi : \Gamma \to [0, 1]$ is:

$$\psi(\gamma^j) = \Pr(\gamma = \gamma^j) = \psi_1(\gamma_1^{j_1})\psi_2(\gamma_2^{j_2})...\psi_n(\gamma_n^{j_n}) \tag{7.6}$$

with $j = 1, ..., n_\gamma$ and $n_\gamma = n_{\gamma_1} n_{\gamma_2} ... n_{\gamma_n}$.

Then, the overall uncertainty is:

$$w = \mathrm{diag}(\lambda)x + \gamma. \tag{7.7}$$

Further, the following assumption is made:

Assumption 7.3. $y_{\min} < 0 < y_{\max}$ *and* $u_{\min} < 0 < u_{\max}$.

As in [35], first a H_∞ control problem is defined:

Problem 7.1 (H_∞ control problem):
Design a state-feedback control law:

$$u = k(x) \tag{7.8}$$

guaranteeing that the closed-loop system (7.1)–(7.8) with input $w = \mathrm{diag}(\lambda)x + \gamma$, $\lambda \in \Lambda$, $\gamma \in \Gamma$, and output y has a finite l_2-gain $\leq \alpha$ in a bounded positively invariant set Ω, that is, $\forall x(t) \in \Omega$:

i. $x(t+i) \in \Omega, \forall i > 0$.
ii. $u_{\min} \leq k(x(t+i)) \leq u_{\max}$ and $y_{\min} \leq h(x(t+i), k(x(t+i)), w(t+i)) \leq y_{\max}$, $\forall i \geq 0$.
iii. There exists a positive definite function $\beta(x(t))$, such that $\forall T \geq 0$:

$$\sum_{i=0}^{T} \|y(t+i)\|^2 \leq \alpha^2 \sum_{i=0}^{T} \|w(t+i)\|^2 + \beta(x(t)) \tag{7.9}$$

for any non-zero w.

The following assumption is also made [35]:

Assumption 7.4. *There exists an auxiliary control law $u = k_a(x)$ that solves the H_∞ control problem, with a domain of attraction $\Omega_a(k_a, \alpha)$, whose boundary is assumed to be a level curve of a positive function $V_{k_a}(x)$ such that:*

$$V_{k_a}(f(x, k_a(x), w)) - V_{k_a}(x) < -\frac{1}{2}(\|y\|^2 - \alpha^2\|w\|^2)$$
$$\forall x \in \Omega_a(k_a, \alpha), \ \forall \lambda \in \Lambda, \ \forall \gamma \in \Gamma \tag{7.10}$$

and $V_{k_a}(0) = 0$.

Denote with:

$$K = \{k_0, k_1, \dots, k_{N-1}\} \triangleq \{k_0(x_{t|t}), k_1(x_{t+1|t}), \dots, k_{N-1}(x_{t+N-1|t})\} \qquad (7.11)$$

a vector of feedback control policies. It is supposed that a full measurement x of the state is available at the current time t. We consider the following feedback stochastic NMPC problem [22]:

Problem 7.2 (Constrained feedback stochastic NMPC problem):
Suppose that Assumptions 7.1–7.4 hold. For the current x, the feedback stochastic NMPC solves the following optimization problem:

$$V_E^*(x) = \min_K \underset{\lambda\in\Lambda,\, \gamma\in\Gamma}{E} \{J(K,x,\lambda,\gamma)\} \qquad (7.12)$$

subject to $x_{t|t} = x$ and:

$$y_{\min} \le y_{t+i|t} \le y_{\max},\ \forall\lambda \in \Lambda,\ \forall\gamma\in\Gamma,\ i=1,\dots,N \qquad (7.13)$$

$$u_{\min} \le u_{t+i} \le u_{\max},\ i=0,1,\dots,N-1 \qquad (7.14)$$

$$x_{t+N|t} \in \Omega_a(k_a,\alpha),\ \forall\lambda \in \Lambda,\ \forall\gamma\in\Gamma \qquad (7.15)$$

$$u_{t+i} = k_i(x_{t+i|t}),\ i=0,1,\dots,N-1 \qquad (7.16)$$

$$x_{t+i+1|t} = f(x_{t+i|t}, u_{t+i}, w_{t+i}),\ i\ge 0 \qquad (7.17)$$

$$y_{t+i|t} = h(x_{t+i|t}, u_{t+i}, w_{t+i}),\ i\ge 0 \qquad (7.18)$$

and the cost function given by:

$$J(K,x,\lambda,\gamma) = \frac{1}{2}\sum_{i=0}^{N-1}\left[\|y_{t+i|t}\|^2 - \alpha^2\|w_{t+i}\|^2\right] + V_{k_a}(x_{t+N|t}). \qquad (7.19)$$

Here, N is a finite horizon, α is the l_2-gain which is interpreted as the uncertainty attenuation level, and $E\{.\}$ means mathematical expectation. It is supposed that $N \ll N_\gamma$, and by Assumption 7.2 it can be accepted that $\gamma_{t+i} = const,\ i=0,1,\dots,N-1$. Then by Assumption 7.2, the expectation can be expressed:

$$\underset{\lambda\in\Lambda,\, \gamma\in\Gamma}{E}\{J(K,x,\lambda,\gamma)\} = \sum_{i=1}^{n_\lambda}\sum_{j=1}^{n_\gamma} J(K,x,\lambda^i,\gamma^j)\varphi(\lambda^i)\psi(\gamma^j) \qquad (7.20)$$

An auxiliary control law $k_a(x)$ is typically obtained by solving the H_∞ control problem for the linearized system [35]. Thus, a practical way to compute a nonlinear control $k_a(x)$ satisfying Assumption 7.4 for nonlinear input-affine systems is suggested in [35].

An optimal solution to the feedback stochastic NMPC problem (7.12)–(7.19) is denoted $K^* = \{k_0^*, k_1^*, \dots, k_{N-1}^*\}$ and the control input is chosen according to the receding horizon policy $u(x_{t|t}) = k_0^*(x_{t|t})$.

The following assumption is made:

Assumption 7.5. *Each feedback control policy $k_i(x_{t+i|t})$, $i = 0, ..., N-1$ has the form:*

$$k_i(x_{t+i|t}) = \eta_i k_a(x_{t+i|t}) + r_i(\xi_i, x_{t+i|t}) = g_i(p_i, x_{t+i|t}), \tag{7.21}$$

where $p_i = [\eta_i^T \; \xi_i^T]^T \in \mathbb{R}^{n_i}$ are the parameters that need to be optimized, $k_a(x_{t+i|t})$ is an auxiliary control law that satisfies Assumption 7.4, and $r_i(\xi_i, x_{t+i|t})$ is a parameterized continuous function with $r_i(\xi_i, 0) = 0$.

In general, the parameterization of the form (7.21) would lead to an approximate solution to the feedback stochastic NMPC problem (7.12)–(7.19). Denote with P the whole set of parameters that need to be determined, i.e.:

$$P = [p_0^T \; p_1^T \; ... \; p_{N-1}^T]^T \in \mathbb{R}^{n_p}, \; n_p = \sum_{i=0}^{N-1} n_i. \tag{7.22}$$

Then, the expected value (7.20) of the cost function is:

$$V_E(P, x) = \underset{\lambda \in \Lambda, \gamma \in \Gamma}{E} \{J(P, x, \lambda, \gamma)\}. \tag{7.23}$$

It should be noted that the argument K in the cost function (7.19) is now replaced with the argument P.

Using the ideas of direct single shooting to eliminate the equality constraints, the optimization problem (7.12)–(7.19) can be formulated in a compact form as follows [22]:

Problem 7.3:

$$V_E^*(x) = \min_P \sum_{i=1}^{n_\lambda} \sum_{j=1}^{n_\gamma} J(P, x, \lambda^i, \gamma^j) \varphi(\lambda^i) \psi(\gamma^j) \tag{7.24}$$

$$\text{subject to}: \; G(P, x, \lambda, \gamma) \le 0, \; \forall \lambda \in \Lambda, \forall \gamma \in \Gamma. \tag{7.25}$$

Problem 7.3 defines a multi-parametric Nonlinear Programming (mp-NLP) problem, since it is NLP in P parameterized by x. We remark that the constraints function $G(P, x, \lambda, \gamma)$ in (7.25) is implicitly defined by (7.13)–(7.18). Also, since Λ and Γ are discrete sets, (7.25) represents a finite number of constraints. It should be noted that the number of constraints (7.25) increases rapidly with the increase of the horizon and the sizes n_λ and n_γ of the uncertainty sets Λ and Γ. Thus, as the horizon increases from N_1 to N_2 and the sizes of the sets Λ and Γ increase from n_λ^1 to n_λ^2, and from n_γ^1 to n_γ^2, respectively, the number of constraints will increase $(N_2 n_\lambda^2 n_\gamma^2)/(N_1 n_\lambda^1 n_\gamma^1)$ times. This leads to a stronger motivation for an explicit approach, where the computational complexity is handled in off-line.

Define the set of N-step robustly feasible initial states:

$$X_f = \{x \in \mathbb{R}^n \,|\, G(P, x, \lambda, \gamma) \le 0, \forall \lambda \in \Lambda, \forall \gamma \in \Gamma \text{ for some } P \in \mathbb{R}^{n_p}\}. \tag{7.26}$$

In parametric programming problems one seeks the solution $P^*(x)$ as an explicit function of the parameters x in some set $X \subseteq X_f \subseteq \mathbb{R}^n$ [14].

7.2.2 Approximate mp-NLP Approach to Explicit Feedback Stochastic NMPC

The approximate mp-NLP approach [22] to explicit feedback stochastic NMPC is similar to the approach to explicit feedback min-max NMPC (described in Section 6.3.2). Thus, we restrict our attention to a hyper-rectangle $X \subset \mathbb{R}^n$ where we seek to approximate the optimal sequence of control policies $K^* = \{k_0^*, k_1^*, ..., k_{N-1}^*\}$. The associated optimal control input is

$$u_{t+i}^* = k_i^*(x_{t+i|t}) = g_i(p_i^*, x_{t+i|t}),\ i = 0, 1, ..., N-1, \tag{7.27}$$

where $P^* = [p_0^{*T}\ p_1^{*T}\ ...\ p_{N-1}^{*T}]^T$ is determined by solving Problem 7.3. We require that the state space partition is orthogonal and can be represented as a $k-d$ tree. The main idea of the approximate mp-NLP approach [22] is to construct a *piecewise nonlinear* (PWNL) approximation $\hat{K} = \{\hat{k}_0, \hat{k}_1, ..., \hat{k}_{N-1}\}$ to the optimal feedback $K^* = \{k_0^*, k_1^*, ..., k_{N-1}^*\}$ on X, where the constituent *nonlinear* control functions $K_{X_i} = \{k_{0,X_i}, k_{1,X_i}, ..., k_{N-1,X_i}\}$ are defined on hyper-rectangles X_i covering X. Let $K_{X_0} = \{k_{0,X_0}, ..., k_{N-1,X_0}\}$ be an approximation to the optimal solution $K^* = \{k_0^*, ..., k_{N-1}^*\}$, valid in the whole hyper-rectangle X_0. Denote with $P_{X_0} = [p_{0,X_0}^T \cdots p_{N-1,X_0}^T]^T$ the parameters of K_{X_0}. The corresponding approximate value of the control input is

$$\hat{u}_{t+i} = k_{i,X_0}(x_{t+i|t}) = g_i(p_{i,X_0}, x_{t+i|t}),\ i = 0, 1, ..., N-1. \tag{7.28}$$

Let $\hat{V}_E(P_{X_0}, x)$ be the cost function value due to initial state $x = x_{t|t}$ and control function K_{X_0}, i.e.

$$\hat{V}_E(P_{X_0}, x) = \underset{\lambda \in \Lambda,\, \gamma \in \Gamma}{E} \{J(P_{X_0}, x, \lambda, \gamma)\}. \tag{7.29}$$

Then, the approximate control function K_{X_0} is determined by applying the following procedure [22]:

Procedure 7.1 (Computation of explicit approximate solution). *Suppose Assumptions 7.1–7.5 hold. Consider any hyper-rectangle $X_0 \subseteq X_f$ with a set of points $V_0 = \{v_0, v_1, v_2, ..., v_{N_1}\} \subseteq X_0$. Compute the parameters $P_{X_0} = [p_{0,X_0}^T \cdots p_{N-1,X_0}^T]^T$ of the control function $K_{X_0} = \{k_{0,X_0}, ..., k_{N-1,X_0}\}$ by solving the following NLP:*

$$\min_{P_{X_0}} \sum_{i=0}^{N_1} \left(\hat{V}_E(P_{X_0}, v_i) - V_E^*(v_i) + \mu \|g_0(p_{0,X_0}, v_i) - g_0(p_0^{*i}, v_i)\|^2\right) \tag{7.30}$$

$$\textit{subject to}\ \ G(P_{X_0}, v_i, \lambda, \gamma) \leq 0, \forall v_i \in V_0, \forall \lambda \in \Lambda, \forall \gamma \in \Gamma. \tag{7.31}$$

We remark that the optimal parameters p_0^{*i} in the feedback function $k_0^*(v_i) = g_0(p_0^{*i}, v_i)$ in (7.30) are determined by solving Problem 7.3 for $x = v_i$, and the parameter $\mu > 0$ is a weighting coefficient.

Note that the control function $K_{X_0} = \{k_{0,X_0}, \ldots, k_{N-1,X_0}\}$, computed with Procedure 7.1, satisfies the constraints in Problem 7.3 only for the discrete set of points V_0 in the hyper-rectangle X_0 and for the discrete sets of values Λ and Γ of the internal and external uncertainties.

Suppose that the parameter vector P_{X_0} of the control function K_{X_0}, valid in X_0, has been determined by applying Procedure 7.1. Then, for the cost function approximation error in X_0 we have:

$$\varepsilon(x) = \hat{V}_E(P_{X_0}, x) - V_E^*(x) \leq \varepsilon_0,\ x \in X_0. \tag{7.32}$$

An estimate $\hat{\varepsilon}_0$ of the error bound ε_0 is computed as:

$$\hat{\varepsilon}_0 = \max_{i \in \{0,1,2,\ldots,N_1\}} (\hat{V}_E(P_{X_0}, v_i) - V_E^*(v_i)). \tag{7.33}$$

If $\hat{\varepsilon}_0 > \bar{\varepsilon}$, where $\bar{\varepsilon} > 0$ is the specified tolerance of the cost function approximation error, the region X_0 is divided and the procedure is repeated for the new regions.

The approximate mp-NLP algorithm for design of explicit feedback stochastic NMPC represents a slight modification of the algorithm, described in Section 6.3.2.4.

It should be noted that in case of non-convexity of Problem 7.3, it can not be guaranteed that the approximation error $\varepsilon(x)$ associated to the explicit feedback stochastic NMPC will satisfy the requirement $\varepsilon(x) \leq \bar{\varepsilon}$ for all $x \in X$. The non-convexity may also imply that the constraints (7.25) are violated at some points of the state space. In this respect, the described computational method does not necessarily lead to guaranteed properties, but when combined with verification and analysis methods gives a practical tool for development and implementation of explicit feedback stochastic NMPC. The possibility for implementation verification is a significant advantage of the explicit NMPC in comparison to NMPC based on real-time optimization.

7.2.3 Application 1: Stochastic MPC of the Cart and Spring System

Consider the cart and spring system, described in Section 6.3.4. The damping factor h_d is uncertain, but it is known that $h_d = \bar{h}_d + \lambda$. Here, $\bar{h}_d = 1.1$ and λ is a stochastic parameter. It is supposed that $-0.5 \leq \lambda \leq 0.5$ and the following discrete set of values is considered $\lambda \in \Lambda = \{-0.5, 0, 0.5\}$ with the corresponding values of the probability mass function $\varphi(-0.5) = 0.2$, $\varphi(0) = 0.6$, $\varphi(0.5) = 0.2$. The external uncertainty for this system is $\gamma = 0$. Recall that the system is described by the following nonlinear discrete-time model [35]:

$$x_1(t+1) = x_1(t) + T_s x_2(t) \tag{7.34}$$

$$x_2(t+1) = x_2(t) - T_s \frac{\rho_0}{M} e^{-x_1(t)} x_1(t) - T_s \frac{\overline{h}_d}{M} x_2(t) + T_s \frac{u(t)}{M} + T_s w(t), \tag{7.35}$$

where x_2 is the carriage velocity, $w(t) = -\frac{\lambda}{M} x_2(t)$ is a state dependent (internal) uncertainty, $T_s = 0.4$ is the sampling time, $M = 1$ and $\rho_0 = 0.33$. The following input and state constraints are imposed on the system:

$$-4 \leq u \leq 4, \; -1.3 \leq x_2 \leq 1.3. \tag{7.36}$$

The horizon is $N = 15$ and the terminal constraint is:

$$x_{t+N|t} \in \Omega_a, \; \Omega_a = \{x \in \mathbb{R}^n \,|\, x^T \Sigma x \leq \delta\}, \tag{7.37}$$

where $\delta = 0.001$ [35] and $\Sigma = \begin{bmatrix} 1.3 & 1.9 \\ 1.9 & 3.0 \end{bmatrix}$.

The mp-NLP approach described in Section 7.2.2 is applied to design an explicit feedback stochastic NMPC controller for the cart [22]. The NMPC minimizes the mathematical expectation (7.20) of the cost function (7.19) subject to the system equations (7.34)–(7.35) and the constraints (7.36)–(7.37). In (7.19), it is chosen $\alpha = 1$ and the terminal penalty is given by $V_{k_a} = x^T \Sigma x$ [35]. Like in Section 6.3.4, the feedback functions $k_i(x_{t+i|t})$, $i = 0, \ldots, N-1$ have the form:

$$k_i(p_i, x_{t+i|t}) = \eta_i k_a(x_{t+i|t}) + \xi_{i,1} x_{1,t+i|t}^2 + \xi_{i,2} x_{2,t+i|t}^2, \tag{7.38}$$

where $p_i = [\eta_i \; \xi_{i,1} \; \xi_{i,2}]^T$ are the parameters that need to be optimized and $k_a(x_{t+i|t})$ is the auxiliary control law. The control law $k_a(x_{t+i|t})$ is determined in the way described in Section 6.3.4 (respectively in [35]).

In [23], a condition on the approximation tolerance has been derived such that the asymptotic stability of the nonlinear system in closed-loop with the approximate explicit NMPC is guaranteed. According to this condition, the tolerance is chosen to be dependent on the state, which would lead to a state space partition with less complexity in comparison to that corresponding to an uniform tolerance. In [22], a similar approach is applied and the approximation tolerance is chosen to be dependent on the state as $\bar{\varepsilon}(X_0) = \max(\bar{\varepsilon}_a, \bar{\varepsilon}_r \min_{x \in X_0} V_E^*(x))$, where $\bar{\varepsilon}_a = 0.005$ and $\bar{\varepsilon}_r = 0.03$ are the absolute and the relative tolerances.

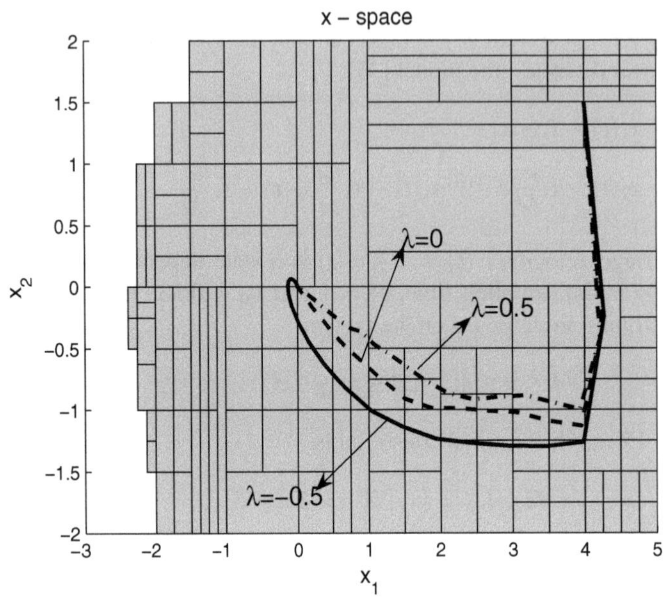

Fig. 7.1 State space partition of the explicit approximate feedback stochastic NMPC and the state trajectories corresponding to $\lambda = -0.5$, $\lambda = 0$, $\lambda = 0.5$.

The state space partition of the explicit approximate feedback stochastic NMPC controller is shown in Fig. 7.1. The partition has 150 regions and 11 levels in a binary search tree representation. Totally, 27 arithmetic operations are needed in real-time to compute the control input by traversing the binary search tree (11 comparisons, 10 multiplications, 5 additions and 1 exponential).

The performance of the closed-loop system was simulated for initial state $x(0) = [4\ 1.5]^T$ and for the three values of the stochastic parameter λ. The response is depicted in the state space (Fig. 7.1), as well as trajectories in time (Fig. 7.2 and Fig. 7.3). In Fig. 7.2 and Fig. 7.3, the control and state trajectories obtained with the explicit min-max NMPC controller (designed in Section 6.3.4) are given for comparison. The cost function values corresponding to the closed-loop trajectories associated to the explicit stochastic NMPC and to the explicit min-max NMPC are $\hat{V}_E = 121.64$ and $\hat{V}_{\min-\max} = 122.57$ (for $h_d = 0.6$), and $\hat{V}_E = 141.59$ and $\hat{V}_{\min-\max} = 141.70$ (for $h_d = 1.6$). Therefore, the explicit min-max NMPC appears to be slightly more conservative, since it is characterized with larger values of the cost function in comparison to the explicit stochastic NMPC. It can be seen that the explicit feedback stochastic NMPC controller brings the cart to the equilibrium despite of the presence of stochastic uncertainty, and the constraints imposed on the system are satisfied.

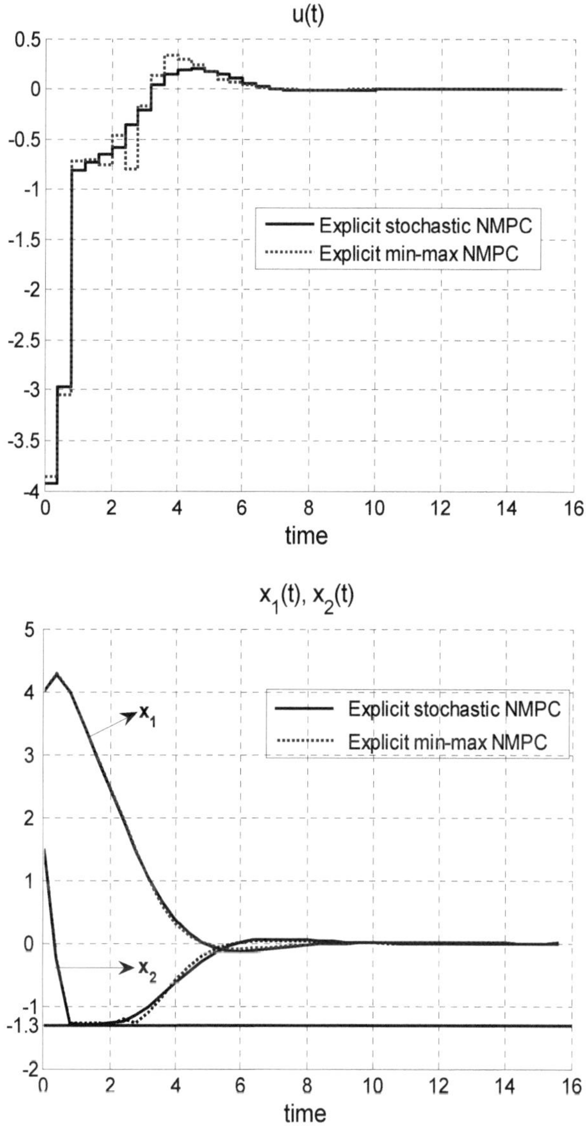

Fig. 7.2 Control input and state trajectory for $h_d = 0.6$.

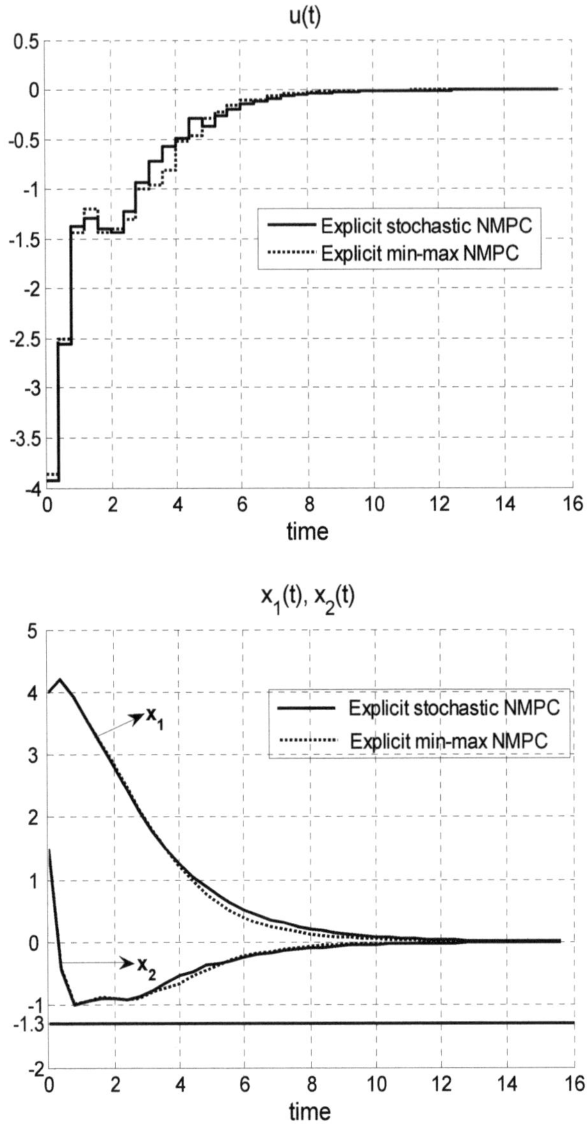

Fig. 7.3 Control input and state trajectory for $h_d = 1.6$.

7.3 Explicit Stochastic NMPC Based on Gaussian Process Models

In this section, the approximate mp-NLP approach [19, 20] to explicit *open-loop* stochastic NMPC based on Gaussian process models (abbreviated as GP-NMPC) is presented. The approach constructs a piecewise linear (PWL) approximation to the optimal control sequence, defined on an orthogonal partition of the state space.

7.3.1 Modeling of Dynamic Systems with Gaussian Processes

A Gaussian process is an example of the use of a flexible, probabilistic, nonparametric model which directly provides us with quantification of the uncertainty of predictions. Its use and properties for modeling are reviewed in [41].

A Gaussian process is a collection of random variables which have a joint multivariate Gaussian distribution. Assuming a relationship of the form $y = f(z)$ between an input $z \in \mathbb{R}^D$ and output $y \in \mathbb{R}$, we have $y(1), y(2), \dots, y(M) \sim \mathscr{N}(0, \mathbf{K})$, where $K_{pq} = \mathrm{Cov}(y(p), y(q)) = C(z(p), z(q))$ gives the covariance between the output points $y(p)$ and $y(q)$ corresponding to the input points $z(p)$ and $z(q)$. Thus, the mean $\mu(z)$ (usually assumed to be zero) and the covariance function $C(z(p), z(q))$ fully specify the Gaussian process. Note that the covariance function $C(z(p), z(q))$ can be any function with the property that it generates a positive definite covariance matrix. A common choice is the Gaussian covariance function [45, 41]:

$$C(z(p), z(q)) = v_1 \exp\left[-\frac{1}{2}\sum_{i=1}^{D} w_i (z_i(p) - z_i(q))^2\right] + v_0 \alpha_{pq} \tag{7.39}$$

where $\Theta = [w_1, \dots, w_D, v_0, v_1]$ is a vector of parameters called hyperparameters and z_i denotes the i-th component of the D-dimensional input vector z. The hyperparameter v_1 controls the magnitude of the covariance and the hyperparameters w_i represent the relative importance of each component z_i of vector z. The part $v_0 \alpha_{pq}$ represents the covariance between outputs due to white noise, where α_{pq} is the Kronecker operator and v_0 is the white noise variance (when assuming different kinds of noise the covariance function should be changed appropriately, e.g. [15]). For a given problem, the hyperparameters are learned (identified) using the data at hand. After the learning, one can use the w parameters as indicators of 'how important' the corresponding input components (dimensions) are: if w_i is zero or near zero it means that the inputs in dimension i contain little information and could possibly be removed.

Consider a set of M D-dimensional input vectors $\mathbf{Z} = [z(1), z(2), \dots, z(M)]^T$ and a vector of output data $Y = [y(1), y(2), \dots, y(M)]^T$. Based on the data $(\mathbf{Z}, Y)$, and given a new input vector z^*, we wish to estimate the probability distribution of the corresponding output y^*. Unlike other models, there is no model parameter determination as such, within a fixed model structure. With this model, most of the effort consists in *tuning* the parameters of the covariance function. This is done by maximizing the log-likelihood with the vector of hyperparameters Θ:

$$\mathscr{L}(\Theta) = \log(p(Y\,|\,\mathbf{Z})) = -\frac{1}{2}\log(|\mathbf{K}|) - \frac{1}{2}Y^T\mathbf{K}^{-1}Y - \frac{M}{2}\log(2\pi) \tag{7.40}$$

where $\mathbf{K}$ is the $M \times M$ training covariance matrix with determinant $|\mathbf{K}|$ and the hyperparameters distribution $p(\Theta\,|\,Y,\mathbf{Z})$ is approximated with their most likely values. The optimization requires the computation of the derivative of $\mathscr{L}$ with respect to each of the parameters:

$$\frac{\partial \mathscr{L}(\Theta)}{\partial \theta_i} = -\frac{1}{2}\text{trace}\left(\mathbf{K}^{-1}\frac{\partial \mathbf{K}}{\partial \theta_i}\right) + \frac{1}{2}Y^T\mathbf{K}^{-1}\frac{\partial \mathbf{K}}{\partial \theta_i}\mathbf{K}^{-1}Y \tag{7.41}$$

Here, it involves the computation of the inverse of the $M \times M$ covariance matrix $\mathbf{K}$ at every iteration, which can be computationally demanding for large M. The reader is referred to [41] for a detailed description of the parameter optimization methods.

Given that the hyperparameters are known, we can estimate the probability distribution of the corresponding output y^* at some new input vector z^*:

$$p(y^*\,|\,Y,\mathbf{Z},z^*) = \frac{p(Y,y^*,\mathbf{Z},z^*)}{p(Y\,|\,\mathbf{Z},z^*)} \tag{7.42}$$

It can be shown that this distribution is Gaussian with mean and variance [45]:

$$\mu(z^*) = k(z^*)^T\mathbf{K}^{-1}Y \tag{7.43}$$

$$\sigma^2(z^*) = k_0(z^*) - k(z^*)^T\mathbf{K}^{-1}k(z^*) \tag{7.44}$$

where $k(z^*) = [C(z(1),z^*),\dots,C(z(M),z^*)]^T$ is the $M \times 1$ vector of covariances between the test input and the training inputs and $k_0(z^*) = C(z^*,z^*)$ is the autocovariance of the test input. The vector $k(z^*)^T\mathbf{K}^{-1}$ in (7.43) can be interpreted as a vector of smoothing terms which weights the training outputs Y to make a prediction at the test point z^*. If the new input is far away from the data points, the term $k(z^*)^T\mathbf{K}^{-1}k(z^*)$ in (7.44) will be small, so that the predicted variance $\sigma^2(z^*)$ will be large. Thus, from the system identification point of view equation (7.43) provides the model prediction and equation (7.44) its confidence.

Gaussian processes can be used to model static nonlinearities and can therefore be used for modeling of dynamic systems if delayed input and output signals are used as regressors [26]. In such cases an autoregressive model is considered, such that the current predicted output depends on previous estimated outputs, as well as on previous control inputs:

$$z(t) = [\hat{y}(t-1),\hat{y}(t-2),\dots,\hat{y}(t-L),u(t-1),u(t-2),\dots,u(t-L)]^T$$

$$\hat{y}(t) = f(z(t)) + \eta(t) \tag{7.45}$$

where t denotes consecutive number of data sample, L is a given lag, and $\eta(t)$ is the prediction error. The quality of the predictions with a Gaussian process model is assessed by computing the average squared error (ASE):

$$\mathrm{ASE} = \frac{1}{M}\sum_{i=1}^{M}[\mu(\hat{y}(i)) - y(i)]^2 \tag{7.46}$$

and by the log predictive density error (LD) [26]:

$$\mathrm{LD} = \frac{1}{2}\log(2\pi) + \frac{1}{2M}\sum_{i=1}^{M}\left(\log[\sigma^2(\hat{y}(i))] + \frac{[\mu(\hat{y}(i)) - y(i)]^2}{\sigma^2(\hat{y}(i))}\right) \tag{7.47}$$

In (7.46), (7.47), $\mu(\hat{y}(i))$ and $\sigma^2(\hat{y}(i))$ are the prediction mean and variance, $y(i)$ is the system's output and M is the number of the training points.

The Gaussian process model now not only describes the dynamic characteristics of the non-linear system, but at the same time provides information about the confidence in the predictions. The Gaussian process can highlight areas of the input space where prediction quality is poor, due to the lack of data, by indicating the higher variance around the predicted mean. The Gaussian process modelling approach in [26] has been applied to model the dynamics of various systems e.g. [2, 29, 20, 28, 39].

7.3.2 *Formulation of the Stochastic GP-NMPC Problem as an mp-NLP Problem*

Consider a stochastic nonlinear discrete-time system:

$$x(t+1) = f(x(t), u(t)) + \xi(t) \tag{7.48}$$

where $x(t) \in \mathbb{R}^n$ and $u(t) \in \mathbb{R}^m$ are the state and input variables, $\xi(t) \in \mathbb{R}^n$ are Gaussian disturbances, and $f : \mathbb{R}^n \times \mathbb{R}^m \to \mathbb{R}^n$ is a nonlinear continuous function. Suppose that a Gaussian process model of the system (7.48) is obtained by applying the approach described in Section 7.3.1. Suppose the initial state $x(t) = x_{t|t}$ and the control inputs $u(t+k) = u_{t+k}$, $k = 0, 1, \ldots, N-1$ are given. Then, the probability distribution of the predicted states $x_{t+k+1|t}$, $k = 0, 1, \ldots, N-1$ which correspond to the given initial state $x_{t|t}$ and control inputs u_{t+k}, $k = 0, 1, \ldots, N-1$ can be obtained [16]:

$$x_{t+k+1|t}\,|\,x_{t+k|t}, u_{t+k} \sim \mathcal{N}(\mu(x_{t+k+1|t}), \sigma^2(x_{t+k+1|t})),\ k = 0, 1, \ldots, N-1 \tag{7.49}$$

The 95% confidence interval of the random variable $x_{t+k+1|t}$ is $[\mu(x_{t+k+1|t}) - 2\sigma(x_{t+k+1|t});\ \mu(x_{t+k+1|t}) + 2\sigma(x_{t+k+1|t})]$, where $\sigma(x_{t+k+1|t})$ is the standard deviation.

In [19, 20], a reference tracking NMPC problem based on a Gaussian process model (GP-NMPC) is considered, where the goal is to have the state vector $x(t)$ track the reference signal $r(t) \in \mathbb{R}^n$. In the problem formulation, the type of the cost function is like the one used in [3]. Suppose that a full measurement of the state $x(t)$ is available at the current time t. For the current $x(t)$, the reference tracking GP-NMPC solves the following optimization problem [19, 20]:

Problem 7.4:

$$V^*(x(t),r(t),u(t-1)) = \min_U J(U,x(t),r(t),u(t-1)) \tag{7.50}$$

subject to $x_{t|t} = x(t)$ and:

$$\mu(x_{t+k|t}) - 2\sigma(x_{t+k|t}) \geq x_{\min},\ k = 1, \ldots, N \tag{7.51}$$

$$\mu(x_{t+k|t}) + 2\sigma(x_{t+k|t}) \leq x_{\max},\ k = 1, \ldots, N \tag{7.52}$$

$$u_{\min} \leq u_{t+k} \leq u_{\max},\ k = 0, 1, \ldots, N-1 \tag{7.53}$$

$$\Delta u_{\min} \leq \Delta u_{t+k} \leq \Delta u_{\max},\ k = 0, 1, \ldots, N-1 \tag{7.54}$$

$$\max\{\|\mu(x_{t+N|t}) - 2\sigma(x_{t+N|t}) - r(t)\|, \\ \|\mu(x_{t+N|t}) + 2\sigma(x_{t+N|t}) - r(t)\|\} \leq \delta \tag{7.55}$$

$$\Delta u_{t+k} = u_{t+k} - u_{t+k-1},\ k = 0, 1, \ldots, N-1 \tag{7.56}$$

$$x_{t+k+1|t} \,|\, x_{t+k|t}, u_{t+k} \sim \mathcal{N}(\mu(x_{t+k+1|t}), \sigma^2(x_{t+k+1|t})) \\ k = 0, 1, \ldots, N-1 \tag{7.57}$$

with $U = [u_t, u_{t+1}, \ldots, u_{t+N-1}]$ and the cost function given by:

$$J(U,x(t),r(t),u(t-1)) = \sum_{k=0}^{N-1} [\|\mu(x_{t+k|t}) - r(t)\|_Q^2 + \|\Delta u_{t+k}\|_R^2] \\ + \|\mu(x_{t+N|t}) - r(t)\|_P^2 \tag{7.58}$$

Here, N is a finite horizon and $P, Q, R \succ 0$. From a stability point of view it is desirable to choose δ in the terminal constraint (7.55) sufficiently small [37]. If the horizon N is large and the Gaussian process model has a small prediction uncertainty, then it is more likely that the choice of a small δ will be possible.

It should be noted that a more general stochastic MPC problem is formulated in [31, 7, 8, 5], where a probabilistic formulation of the cost includes the probabilistic bounds of the predicted variable. The stochastic MPC problem considered here (Problem 7.4) is of a more special form since the cost function (7.58) includes the mean value of the random variable. However, the approximate approach to the explicit solution of Problem 7.4 (which is based on the approximate mp-NLP algorithms, given in Section 1.1.5) can be easily extended to the more general case of stochastic MPC problem formulation where the optimization is performed on the expected value of the cost function.

We introduce a parameter vector:

$$\tilde{x}(t) = [x(t), r(t), u(t-1)] \in \mathbb{R}^{\tilde{n}},\ \tilde{n} = 2n + m \tag{7.59}$$

Let $\tilde{x}$ be the value of the parameter vector at the current time t. Using a direct single shooting strategy, the equality constraints are eliminated and the optimization Problem 7.4 can be formulated in a compact form as follows [19, 20]:

Problem 7.5:

$$V^*(\tilde{x}) = \min_U J(U,\tilde{x}) \text{ subject to } G(U,\tilde{x}) \leq 0 \tag{7.60}$$

The GP-NMPC problem defines an mp-NLP, since it is NLP in U parameterized by $\tilde{x}$. An optimal solution to this problem is denoted $U^* = [u_t^*, u_{t+1}^*, ..., u_{t+N-1}^*]$ and the control input is chosen according to the receding horizon policy $u(t) = u_t^*$. Define the set of feasible parameter vectors as follows:

$$X_f = \{\tilde{x} \in \mathbb{R}^{\tilde{n}} \,|\, G(U,\tilde{x}) \leq 0 \text{ for some } U \in \mathbb{R}^{Nm}\} \tag{7.61}$$

If δ in (7.55) is chosen such that the Problem 7.4 is feasible, then X_f is a non-empty set. In parametric programming problems one seeks the solution $U^*(\tilde{x})$ as an explicit function of the parameters $\tilde{x}$ in some set $X \subseteq X_f \subseteq \mathbb{R}^{\tilde{n}}$ [14]. In case the Problem 7.5 is convex, its approximate solution can be found by applying the approximate mp-NLP approach, described in Section 1.1.5.1. Otherwise, the approximate mp-NLP approach from Section 1.1.5.2 should be used, where in addition to the set of vertices of a given hyper-rectangle in the parameter space, the optimal solution is also computed at several interior points and global optimization methods are applied.

7.3.3 Application 2: Reference Tracking Control of a Combustion Plant

Energy production is one of the largest sources of air pollution and CO_2. Therefore a rational and ecological use of energy is the main task of the thermoelectric power plants. A feasible method to reduce the NO_x, CO, CO_2 emissions and to increase the efficiency is to improve the control strategies of existing power plants, i.e. to optimize the combustion process [11]. The objectives for the improvement of the power plant combustion process are energy saving, pollution reduction, longer plant lifetime, less downtime and maintenance effort, increased safety in operation, i.e. overall cost reduction. These goals can be achieved through application of modern control algorithms with low on-line computational complexity and high reliability of the implementation. Feedback combustion control is possible since continuous flue gases analyzers are available [11]. For control purposes it would be ideal to measure all flue gases components. But the price for such a realization would currently be too high in comparison with the savings achieved. Therefore the control of the oxygen fraction in the flue gases, measured on-line by the well known in-situ ZrO_2 analyzers, is often the best solution [10]. Based on that, different algorithms for combustion control have been studied in [42, 9, 4, 10, 25] and more recent in e.g. [40, 13]. It should be noted that these methods assume that the combustion model is known exactly. However, the mathematical models are only an approximation of the real process and they usually contain some amount of uncertainty (unknown additive disturbances and/or uncertain model parameters). In order to achieve a robust performance of the control system it would be necessary to take into account the uncertainty when designing the controller.

In [20], a Gaussian process model of a combustion plant (a steam boiler PK 401 at Cinkarna Celje Company, Celje, Slovenia) is obtained. Then, the approximate mp-NLP approach (described in Section 1.1.5.2) is applied to design an explicit reference tracking GP-NMPC controller that brings the air factor of the combustion plant to its optimal value with every change of the load factor. Thus, an efficient on-line optimization of the combustion plant is achieved where both the economic and the environmental aspects are taken into account. Because of the operation safety of the considered combustion plant and because interrupts in plant operation are not favored by company management, the results obtained in [20] are based on simulation data to show the potential use of the approximate mp-NLP approach to the optimal control of industrial combustion plants.

7.3.3.1 Optimal Operation of Combustion Plants

Fuel composition can be expressed with percentage of carbon C, hydrogen H, oxygen O, nitrogen N, sulphur S, ash A and water W [12]:

$$\mathrm{C+H+O+N+S+A+W} = 100\% \tag{7.62}$$

Composition of the air is usually expressed only with the percentage of oxygen O_2 and nitrogen N_2:

$$\mathrm{O_2 + N_2} = 21\% + 79\% = 100\% \tag{7.63}$$

The combustion process is schematically shown in Fig. 7.4. The limited fuel sources, considerable increase in the fuel prices and the enormous environment pollution require decreasing the fuel use, the heat losses and the amount of harmful flue gases emissions, i.e. to optimize the combustion process [9]. It has been shown in [9] that in order to achieve an optimal operation of the combustion plants, it is necessary to optimize the air factor λ defined as:

$$\lambda = \frac{V_{\mathrm{air}}}{V_{\mathrm{air,stoich}}} \tag{7.64}$$

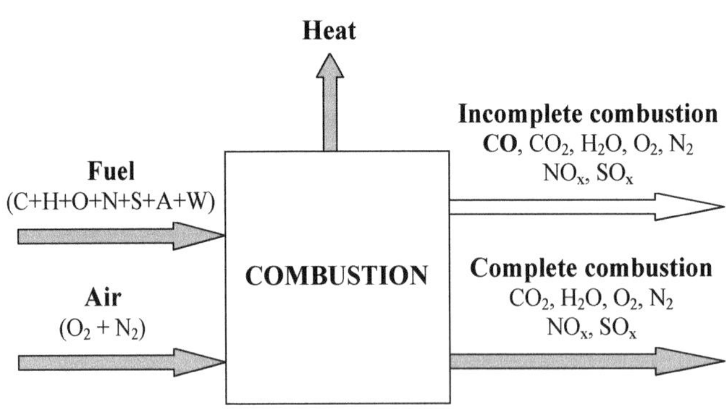

Fig. 7.4 Input and output flows of the combustion process.

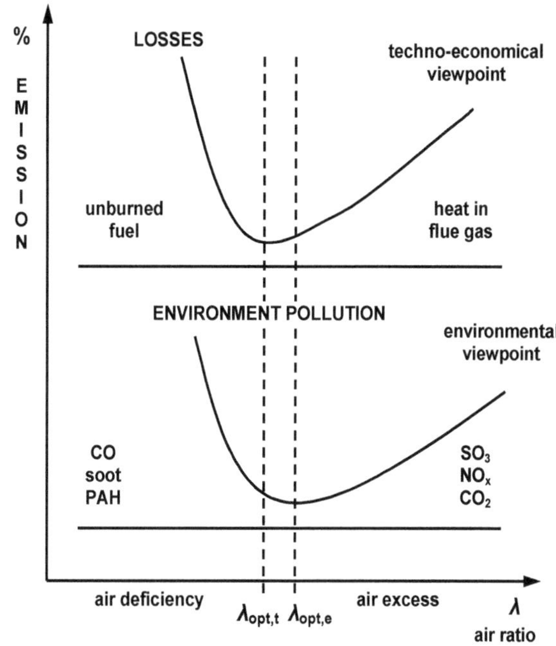

Fig. 7.5 Techno-economical and environmental viewpoints of combustion [9].

where V_{air} is the volume of the air which goes into the combustion chamber and $V_{\text{air,stoich}}$ is the stoichiometrically required volume of the air necessary for complete combustion of 1 kg fuel. The combustion plant is working with air deficiency when $\lambda < 1$, and with air excess when $\lambda > 1$. Fig. 7.5 from [9] shows the aspects of the optimal combustion of fuel. From techno-economical viewpoint, the losses of the combustion can be reduced in two ways: 1) by reducing the quantity of the unburned fuel and 2) by reducing the quantity of the flue gases, i.e. of the heat losses. This leads to the optimal value $\lambda_{\text{opt,t}}$ of the air factor (cf. Fig. 7.5). From environmental viewpoint, it is desired to minimize the quantity of the harmful emissions and the corresponding optimal value of the air factor is $\lambda_{\text{opt,e}}$ (cf. Fig. 7.5). By taking into account both the techno-economical and the environmental aspects of combustion operation, it follows that the value λ of the air factor should be kept within the interval $[\lambda_{\text{opt,t}}; \lambda_{\text{opt,e}}]$.

It has been also shown in practice that the optimal air factor λ_{opt} depends on the load factor β defined as:

$$\beta = \frac{\Phi_{\text{fuel}}}{\Phi_{\text{fuel,max}}} \tag{7.65}$$

where Φ_{fuel} and $\Phi_{\text{fuel,max}}$ are respectively the current and the maximal allowed fuel flowrate. The relation $\lambda_{\text{opt}} = f(\beta)$ is shown in Fig. 7.6, where it can be seen that the optimal operation of the combustion plant is achieved with an air excess.

Therefore, the goal is to apply control algorithms that will maintain the air factor on its optimal value with every change of the load factor. Due to the importance of

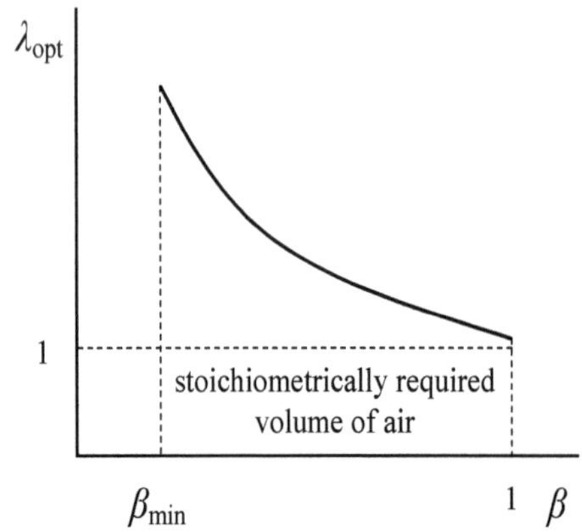

Fig. 7.6 The dependence of the optimal air factor on the load factor [9].

the described issue from economic and also environmental aspect, the combustion control is the field of constant development and research. This is also the driver for the development of the modeling and control approaches presented in the next sections.

7.3.3.2 Gaussian Process Model of a Combustion Plant

In [20], the system under investigation is a process of combustion in a steam boiler PK 401 at Cinkarna Celje Company, Celje, Slovenia. It was not possible to perform experiments on this plant during its operation because of plant safety and because interrupts in plant operation are not favored by company management. Therefore, the Gaussian process model identification was based on simulation data generated by adding a Gaussian disturbance to the analytical model developed in [12].

The fuel composition is expressed with the percentages of carbon C, hydrogen H, oxygen O, nitrogen N, sulphur S, ash A and water H_2O (denoted respectively with $x_C^{\text{fuel}}, x_H^{\text{fuel}}, x_O^{\text{fuel}}, x_N^{\text{fuel}}, x_S^{\text{fuel}}, x_A^{\text{fuel}}, x_{H_2O}^{\text{fuel}}$):

$$x_C^{\text{fuel}} + x_H^{\text{fuel}} + x_O^{\text{fuel}} + x_N^{\text{fuel}} + x_S^{\text{fuel}} + x_A^{\text{fuel}} + x_{H_2O}^{\text{fuel}} = 100\% \tag{7.66}$$

The composition of the air is assumed to be 21% oxygen and 79% nitrogen. The equations of the developed analytical model [12] are based on the stoichiometric chemical reactions of combustion:

$$C + O_2 \rightarrow CO_2 + Q_1 \tag{7.67}$$

$$C + \frac{1}{2}O_2 \rightarrow CO + Q_2 \tag{7.68}$$

$$2H_2 + O_2 \rightarrow 2H_2O + Q_3 \tag{7.69}$$

$$S + O_2 \rightarrow SO_2 + Q_4 \tag{7.70}$$

where Q_1, Q_2, Q_3, Q_4 are the heats of the reactions. The composition of the flue gases, resulting from the combustion process, is expressed in the following way [12]:

$$x_{O_2} + x_{CO} + x_{CO_2} + x_{SO_2} + x_{N_2} + x_{H_2O} = 100\% \tag{7.71}$$

where x_{O_2}, x_{CO}, x_{CO_2}, x_{SO_2}, x_{N_2} and x_{H_2O} are the volume percentages of oxygen, carbon monoxide, carbon dioxide, sulphur dioxide, nitrogen and water. Then, the volume balances for the separate components of the flue gases are described by the following equations [12]:

$$\frac{dx_{O_2}}{dt} = \frac{1}{V_k}\{-x_{O_2}[\Phi_{\text{air}} + \Phi_{\text{fuel}}(V_d - V_o)] + 21\Phi_{\text{air}} - 100V_o\Phi_{\text{fuel}}\} \tag{7.72}$$

$$\frac{dx_{CO}}{dt} = \frac{1}{V_k}\{-x_{CO}[\Phi_{\text{air}} + \Phi_{\text{fuel}}(V_d - V_o)] + 1.866ax_{\text{C}}^{\text{fuel}}\Phi_{\text{fuel}}\} \tag{7.73}$$

$$\frac{dx_{CO_2}}{dt} = \frac{1}{V_k}\{-x_{CO_2}[\Phi_{\text{air}} + \Phi_{\text{fuel}}(V_d - V_o)] + 1.866(1-a)x_{\text{C}}^{\text{fuel}}\Phi_{\text{fuel}}\} \tag{7.74}$$

$$\frac{dx_{SO_2}}{dt} = \frac{1}{V_k}\{-x_{SO_2}[\Phi_{\text{air}} + \Phi_{\text{fuel}}(V_d - V_o)] + 0.699x_{\text{S}}^{\text{fuel}}\Phi_{\text{fuel}}\} \tag{7.75}$$

$$\frac{dx_{N_2}}{dt} = \frac{1}{V_k}\{-x_{N_2}[\Phi_{\text{air}} + \Phi_{\text{fuel}}(V_d - V_o)] + 79\Phi_{\text{air}} + 0.8x_{\text{N}}^{\text{fuel}}\Phi_{\text{fuel}}\} \tag{7.76}$$

$$\begin{aligned}\frac{dx_{H_2O}}{dt} = \frac{1}{V_k}\{&-x_{H_2O}[\Phi_{\text{air}} + \Phi_{\text{fuel}}(V_d - V_o)] + 11.117x_{\text{H}}^{\text{fuel}}\Phi_{\text{fuel}} \\ &+1.244x_{\text{H}_2\text{O}}^{\text{fuel}}\Phi_{\text{fuel}}\}\end{aligned} \tag{7.77}$$

In (7.72)–(7.77), V_k is the volume of the combustion chamber [m^3], Φ_{fuel} is the normalized total flow of fuel [kg s^{-1}], Φ_{air} is the normalized total flow of air [$\text{N m}^3\ \text{s}^{-1}$], V_o is the theoretically required oxygen volume for the combustion of one unit of fuel [$\text{N m}^3\ \text{kg}^{-1}$], V_d is the theoretically obtained gas volume from one unit of fuel [$\text{N m}^3\ \text{kg}^{-1}$], a is the relative portion of carbon converted into CO.

The model (7.72)–(7.77) enables the simulation of the six flue gases components. However, for control design purposes only its O_2-part (equation (7.72)) named also O_2-model is used [9, 10]. The input to the O_2-model is the angular position of the damper, which is used to control the air flow Φ_{air}. The model output is the oxygen concentration in the flue gases. As the damper is a part of the closed-loop, it has to be modeled and added to the O_2-model (7.72). The dependence of the air flow Φ_{air} on the angle ϕ of the damper is given by the following relation [9]:

$$\Phi_{\text{air}} = \frac{\Phi_{\text{air,max}}}{2}\exp\left(\frac{3(\phi-45)}{45}\right),\ 0^\circ \le \phi \le 45^\circ \tag{7.78}$$

$$\Phi_{\text{air}} = \frac{\Phi_{\text{air,max}}}{2}\left(2-\exp\left(\frac{-3(\phi-45)}{45}\right)\right),\ 45^\circ \le \phi \le 90^\circ \tag{7.79}$$

where $\Phi_{\text{air,max}}$ is the maximum flow of air.

The O_2-model (7.72) is a deterministic model, which does not take into account the stochastic disturbances (e.g. change in the fuel composition, change of the humidity of the air flow) that may influence the combustion process. In order to consider the stochastic nature of plant operation, the dynamics of x_{O_2} is represented by the following stochastic discrete-time model [20]:

$$x_{O_2}(t+1) = f(x_{O_2}(t), \Phi_{\text{fuel}}(t), \phi(t)) + \xi(t) \tag{7.80}$$

Here, $\xi(t) \in \mathbb{R}$ is a Gaussian disturbance which represents the additive effect of the unmeasured stochastic disturbances. The sampling time, determined according to system dynamics, was selected to be $T_s = 1\ [s]$.

In [20], the signals ϕ and Φ_{fuel} for identification were generated by random number generators with normal distributions. The signal x_{O_2} was computed from the O_2-model (7.72) and a Gaussian disturbance ξ with zero mean and variance 0.05 was added to it. The ϕ signal blocking was $T_\phi = 5T_s$, i.e. it is kept constant for 5 time instants. The Φ_{fuel} signal blocking was $T_{\Phi_{\text{fuel}}} = 100T_s$. The number M of the signals samples used for the identification determines the dimension of the covariance matrix. In our case, $M = 1000$. Based on the generated data set, the discrete-time system (7.80) is approximated with a Gaussian process with the following hyperparameters [20]:

$$\Theta = [w_1, w_2, w_3, v_0, v_1] = [0.01346, 0.02847, 0.00036, 0.21984, 55.56554] \tag{7.81}$$

The maximum likelihood framework was used to determine the hyperparameters. The optimization method applied for identification of the Gaussian process model was the conjugate gradient method with line searches [16]. The response of the Gaussian process model to the identification signal is shown in Fig. 7.7. The associated average squared error and log density error are respectively ASE=0.6051 and LD=143.4835.

In [20], the signals ϕ and Φ_{fuel} for validation were generated by random number generator with normal distribution and rate of change that is different from the one used for the identification signals. The mean and the variance of x_{O_2} predicted with the identified Gaussian process model are obtained by iterative one-step ahead predictions, where at each step the predicted mean of x_{O_2} is fed back to the input. The response of the Gaussian process model to the validation signals is shown in Fig. 7.8. The associated prediction errors are ASE=0.9177 and LD=188.8626.

7.3.3.3 Design and Performance of Explicit Stochastic Reference Tracking Controller for the Combustion Plant

In [20], an explicit stochastic reference tracking GP-NMPC controller for the combustion plant considered in Section 7.3.3.2 is designed. The block-scheme of the control system is shown in Fig. 7.9. The controller brings the air factor (respectively the concentration of oxygen in the flue gases) on its optimal value with every change of the load factor and thus an optimal operation of the combustion plant is achieved.

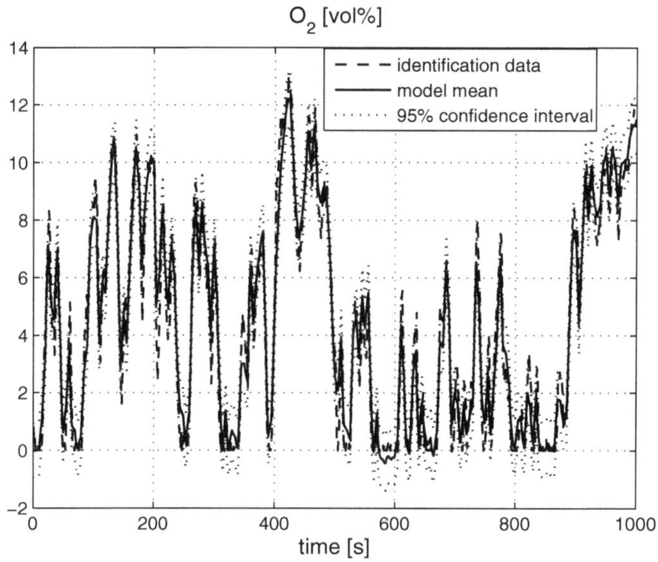

Fig. 7.7 Response of the Gaussian process model to the excitation signal used for identification.

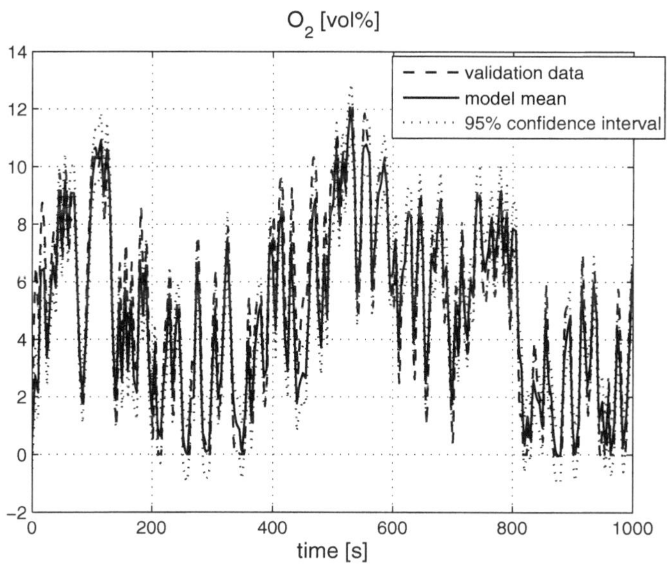

Fig. 7.8 Response of the Gaussian process model to the excitation signal used for validation.

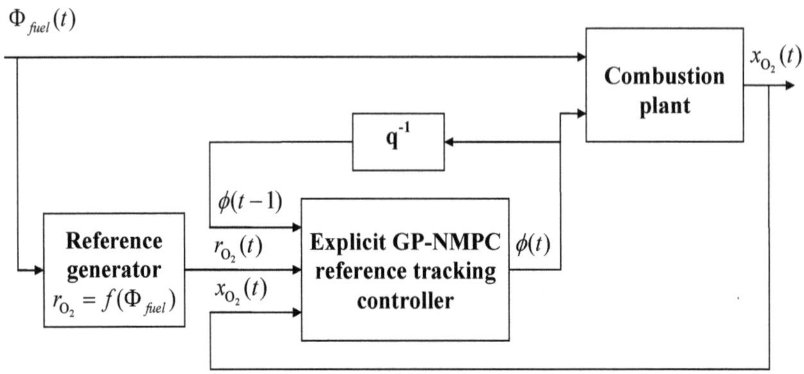

Fig. 7.9 Block scheme of the control system.

Table 7.1 Reference values for the percentage of O_2 in the flue gases

Φ_{fuel} [kg s^{-1}]	0.7	0.8	0.9	1.0	1.1	1.2	1.3
r_{O_2} [vol%]	4.5	4.1	3.7	3.4	3.2	3.0	2.8

The control input is $u = \phi$ (the angle of the damper for the air flow), the state variable is $x = x_{O_2}$ (the percentage of O_2 in the flue gases), and the reference signal is $r = r_{O_2}$ (the required percentage of O_2 in the flue gases). For this particular combustion plant, the reference values r_{O_2} corresponding to different values of the fuel flowrate Φ_{fuel} have been obtained by experiments and are given in Table 7.1 [9]. In case the fuel flowrate Φ_{fuel} does not take a value from this table, then the reference value r_{O_2} is computed through linear interpolation between the neighboring points in the table.

The mp-NLP approach described in Section 1.1.5.2 is applied to design an explicit stochastic reference tracking GP-NMPC controller for the combustion plant based on its Gaussian process model obtained in Section 7.3.3.2:

$$x_{O_2}(t+1)\,|\,x_{O_2}(t),\,\Phi_{\text{fuel}}(t),\,\phi(t) \sim \mathscr{N}(\mu(x_{O_2}(t+1)),\sigma^2(x_{O_2}(t+1))) \tag{7.82}$$

The following control input and rate constraints are imposed on the plant:

$$30^\circ \le \phi \le 60^\circ,\ -3^\circ \le \Delta\phi \le 3^\circ \tag{7.83}$$

The prediction horizon is $N = 10$ and the terminal constraint is:

$$\left|\mu(x_{O_2}(t+N)) - r_{O_2}(t)\right| \le 0.001 \tag{7.84}$$

The weighting matrices in the cost function (7.58) are $Q = 20$, $R = 1$, $P = 20$. The GP-NMPC minimizes the cost function (7.58) subject to the Gaussian process model (7.82) and the constraints (7.83)–(7.84). The parameter vector is $\tilde{x}(t) = [x_{O_2}(t), \Phi_{\text{fuel}}(t), \phi(t-1)] \in \mathbb{R}^3$, which leads to a 3-dimensional parameter space to be partitioned. The latter is defined by $X = [0; 7] \times [0.7; 1.3] \times [30; 60]$. The cost function approximation tolerance is chosen as $\bar{\varepsilon}(X_0) = \max(\bar{\varepsilon}_a, \bar{\varepsilon}_r \min_{\tilde{x} \in X_0} V^*(\tilde{x}))$, where $\bar{\varepsilon}_a = 0.005$ and $\bar{\varepsilon}_r = 0.1$ are the absolute and the relative tolerances, respectively. The partition of the explicit GP-NMPC controller is shown in Fig. 7.10. It has 513 regions and 12 levels of a binary search tree representation. Totally, 18 arithmetic operations are needed in real-time to compute the control input by traversing the binary search tree (12 comparisons, 3 multiplications and 3 additions).

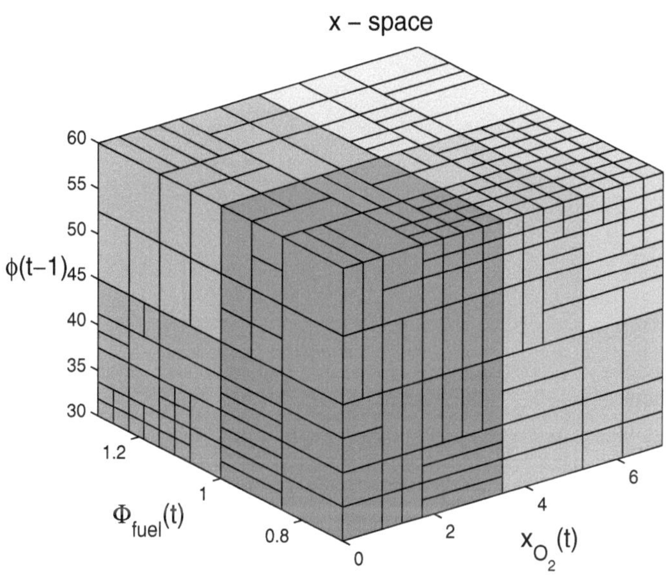

Fig. 7.10 Parameter space partition of the explicit approximate GP-NMPC controller.

The performance of the closed-loop system was simulated for the following change in the fuel flowrate:

$$\begin{aligned} \Phi_{\text{fuel}}(t) &= 1.1\,[\text{kg s}^{-1}], t \in [0; 50]; \\ \Phi_{\text{fuel}}(t) &= 1.25\,[\text{kg s}^{-1}], t \in [51; 100]; \\ \Phi_{\text{fuel}}(t) &= 1.05\,[\text{kg s}^{-1}], t \in [101; 150] \end{aligned} \quad (7.85)$$

and initial conditions for the state and control variable $x_{O_2}(0) = 3.3\,[\text{vol\%}]$ and $\phi(0) = 46^\circ$, respectively. The resulting closed-loop response is depicted in Fig. 7.11 and Fig. 7.12.

The results show that the exact and the approximate solutions are almost indistinguishable.

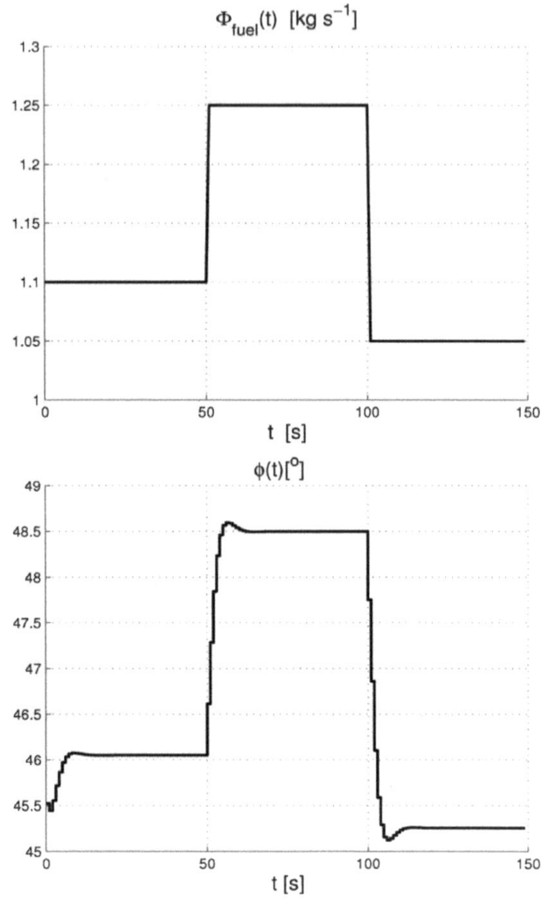

Fig. 7.11 Top: Change of the fuel flowrate. Bottom: The control input with the approximate explicit GP-NMPC (the solid curve) and with the exact GP-NMPC (the dotted curve).

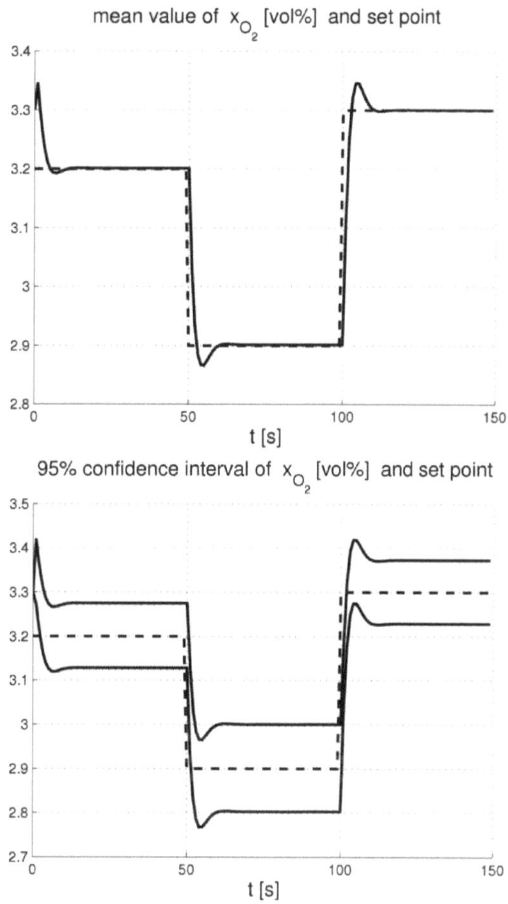

Fig. 7.12 Top: The mean value of the state variable predicted with the Gaussian process model. Bottom: The 95% confidence interval of the state variable predicted with the Gaussian process model. The solid curves are with the approximate explicit GP-NMPC, the dotted curves are with the exact GP-NMPC and the dashed curve is the set point.

References

1. Arellano-Garcia, H., Wendt, M., Barz, T., Wozny, G.: Close-loop stochastic dynamic optimization under probabilistic output constraints. In: Findeisen, R., Allgöwer, F., Biegler, L. (eds.) Assessment and Future Directions of Nonlinear Model Predictive Control. LNCIS, vol. 358, pp. 305–315. Springer, Heidelberg (2007)
2. Ažman, K., Kocijan, J.: Application of Gaussian processes for black-box modelling of biosystems. ISA Transactions 46, 443–457 (2007)
3. Bemporad, A., Morari, M., Dua, V., Pistikopoulos, E.N.: The explicit linear quadratic regulator for constrained systems. Automatica 38, 3–20 (2002)
4. Bitenc, A., Čretnik, J., Petrovčič, J., Strmčnik, S.: Design and application of an industrial controller. Computing & Control Engineering Journal 3, 29–34 (1992)
5. Cannon, M., Couchman, P., Kouvaritakis, B.: MPC for stochastic systems. In: Findeisen, R., Allgöwer, F., Biegler, L. (eds.) Assessment and Future Directions of Nonlinear Model Predictive Control. LNCIS, vol. 358, pp. 255–268. Springer, Heidelberg (2007)

6. Cannon, M., Ng, D., Kouvaritakis, B.: Successive Linearization NMPC for a Class of Stochastic Nonlinear Systems. In: Magni, L., Raimondo, D.M., Allgöwer, F. (eds.) Nonlinear Model Predictive Control: Towards New Challenging Applications. LNCIS, vol. 384, pp. 249–262. Springer, Heidelberg (2009)
7. Couchman, P., Cannon, M., Kouvaritakis, B.: Stochastic MPC with inequality stability constraints. Automatica 42, 2169–2174 (2006)
8. Couchman, P., Kouvaritakis, B., Cannon, M.: LTV models in MPC for sustainable development. International Journal of Control 79, 63–73 (2006)
9. Čretnik, J.: Modern automatic combustion control. Ph.D. thesis, Faculty of Electrical and Computing Engineering, Ljubljana (1992) (in Slovenian)
10. Čretnik, J.: Modern combustion control. In: Proceedings of the Conference on Modeling and Simulation, Barcelona, pp. 804–808 (1994)
11. Čretnik, J., Gumprecht, F.: In-situ oxygen and cumbustibles measurement. In: Proceedings of the Symposium Power Plants, Vrnjacka banja, Serbia (2006)
12. Čretnik, J., Strmčnik, S., Zupančič, B.: A model for combustion of fuel in the boiler. In: Proceedings of the 3rd Symposium on Simulationstechnik, Bad Münster, Germany, pp. 469–473 (1985)
13. Dong, X.C., Wang, H.B., Zhao, X.X.: Model reference neural network control for boiler combustion system. In: Proceedings of the 2005 International Conference on Machine Learning and Cybernetics, ICMLC 2005, pp. 4694–4698 (2005)
14. Fiacco, A.V.: Introduction to sensitivity and stability analysis in nonlinear programming. Academic Press, Orlando (1983)
15. Gibbs, M.N.: Bayesian Gaussian processes for regression and classification. Ph.D. thesis. Cambridge University, Cambridge, UK (1997)
16. Girard, A., Murray-Smith, R.: Gaussian Processes: Prediction at a Noisy Input and Application to Iterative Multiple-Step Ahead Forecasting of Time-Series. In: Murray-Smith, R., Shorten, R. (eds.) Switching and Learning in Feedback Systems. LNCS, vol. 3355, pp. 158–184. Springer, Heidelberg (2005)
17. Girard, A., Rasmussen, C.E., Quinonero Candela, J., Murray-Smith, R.: Gaussian process priors with uncertain inputs & application to multiple-step ahead time series forecasting. In: Proceedings of NIPS 15. MIT Press, Vancouver (2003)
18. Goodwin, G.C., Østergaard, J., Quevedo, D.E., Feuer, A.: A Vector Quantization Approach to Scenario Generation for Stochastic NMPC. In: Magni, L., Raimondo, D.M., Allgöwer, F. (eds.) Nonlinear Model Predictive Control: Towards New Challenging Applications. LNCIS, vol. 384, pp. 235–248. Springer, Heidelberg (2009)
19. Grancharova, A., Kocijan, J., Johansen, T.A.: Explicit stochastic nonlinear predictive control based on Gaussian process models. In: Proceedings of the European Control Conference, Kos, Greece, pp. 2340–2347 (2007)
20. Grancharova, A., Kocijan, J., Johansen, T.A.: Explicit stochastic predictive control of combustion plants based on Gaussian process models. Automatica 44, 1621–1631 (2008)
21. Grancharova, A., Johansen, T.A.: Computation, approximation and stability of explicit feedback min-max nonlinear model predictive control. Automatica 45, 1134–1143 (2009)
22. Grancharova, A., Johansen, T.A.: A computational approach to explicit feedback stochastic nonlinear model predictive control. In: Proceedings of the IEEE Conference on Decision and Control, Atlanta, USA, pp. 6083–6088 (2010)
23. Johansen, T.A.: Approximate explicit receding horizon control of constrained nonlinear systems. Automatica 40, 293–300 (2004)

24. Kantas, N., Maciejowski, J.M., Lecchini-Visintini, A.: Sequential Monte Carlo for Model Predictive Control. In: Magni, L., Raimondo, D.M., Allgöwer, F. (eds.) Nonlinear Model Predictive Control: Towards New Challenging Applications. LNCIS, vol. 384, pp. 263–274. Springer, Heidelberg (2009)
25. Kocijan, J.: An approach to multivariable combustion control design. Journal on Process control 7, 291–301 (1997)
26. Kocijan, J., Girard, A., Banko, B., Murray-Smith, R.: Dynamic systems identification with Gaussian processes. Mathematical and Computer Modelling of Dynamic Systems 11, 411–424 (2005)
27. Kocijan, J.: Control algorithms based on Gaussian process models: A state-of-the-art survey. In: Proceedings of the Special International Conference on Complex Systems: Synergy of Control Communications and Computing, Ohrid, Republic of Macedonia, pp. 69–80 (2011)
28. Kocijan, J., Grancharova, A.: Gaussian process modelling case study with multiple outputs. Comptes Rendus de l'Académie Bulgare des Sciences 63, 601–607 (2010)
29. Kocijan, J., Likar, B.: Gas-liquid separator modelling and simulation with Gaussian-process models. Simulation Modelling Practice and Theory 16, 910–922 (2008)
30. Kocijan, J., Murray-Smith, R.: Nonlinear Predictive Control with a Gaussian Process Model. In: Murray-Smith, R., Shorten, R. (eds.) Switching and Learning in Feedback Systems. LNCS, vol. 3355, pp. 185–200. Springer, Heidelberg (2005)
31. Kouvaritakis, B., Cannon, M., Couchman, P.: MPC as a tool for sustainable development integrated policy assessment. IEEE Transactions on Automatic Control 51, 145–149 (2006)
32. Lee, J.H., Cooley, B.L.: Optimal feedback control strategies for state-space systems with stochastic parameters. IEEE Transactions on Automatic Control 43, 1469–1475 (1998)
33. Likar, B., Kocijan, J.: Predictive control of a gas-liquid separation plant based on a Gaussian process model. Computers & Chemical Engineering 31, 142–152 (2007)
34. Limon, D., Alamo, T., Salas, F., Camacho, E.F.: Input-to-state stability of min-max MPC controllers for nonlinear systems with bounded uncertainties. Automatica 42, 797–803 (2006)
35. Magni, L., de Nicolao, G., Scattolini, R., Allgöwer, F.: Robust model predictive control for nonlinear discrete-time systems. International Journal of Robust and Nonlinear Control 13, 229–246 (2003)
36. Magni, L., Scattolini, R.: Robustness and robust design of MPC for nonlinear discrete time systems. In: Findeisen, R., Allgöwer, F., Biegler, L. (eds.) Assessment and Future Directions of Nonlinear Model Predictive Control. LNCIS, vol. 358, pp. 239–254. Springer, Heidelberg (2007)
37. Mayne, D.Q., Rawlings, J.B., Rao, C.V., Scokaert, P.O.M.: Constrained model predictive control: Stability and optimality. Automatica 36, 789–814 (2000)
38. Murray-Smith, R., Sbarbaro, D., Rasmussen, C.E., Girard, A.: Adaptive, cautious, predictive control with Gaussian process priors. In: Proceedings of the 13th IFAC Symposium on System Identification, Rotterdam (2003)
39. Petelin, D., Kocijan, J., Grancharova, A.: On-line Gaussian process model for the prediciton of the ozone concentration in the air. Comptes Rendus de l' Académie Bulgare des Sciences 64, 117–124 (2011)
40. Rangaswamy, T.R., Shanmugan, J., Mohammed, K.P., Thyagarajan, T.: Cascaded fuzzy controller for combustion control of utility boiler. Advances in Modelling and Analysis C 60, 37–56 (2005)
41. Rasmussen, C.E., Williams, C.K.I.: Gaussian processes for machine learning. MIT Press, Cambridge (2006)

42. Slevin, M.J.: Boiler oxygen trim control. In: Heating, Piping, Air Conditioning, pp. 95–98 (1984)
43. Solak, E., Murray-Smith, R., Leithead, W.E., Leith, D.J., Rasmussen, C.E.: Derivative observations in Gaussian process models of dynamic systems. In: Proceedings of NIPS 15. MIT Press, Vancouver (2003)
44. van Hessem, D., Scherer, C.W., Bosgra, O.H.: LMI-based closed-loop economic optimization of stochastic process operation under state and input constraints. In: Proceedings of the IEEE Conference on Decision and Control, pp. 4228–4233 (2001)
45. Williams, C.K.I.: Prediction with Gaussian processes: From linear regression and beyond. In: Jordan, M.I. (ed.) Learning and Inference in Graphical Models, pp. 599–621. Kluwer Academic Press (1998)
46. Yan, J., Bitmead, R.R.: Incorporating state estimation into model predictive control and its application to network traffic control. Automatica 41, 595–604 (2005)

Chapter 8
Explicit NMPC Based on Neural Network Models

Abstract. This chapter considers an approximate mp-NLP approach to explicit solution of deterministic NMPC problems for constrained nonlinear systems described by neural network NARX models. The approach builds an orthogonal search tree structure of the *regressor* space partition and consists in constructing a piecewise linear (PWL) approximation to the optimal control sequence. A dual-mode control strategy is proposed in order to achieve an offset-free closed-loop response in the presence of bounded disturbances and/or model errors. It consists in using the explicit NMPC (based on NARX model) when the output variable is far from the origin and applying an LQR in a neighborhood of the origin. The LQR design is based on an augmented linear ARX model which takes into account the integral regulation error. The approximate mp-NLP approach and the dual-mode approach are applied to design an explicit output-feedback NMPC for regulation of a pH maintaining system.

8.1 Introduction

The NMPC algorithms are based on various nonlinear models. Often these models are developed as *first-principles* models, but other approaches, like *black-box* identification approaches are also popular. In Chapters 3, 4, 5, 6 and in Section 7.2, approaches to explicit solution of NMPC problems based on *first-principles* models were presented, which assume that the state variables can be measured.

Alternatively, there exists a number of NMPC approaches based on various *black-box* models e.g. based on neural network models (e.g. [16, 22]), fuzzy models (e.g. [12]), local model networks (e.g. [18]), Gaussian process models (e.g. [11]). The common feature of these NMPC approaches is that an on-line optimization needs to be performed in order to compute the optimal control input. Consequently, the computation is time consuming and the real-time NMPC implementation is limited to processes where the sampling time is sufficient to support the computational needs. However, the on-line computational complexity can be circumvented with an explicit approach to NMPC, where the only computation performed on-line would

A. Grancharova & T.A. Johansen: Explicit Nonlinear Model Predictive Ctrl., LNCIS 429, pp. 187–207.
springerlink.com

be a simple function evaluation. Thus, in Section 7.3, an approach for off-line computation of explicit stochastic NMPC controller for constrained nonlinear systems based on a stochastic *black-box* model (Gaussian process model) was described.

In this chapter, the approximate mp-NLP approach [5, 6] to explicit solution of deterministic NMPC problems for constrained nonlinear systems described by neural network NARX models [2] is considered. The NMPC problem based on neural network model will be referred to as NN-NMPC problem. The approach builds an orthogonal search tree structure of the *regressor* space partition and consists in constructing a piecewise linear (PWL) approximation to the optimal control sequence. A dual-mode control strategy is proposed in order to achieve an offset-free closed-loop response in the presence of bounded disturbances and/or model errors. It is similar to the dual-mode receding horizon control concept developed in [15] (based on *state space* models), however here *black-box* models are considered and an explicit solution of the NMPC problem is sought. Thus, the suggested strategy consists in using the explicit NMPC (based on NARX model) when the output variable is far from the origin and applying an LQR in a neighborhood of the origin. The LQR design is based on an augmented linear ARX model which takes into account the integral regulation error. The main motivations behind the dual-mode control strategy are the following. First, it may be beneficial to use a separate linear model in a neighborhood of the equilibrium, since the nonlinear *black-box* model may not have accurate linearizations unlike a *first-principles* model, and the requirement for accurate control is highest at the equilibrium. Second, it leads to a reduced complexity of the explicit NMPC compared to augmenting the nonlinear model with an integrator to achieve an integral action directly in the NMPC.

8.2 Formulation of the NN-NMPC Problem as an mp-NLP Problem

8.2.1 Modeling of Dynamic Systems with Neural Networks

The black-box identification of nonlinear systems is an area which is quite diverse. It covers topics from mathematical approximation theory, estimation theory, nonparametric regression and concepts like neural networks, fuzzy models, wavelets etc. A unified overview of this topic is given in [20].

Consider a nonlinear dynamical system with input $u \in \mathbb{R}^m$ and output $y \in \mathbb{R}^p$ and let $U = [u(1), u(2), \ldots, u(M)]$ and $Y = [y(1), y(2), \ldots, y(M)]$ be sets of observed values of u and y to the number of M. Based on these data, the dynamics of the system can be described with a neural network NARX model [2], where the future predicted output $y(i+1)$ depends on previous estimated outputs, as well as on previous control inputs:

$$y(i+1) = f(z(i), \theta) \tag{8.1}$$

$$z(i) = [y(i), y(i-1), \ldots, y(i-L), u(i), u(i-1), \ldots, u(i-L)] \tag{8.2}$$

Here, L is a given lag, i denotes the consecutive index of data samples ($i \geq L$), $z(i)$ is the *regressor* vector, f is the function realized by the black-box model, and θ is a finite-dimensional vector of parameters. Thus, the function f is a concatenation of two mappings: one that takes the increasing number of the past values of the observed inputs and outputs and maps them into the finite dimensional *regressor* vector and one that takes this vector to the space of the outputs. The nonlinear mapping from the *regressor* space to the output space can be of various kinds. In our case we will use neural network with sigmoid basis functions in the hidden layer and linear basis functions in the output layer. This form of neural network is called Multilayer Perceptron (MLP), which is probably the most frequently considered member of the neural network family (e.g. [16]) and can be used as an universal approximator. This particular choice was subjective. Any other choice of *regressor* vector composition or any other choice of mapping is possible until it enables satisfactory description of the modeled dynamic system. It should be noted that the results given in [5, 6] are not limited to MLP approach only.

The parameters of the MLP are the weights of its units. After the structure (number of layers and units) is determined, the model parameters are obtained with optimization, based on a chosen cost function. This cost function is most frequently a least squares combination of errors between estimated and measured output signals:

$$E = \frac{1}{2M} \sum_{i=1}^{M} \|y(i) - \hat{y}(i|\theta)\|^2 \tag{8.3}$$

where $\hat{y}(i|\theta)$ is estimated output signal, θ is a vector containing the weights, and M is the number of measured output signals $y(i)$. The quality of prediction can be assessed with evaluation of residuals, estimation of the average prediction error or visualization of the network model's ability to predict. The reader is referred to [16] for more details.

8.2.2 Formulation of the NN-NMPC Problem

Consider the discrete-time nonlinear system:

$$x(t+1) = h(x(t), u(t)) \tag{8.4}$$

$$y(t) = g(x(t), u(t)) \tag{8.5}$$

where $x(t) \in \mathbb{R}^n$, $u(t) \in \mathbb{R}^m$, and $y(t) \in \mathbb{R}^p$ are the state, input and output vectors, and $h : \mathbb{R}^n \times \mathbb{R}^m \rightarrow \mathbb{R}^n$ and $g : \mathbb{R}^n \times \mathbb{R}^m \rightarrow \mathbb{R}^p$ are nonlinear functions. The following input and output constraints are imposed on the system (8.4)–(8.5):

$$u_{\min} \leq u(t) \leq u_{\max}\ ,\ y_{\min} \leq y(t) \leq y_{\max} \tag{8.6}$$

Assume that the dynamics of the nonlinear system (8.4)–(8.5) is approximated with an MLP neural network with NARX structure of the form (8.1)–(8.2). Then for $t \geq L$, define a *modified regressor* vector:

$$\tilde{z}(t) = \begin{cases} [y(t), y(t-1), \ldots, y(t-L), u(t-1), \ldots, u(t-L)], \text{ if } L > 0 \\ y(t), \text{ if } L = 0 \end{cases}, \tag{8.7}$$

where $u(t-1), \ldots, u(t-L)$ and $y(t), y(t-1), \ldots, y(t-L)$ are the measured values of the input u and the output y. Thus, $\tilde{z}(t) \in \mathbb{R}^q$ with $q = (L+1)p + Lm$. Then, the NARX model, used to obtain one-step ahead prediction of the output for $t \geq L$, is represented:

$$\hat{y}(t+1|\theta) = f_{NN}(\tilde{z}(t), u(t), \theta), \tag{8.8}$$

where f_{NN} is the function realized by the neural network (NN) and θ contains the network weights. Suppose the initial *regressor* vector $\tilde{z}(t) = \tilde{z}_{t|t}$ is known and the control inputs $u(t+k) = u_{t+k}$, $k = 0, 1, \ldots, N-1$ are given. Then, the model (8.8) can be used to obtain the predicted output $y_{t+k+1|t}$, $k = 0, 1, \ldots, N-1$ through iterative one-step ahead predictions, where at each step the predicted output value is fed back to the *regressor* vector:

$$y_{t+k+1|t} = f_{NN}(\tilde{z}_{t+k|t}, u_{t+k}, \theta) \tag{8.9}$$

$$\tilde{z}_{t+k|t} = \begin{cases} [y_{t+k|t}, y_{t+k-1|t}, \ldots, y_{t+k-L|t}, u_{t+k-1}, \ldots, u_{t+k-L}], \text{ if } L > 0 \\ y_{t+k|t}, \text{ if } L = 0 \end{cases} \tag{8.10}$$

The following assumptions are made [5, 6]:

Assumption 8.1. *There exists* $u_{st}^{NN} \in \mathbb{R}^m$ *satisfying* $u_{\min} \leq u_{st}^{NN} \leq u_{\max}$, *and such that* $f_{NN}(\tilde{z}_0, u_{st}^{NN}, \theta) = 0$, *where* $\tilde{z}_0$ *is obtained from (8.10) with* $y_{t+k|t} = y_{t+k-1|t} = \ldots = y_{t+k-L|t} = 0$, $u_{t+k-1} = \ldots = u_{t+k-L} = u_{st}^{NN}$.

Assumption 8.2. $y_{\min} < 0 < y_{\max}$.

Assumption 8.1 means that the point $y = 0$, $u = u_{st}^{NN}$, is an equilibrium point for the NARX model (8.8), and Assumption 8.2 means that it is feasible for (8.6).

We consider the optimal regulation problem where the goal is to steer the output variable y to the origin by minimizing a certain performance criterion. Suppose that a full measurement of the *modified regressor* vector $\tilde{z}(t)$ is available at the current time $t \geq L$. Then, for the current $\tilde{z}(t)$, the regulation NN-NMPC solves the following optimization problem [5, 6]:

Problem 8.1:

$$V^*(\tilde{z}(t)) = \min_U J(U, \tilde{z}(t)) \tag{8.11}$$

subject to $\tilde{z}_{t|t} = \tilde{z}(t)$ and:

$$y_{\min} \leq y_{t+k|t} \leq y_{\max},\ k = 1,...,N \tag{8.12}$$

$$u_{\min} \leq u_{t+k} \leq u_{\max},\ k = 0,1,...,N-1 \tag{8.13}$$

$$\tilde{z}^c_{t+N|t} \in \Omega \tag{8.14}$$

$$y_{t+k+1|t} = f_{NN}(\tilde{z}_{t+k|t}, u_{t+k}, \theta),\ k = 0,1,...,N-1 \tag{8.15}$$

$$\tilde{z}_{t+k|t} = \begin{cases} [y_{t+k|t}, y_{t+k-1|t}, \ldots, y_{t+k-L|t}, u_{t+k-1}, \ldots, u_{t+k-L}],\ \text{if} L > 0 \\ y_{t+k|t},\ \text{if} L = 0, \end{cases} \quad k = 0,1,...,N-1 \tag{8.16}$$

with $U = [u_t, u_{t+1}, \ldots, u_{t+N-1}]$ and the cost function given by:

$$J(U, \tilde{z}(t)) = \sum_{k=0}^{N-1} [\|y_{t+k|t}\|^2_Q + \|u_{t+k} - u^{\text{NN}}_{st}\|^2_R] + \|y_{t+N|t}\|^2_F \tag{8.17}$$

and $\tilde{z}^c = \tilde{z} - [0^T_{(L+1)p}\ u^{NN^T}_{st}\ \ldots\ u^{NN^T}_{st}]^T$, where $0_{(L+1)p}$ is a zero vector with dimension $(L+1)p$. In (8.17), N is a finite horizon and Q, R, $F \succ 0$. In (8.14), Ω is the terminal set defined by $\Omega = \{\tilde{z}^c \in \mathbb{R}^q \,|\, \|\tilde{z}^c\|^2 \leq \delta\}$ with $\delta > 0$. From a stability point of view it is desirable to choose δ as small as possible [14]. If the system is asymptotically stable (or pre-stabilized) and N is large, then it is more likely that the choice of a small δ will be possible.

Let $\tilde{z}$ be the value of the *modified regressor* vector at the current time t. Then, the optimization Problem 8.1 can be formulated in a compact form as follows [5, 6]:

Problem 8.2:

$$V^*(\tilde{z}) = \min_U J(U, \tilde{z}) \ \text{ subject to } \ G(U, \tilde{z}) \leq 0 \tag{8.18}$$

The NN-NMPC problem defines an mp-NLP, since it is an NLP in U parameterized by $\tilde{z}$. We remark that the constraints function $G(U, \tilde{z})$ in (8.18) is implicitly defined by (8.12)–(8.16), and that all equality constraints are eliminated due to the direct single shooting strategy. An optimal solution to this problem is denoted $U^* = [u^*_t, u^*_{t+1}, \ldots, u^*_{t+N-1}]$ and the control input is chosen according to the receding horizon policy $u(t) = u^*_t$. Define the set of N-step feasible initial *regressor* vectors as follows:

$$Z_f = \{\tilde{z} \in \mathbb{R}^q \,|\, G(U, \tilde{z}) \leq 0 \text{ for some } U \in \mathbb{R}^{Nm}\} \tag{8.19}$$

In parametric programming problems one seeks the solution $U^*(\tilde{z})$ as an explicit function of the parameters $\tilde{z}$ in some set $\underline{Z} \subseteq Z_f \subseteq \mathbb{R}^q$ [3].

8.3 Approximate mp-NLP Approach to Explicit NN-NMPC

In [5, 6], an approximate mp-NLP approach is proposed to explicitly solve the *output-feedback* NN-NMPC problem formulated in the previous section. It is similar to the approximate mp-NLP approach to explicit solution of *state-space* NMPC problems. Let $Z \subset \mathbb{R}^q$ be a hyper-rectangle where we seek to approximate the optimal solution $U^*(\tilde{z})$ to Problem 8.2. It is required that the *regressor* space partition is orthogonal and can be represented as a $k-d$ tree. The idea of the approximate mp-NLP approach is to construct a PWL approximation $\widehat{U}(\tilde{z})$ to $U^*(\tilde{z})$ on Z, where the constituent affine functions are defined on hyper-rectangles covering Z. The computation of an affine *regressor* feedback associated to a given region Z_0 includes the following steps [5, 6]. First, a close-to-global solution of Problem 8.2 is computed at a set of points $V_0 = \{v_0, v_1, v_2, \ldots, v_{N_1}\} \subset Z_0$. Then, based on the solutions at these points, a local linear approximation $\widehat{U}_0(\tilde{z}) = K_0\tilde{z} + g_0$ to the close-to-global solution $U^*(\tilde{z})$, valid in the whole hyper-rectangle Z_0, is determined by applying the following procedure [5, 6]:

Procedure 8.1 (Computation of explicit approximate solution). *Consider any hyper-rectangle $Z_0 \subseteq Z$ with a set of points $V_0 = \{v_0, v_1, v_2, \ldots, v_{N_1}\} \subset Z_0$. Compute K_0 and g_0 by solving the following NLP:*

$$\min_{K_0, g_0} \sum_{i=0}^{N_1} (J(K_0 v_i + g_0, v_i) - V^*(v_i) + \alpha \| K_0 v_i + g_0 - U^*(v_i) \|^2) \tag{8.20}$$

$$\textit{subject to} \;\; G(K_0 v_i + g_0, v_i) \leq 0 \,, \; \forall v_i \in V_0 \tag{8.21}$$

In (8.20), $J(K_0 v_i + g_0, v_i)$ is the sub-optimal cost, $V^*(v_i)$ denotes the cost corresponding to the close-to-global solution $U^*(v_i)$, i.e. $V^*(v_i) = J(U^*(v_i), v_i)$, and the parameter α is a weighting coefficient (tuned in an ad-hoc fashion). Note that the computed linear *regressor* feedback $\widehat{U}_0(\tilde{z}) = K_0\tilde{z} + g_0$ satisfies the constraints in Problem 8.2 only at the discrete set of points $V_0 \subset Z_0$. After the feedback $\widehat{U}_0(\tilde{z})$ has been determined, an estimate $\widehat{\varepsilon}_0$ of the maximal cost function approximation error in Z_0 is computed as follows:

$$\widehat{\varepsilon}_0 = \max_{i \in \{0,1,2,\ldots,N_1\}} (J(K_0 v_i + g_0, v_i) - V^*(v_i)) \tag{8.22}$$

If $\widehat{\varepsilon}_0 > \bar{\varepsilon}$, where $\bar{\varepsilon} > 0$ is the specified tolerance of the approximation error, the region Z_0 is divided and the procedure is repeated for the new regions. The approximate PWL *regressor* feedback law is found by applying the approximate mp-NLP algorithm, described in Section 1.1.5.2. The mp-NLP algorithm terminates with a PWL function $\widehat{U}(\tilde{z}) = [\widehat{u}_0(\tilde{z}), \widehat{u}_1(\tilde{z}), \ldots, \widehat{u}_{N-1}(\tilde{z})]$ that is defined on an inner approximation Z_Π of the set $Z \cap Z_f$.

8.4 Design of Explicit Dual-Mode Controller

Generally, it will be difficult to guarantee that the local linearization at a nominal equilibrium point of an NN ARX model is accurate. The inaccuracies of the model may result in a steady-state offset of the explicit NN-NMPC controller. In [5, 6], a dual-mode control strategy is proposed which aims at achieving an offset-free closed-loop response in the presence of bounded disturbances and/or model errors. With this strategy, the control is performed by the explicit NN-NMPC controller when the system is far from equilibrium, and by a Linear Quadratic Regulator (LQR) with integral action when it is close to equilibrium.

8.4.1 Design of LQR with Integral Action in a Neighborhood of the Equilibrium

Consider a linear ARX model ([13]):

$$\begin{aligned} y(t+1) = A_1 y(t) + A_2 y(t-1) + \ldots + A_{l+1} y(t-l) + B_1(u(t) - u^*_{st}) + \\ B_2(u(t-1) - u^*_{st}) + \ldots + B_{l+1}(u(t-l) - u^*_{st}), \end{aligned} \tag{8.23}$$

that is valid in a neighborhood of the equilibrium $y = 0$, $u = u^*_{st}$ of the considered nonlinear dynamical system (8.4)–(8.5). In (8.23), the matrices $A_i \in \mathbb{R}^{p\times p}$ and $B_i \in \mathbb{R}^{p\times m}$, $i = 1, 2, \ldots, l+1$ contain the coefficients of the model, and l is a given lag. To estimate the parameters of the model (8.23), the least squares estimation method or the four-stage instrumental variable method can be applied ([13]). Based on the linear ARX model, an LQR that will regulate the system (8.23) to the origin, is designed. In order to achieve an offset-free performance, the model (8.23) is augmented with the following output $y_{int} \in \mathbb{R}^p$, which takes into account the integral error:

$$y_{int}(t+1) = y_{int}(t) + T_s y(t) \tag{8.24}$$

where T_s is the sampling time. Let $u_e(t) \equiv u(t) - u^*_{st}$. Then, the extended system with input u_e and output $y_e = [y, y_{int}]$ is described by the linear ARX model:

$$\begin{aligned} y_e(t+1) = A^e_1 y_e(t) + A^e_2 y_e(t-1) + \ldots + A^e_{l+1} y_e(t-l) + \\ B^e_1 u_e(t) + B^e_2 u_e(t-1) + \ldots + B^e_{l+1} u_e(t-l), \end{aligned} \tag{8.25}$$

where $A^e_1 = \begin{bmatrix} A_1 & 0_p \\ T_s I_p & I_p \end{bmatrix}$, $A^e_i = \begin{bmatrix} A_i & 0_p \\ 0_p & 0_p \end{bmatrix}$, $i = 2, 3, \ldots, l+1$, $B^e_i = \begin{bmatrix} B_i \\ 0_{p,m} \end{bmatrix}$, $i = 1, 2, \ldots, l+1$. Here, I_p is the p-dimensional identity matrix, 0_p is the p-dimensional square zero matrix, and $0_{p,m}$ is the $p\times m$-dimensional zero matrix. The following *regressor* vector is introduced [5, 6]:

$$\tilde{z}_e(t) = \begin{cases} [y_e(t), y_e(t-1), \ldots, y_e(t-l), u_e(t-1), u_e(t-2), \ldots, u_e(t-l)], \text{ if } l > 0 \\ y_e(t), \text{ if } l = 0 \end{cases} \tag{8.26}$$

Thus, $\tilde{z}_e(t) \in \mathbb{R}^{q_e}$ with $q_e = (l+1)2p + lm$. This vector can also be represented as $\tilde{z}_e(t) = [z_1(t), z_2(t), \ldots, z_{l+l+1}(t)]$, where $z_1(t), \ldots, z_{l+1}(t)$ are the shifted values of y_e and $z_{l+2}(t), \ldots, z_{l+l+1}(t)$ are the shifted values of u_e. The following relations hold [5, 6]:

$$\begin{aligned} & y_e(t+1) = z_1(t+1) \\ z_1(t) &= y_e(t) = z_2(t+1) \\ z_2(t) &= y_e(t-1) = z_3(t+1) \\ &\vdots \\ z_l(t) &= y_e(t-l+1) = z_{l+1}(t+1) \\ z_{l+1}(t) &= y_e(t-l) \end{aligned} \tag{8.27}$$

$$\begin{aligned} & u_e(t) = z_{l+2}(t+1) \\ z_{l+2}(t) &= u_e(t-1) = z_{l+3}(t+1) \\ z_{l+3}(t) &= u_e(t-2) = z_{l+4}(t+1) \\ &\vdots \\ z_{l+l}(t) &= u_e(t-l+1) = z_{l+l+1}(t+1) \\ z_{l+l+1}(t) &= u_e(t-l) \end{aligned} \tag{8.28}$$

Then, the system (8.25) can be represented:

$$\tilde{z}_e(t+1) = \tilde{A}^e \tilde{z}_e(t) + \tilde{B}^e u_e(t) \tag{8.29}$$

For $l > 0$, the matrices $\tilde{A}^e$ and $\tilde{B}^e$ in (8.29) are given by:

$$\tilde{A}^e = \begin{bmatrix} A_1^e & A_2^e & \ldots & A_l^e & A_{l+1}^e & B_2^e & \ldots & B_l^e & B_{l+1}^e \\ I_{2p} & 0_{2p} & \ldots & 0_{2p} & 0_{2p} & 0_{2p,m} & \ldots & 0_{2p,m} & 0_{2p,m} \\ 0_{2p} & I_{2p} & \ldots & 0_{2p} & 0_{2p} & 0_{2p,m} & \ldots & 0_{2p,m} & 0_{2p,m} \\ \vdots & & & & & & & & \\ 0_{2p} & 0_{2p} & \ldots & I_{2p} & 0_{2p} & 0_{2p,m} & \ldots & 0_{2p,m} & 0_{2p,m} \\ 0_{m,2p} & 0_{m,2p} & \ldots & 0_{m,2p} & 0_{m,2p} & 0_m & \ldots & 0_m & 0_m \\ 0_{m,2p} & 0_{m,2p} & \ldots & 0_{m,2p} & 0_{m,2p} & I_m & \ldots & 0_m & 0_m \\ \vdots & & & & & & & & \\ 0_{m,2p} & 0_{m,2p} & \ldots & 0_{m,2p} & 0_{m,2p} & 0_m & \ldots & I_m & 0_m \end{bmatrix} \tag{8.30}$$

$$\tilde{B}^e = [B_1^e \; 0_{2p,m} \; 0_{2p,m} \ldots 0_{2p,m} \; I_m \; 0_m \ldots 0_m]^T \tag{8.31}$$

In (8.30), (8.31), I_{2p} and I_m are identity matrices, 0_{2p} and 0_m are square zero matrices, and $0_{2p,m}$ and $0_{m,2p}$ are zero matrices with dimensions $2p \times m$ and $m \times 2p$ respectively. If $l = 0$, then $\tilde{A}^e = A_1^e$ and $\tilde{B}^e = B_1^e$.

The unconstrained LQR problem for system (8.29) solves the following optimization problem:

$$\min_{\{u_e(t),u_e(t+1),\ldots\}} \sum_{k=0}^{\infty} \left[\|\tilde{z}_e(t+k)\|_{Q_e}^2 + \|u_e(t+k)\|_{R_e}^2 \right] \tag{8.32}$$

where $Q_e, R_e \succ 0$. The solution to (8.32) is the linear feedback control law:

$$u_e(t+k) = -K\tilde{z}_e(t+k),\; k \geq 0, \tag{8.33}$$

where the controller gain K is given by [17]:

$$K = \left(\tilde{B}^{eT} P \tilde{B}^e + R_e\right)^{-1} \tilde{B}^{eT} P \tilde{A}^e \tag{8.34}$$

$$P = \tilde{A}^{eT} P \tilde{A}^e + Q_e - \tilde{A}^{eT} P \tilde{B}^e \left(\tilde{B}^{eT} P \tilde{B}^e + R_e\right)^{-1} \left(\tilde{A}^{eT} P \tilde{B}^e\right)^T \tag{8.35}$$

By taking into account that $u_e(t) \equiv u(t) - u^*_{st}$, it follows from (8.33) that the control input applied to the system is [5, 6]:

$$u(t+k) = -K\tilde{z}_e(t+k) + u^*_{st},\; k \geq 0 \tag{8.36}$$

8.4.2 *Explicit Dual-Mode Controller*

Consider the closed-loop system:

$$\tilde{z}_e(t+k) = (\tilde{A}^e - \tilde{B}^e K)\tilde{z}_e(t+k-1),\; k \geq 0, \tag{8.37}$$

where $\tilde{z}_e(t+k)$ is defined by (8.26) if t is replaced by $t+k$. Assume that $A_{cl} = \tilde{A}^e - \tilde{B}^e K$ is strictly Hurwitz. Let $\Gamma_e = \{\tilde{z}_e \in \mathbb{R}^{q_e} \,|\, \tilde{z}_e^T S \tilde{z}_e \leq \sigma\}$ with $S \succ 0$, $\sigma > 0$, be a positively invariant admissible set for the system (8.37). It means that $\forall \tilde{z}_e(t) \in \Gamma_e$, $\tilde{z}_e(t+k) \in \Gamma_e$, $\forall k > 0$ and:

$$y_{\min} < [\Psi\, 0_{2p}...0_{2p}\, 0_m...0_m]\tilde{z}_e(t+k) < y_{\max},\; k \geq 0 \tag{8.38}$$

$$u_{\min} < -K\tilde{z}_e(t+k) + u^*_{st} < u_{\max},\; k \geq 0 \tag{8.39}$$

where $\Psi = [I_p\; 0_p]$ and I_p, 0_p, 0_{2p}, 0_m are defined above. Γ_e can be determined in a way similar to Lemma 1 in [1]. If S satisfies the Lyapunov equation:

$$A_{cl}^T S A_{cl} - S = -\mu S - Q_e - K^T R_e K \tag{8.40}$$

for some $\mu > 0$, then there exists a constant $\sigma > 0$ such that the set Γ_e is a positively invariant admissible set for the system (8.37). For $l \leq L$, let $\widetilde{\Gamma}_e^1 = \{\xi \in \mathbb{R}^{\tilde{q}_e} \,|\, \xi^T \widetilde{S}_1 \xi \leq \tilde{\sigma}_1\}$ with $\widetilde{S}_1 \succ 0$, $\tilde{\sigma}_1 > 0$ be the orthogonal projection of Γ_e onto $\mathbb{R}^{\tilde{q}_e}$, $\tilde{q}_e = (l+1)p + lm$, by omitting all integrator elements from the *regressor* vector $\tilde{z}_e$. Let $\widetilde{\Omega}^1 = \{\zeta \in \mathbb{R}^{\tilde{q}_e} \,|\, \zeta^T \widetilde{S}_1 \zeta \leq \|\widetilde{S}_1\| \tilde{\delta}_1\}$ be the orthogonal projection of the terminal set Ω onto $\mathbb{R}^{\tilde{q}_e}$, where $\|\widetilde{S}_1\|$ is the induced norm of matrix $\widetilde{S}_1$. Then, it is required $\|\widetilde{S}_1\| \tilde{\delta}_1 < \tilde{\sigma}_1$, so that $\widetilde{\Omega}^1 \subset \widetilde{\Gamma}_e^1$. For $l > L$, let $\widetilde{\Gamma}_e^2 = \{\xi \in \mathbb{R}^q \,|\, \xi^T \widetilde{S}_2 \xi \leq \tilde{\sigma}_2\}$ with $\widetilde{S}_2 \succ 0$, $\tilde{\sigma}_2 > 0$ be the orthogonal projection of Γ_e onto $\mathbb{R}^q$, $q = (L+1)p + Lm$, by omitting all integrator elements and the elements $y(t-L-1),...,y(t-l)$ from $\tilde{z}_e$. Let $\widetilde{\Omega}^2 = \{\zeta \in \mathbb{R}^q \,|\, \zeta^T \widetilde{S}_2 \zeta \leq \|\widetilde{S}_2\| \tilde{\delta}_2\}$, with $\tilde{\delta}_2 > 0$ be a set such that $\Omega \subseteq \widetilde{\Omega}^2$. Similar to above it is required $\|\widetilde{S}_2\| \tilde{\delta}_2 < \tilde{\sigma}_2$, so that $\widetilde{\Omega}^2 \subset \widetilde{\Gamma}_e^2$.

In order to define the dual-mode controller, the *regressor* vector, associated to the system (8.37), is introduced:

$$\tilde{z}_r(t) = \begin{cases} [\Psi y_e(t), \Psi y_e(t-1), ..., \Psi y_e(t-l), \\ u_e(t-1) + u^*_{st}, ..., u_e(t-l) + u^*_{st}] \,, \text{ if } l > 0 \\ \Psi y_e(t) \,, \text{ if } l = 0 \end{cases} \tag{8.41}$$

where Ψ is defined above. Thus, $\tilde{z}_r(t) \in \mathbb{R}^{q_r}$ with $q_r = (l+1)p + lm$. Let $\Gamma_r \in \mathbb{R}^{q_r}$ be the orthogonal projection of Γ_e onto $\mathbb{R}^{q_r}$, specified by (8.41) (note that $q_r < q_e$). Further, for $l = L$, it is required $\Gamma_r \subset Z_\Pi \subset \mathbb{R}^q$. For $l < L$, $\Gamma_r \subset \widetilde{Z}_\Pi \subset \mathbb{R}^{q_r}$, where $\widetilde{Z}_\Pi$ is the orthogonal projection of Z_Π onto $\mathbb{R}^{q_r}$, obtained by omitting the regressors with lag larger than l. For $l > L$, $\widetilde{\Gamma}_r \subset Z_\Pi \subset \mathbb{R}^q$, where $\widetilde{\Gamma}_r$ is the orthogonal projection of Γ_r onto $\mathbb{R}^q$, obtained by omitting the regressors with lag larger than L.

Let $\tilde{z}$, $\tilde{z}_e$, and $\tilde{z}_r$ be the values of the *regressor* vectors (8.7), (8.26), and (8.41) at the current time t. Then, the explicit dual-mode controller is defined as follows:

$$u_d \triangleq \begin{cases} \widehat{u}_0(\tilde{z}) \,, \text{ if } \tilde{z}_r \notin \Gamma_r \\ -K\tilde{z}_e + u^*_{st} \,, \text{ if } \tilde{z}_r \in \Gamma_r \end{cases} \tag{8.42}$$

The expression in the first row of (8.42) means that the control is performed by the explicit NN-NMPC controller when the system is far from equilibrium. The expression in the second row implies that the control will be switched to the LQR when $\tilde{z}_r$ enters the set Γ_r and the LQR will continue controlling the system until $\tilde{z}_r$ leaves this set due to a large disturbance, for example. The integrator output y_{int} is used only when $\tilde{z}_r \in \Gamma_r$. In the case when $\tilde{z}_r \notin \Gamma_r$, y_{int} is set to zero and not used.

If the NN ARX model describes exactly the system dynamics far from the origin (outside the set Γ_r) and the problem (8.18) is convex, then the closed-loop system stability can be ensured by conditions similar to those in [10]. In presence of model errors far from the origin, it would be necessary to apply approaches to explicit *robust* NMPC ([4]). If the problem (8.18) is non-convex, then the closed-loop stability can not be guaranteed, but it can be verified by off-line simulations.

8.5 Application: Regulation of a pH Maintaining System

In [5, 6], the dual-mode approach to explicit output-feedback NMPC, described in the previous two sections, is applied to design an explicit NMPC for regulation of a pH maintaining system. The motivation for this particular example is not to suggest that the mp-NLP approach is particularly suitable for this kind of process, but rather to demonstrate a potential engineering applications of the mp-NLP approach to processes which are modeled with higher order black-box models. Particularly attractive for suggested control method from engineering applications aspect is a benefit to be able to execute the NMPC code in a low-cost PLC type of hardware.

8.5.1 The pH Maintaining System

A simplified schematic diagram of the pH maintaining system taken from [9] is given in Fig. 8.1. The process consists of an acid stream (Q_1), buffer stream (Q_2) and base stream (Q_3) that are mixed in a tank T_1. Prior to mixing, the acid stream enters the tank T_2. The acid and buffer flow rates are assumed to be constant. The effluent pH is the measured variable, which is controlled by manipulating the base flow rate.

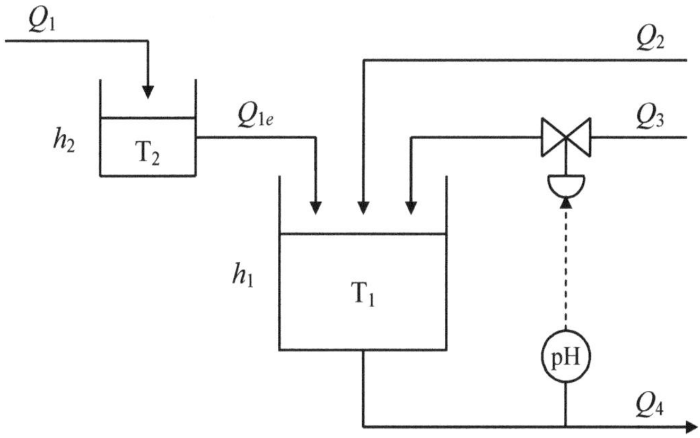

Fig. 8.1 Scheme of the pH maintaining system.

In [9], a dynamic model of the pH maintaining system is derived using conservation equations and equilibrium relations. The model also includes hydraulic relationships for the tank outlet flows. Modeling assumptions include perfect mixing, constant density, and complete solubility of the ions involved. The model is presented briefly according to [9].

The chemical reactions for the system are:

$$H_2CO_3 \longleftrightarrow HCO_3^- + H^+ \tag{8.43}$$

$$HCO_3^- \longleftrightarrow CO_3^= + H^+ \tag{8.44}$$

$$H_2O \longleftrightarrow OH^- + H^+ \tag{8.45}$$

The corresponding equilibrium constants are:

$$K_{a1} = \frac{[HCO_3^-][H^+]}{[H_2CO_3]}, \; K_{a2} = \frac{[CO_3^=][H^+]}{[HCO_3^-]}, \; K_w = [H^+][OH^-] \tag{8.46}$$

The chemical equilibria is modeled by defining two reaction invariants for each of the streams Q_i, $i \in \{1,2,3,4\}$ [9]:

$$W_{ai} = [\mathrm{H}^+]_i - [\mathrm{OH}^-]_i - [\mathrm{HCO}_3^-]_i - 2[\mathrm{CO}_3^=]_i \tag{8.47}$$
$$W_{bi} = [\mathrm{H_2CO_3}]_i + [\mathrm{HCO}_3^-]_i + [\mathrm{CO}_3^=]_i \tag{8.48}$$

The invariant W_a is a charge related quantity, while W_b represents the concentration of the $\mathrm{CO}_3^=$ ion. The pH can be determined from W_a and W_b using the following relations [9]:

$$W_b \frac{\frac{K_{a1}}{[\mathrm{H}^+]} + \frac{2K_{a1}K_{a2}}{[\mathrm{H}^+]^2}}{1 + \frac{K_{a1}}{[\mathrm{H}^+]} + \frac{K_{a1}K_{a2}}{[\mathrm{H}^+]^2}} + W_a + \frac{K_w}{[\mathrm{H}^+]} - [\mathrm{H}^+] = 0 \tag{8.49}$$
$$\mathrm{pH} = -\log([\mathrm{H}^+]) \tag{8.50}$$

In [9], a simplified model of the pH maintaining system is developed, where the dynamics of the pH transmitter and the flow dynamics of tank T_2 are neglected. The mass balance on tank T_1 yields:

$$A_1 \frac{dh_1}{dt} = Q_{1e} + Q_2 + Q_3 - Q_4 \, , \tag{8.51}$$

where h_1 is the liquid level and A_1 is the cross-sectional area of tank T_1. The exit flow rate Q_4 is modeled as:

$$Q_4 = C_v (h_1 + l)^s \, , \tag{8.52}$$

where C_v is a constant valve coefficient, s is a constant valve exponent, and l is the vertical distance between the bottom of tank T_1 and the outlet for Q_4. By combining mass balances on each of the ionic species in the system, the following differential equations for the effluent reaction invariants W_{a4} and W_{b4} are derived [9]:

$$A_1 h_1 \frac{dW_{a4}}{dt} = Q_{1e}(W_{a1} - W_{a4}) + Q_2(W_{a2} - W_{a4}) + Q_3(W_{a3} - W_{a4}) \tag{8.53}$$
$$A_1 h_1 \frac{dW_{b4}}{dt} = Q_{1e}(W_{b1} - W_{b4}) + Q_2(W_{b2} - W_{b4}) + Q_3(W_{b3} - W_{b4}) \tag{8.54}$$

Based on the above relations, a state space model of the pH maintaining system is obtained by defining the following state, input and output variables:

$$x = [W_{a4} \; W_{b4} \; h_1]^T \, , \; \tilde{u} = Q_3 \, , \; \tilde{y} = \mathrm{pH} \tag{8.55}$$

The state space model has the form [9]:

$$\dot{x} = \tilde{f}(x) + \tilde{g}(x)\tilde{u} \tag{8.56}$$
$$c(x, \tilde{y}) = 0 \, , \tag{8.57}$$

where:

$$\tilde{f}(x) = \begin{bmatrix} \frac{Q_1(W_{a1}-x_1)+Q_2(W_{a2}-x_1)}{A_1 x_3} \\ \frac{Q_1(W_{b1}-x_2)+Q_2(W_{b2}-x_2)}{A_1 x_3} \\ \frac{Q_1 - C_v(x_3+l)^s + Q_2}{A_1} \end{bmatrix} , \; \tilde{g}(x) = \begin{bmatrix} \frac{W_{a3}-x_1}{A_1 x_3} \\ \frac{W_{b3}-x_2}{A_1 x_3} \\ \frac{1}{A_1} \end{bmatrix} \tag{8.58}$$

$$c(x,\tilde{y}) = x_1 + 10^{\tilde{y}-14} - 10^{-\tilde{y}} + \frac{x_2(1+2\times 10^{\tilde{y}-pK_2})}{1+10^{pK_1-\tilde{y}}+10^{\tilde{y}-pK_2}} \tag{8.59}$$

The relation between the constants K_{a1}, K_{a2} in (8.49) and the constants K_1, K_2 in (8.59) is:

$$K_{a1} = 10^{-pK_1}\ ,\ K_{a2} = 10^{-pK_2}\ ,\ p > 0\,. \tag{8.60}$$

The parameters of the model (8.56)–(8.60) are given in [9].

8.5.2 *ARX Model Identification*

8.5.2.1 Neural Network ARX Model Identification

The identification and the validation of the NN model of the pH maintaining system is based on simulation data, generated with the model (8.56)–(8.57), where the liquid level h_1 in tank T_1 is assumed to be constant [5, 6]. Thus, it is presumed that a controller has been already designed to keep the level h_1 on the nominal value $h_1^* = 14$ [cm] by manipulating the exit flow rate Q_4. To get an idea about the system dynamics, necessary for sampling time and regressor vector selection, some preliminary tests were pursued. The process model (8.56)–(8.57) was excited with a combination of step-like signals for estimation of the dominant time constant and settling time. The dominant time constant was estimated in range between 65 [s] and 185 [s] and settling time between 135 [s] and 325 [s]. This 'provisional' dynamics is necessary for the estimation of appropriate sampling time. Based on responses and iterative cut-and-try procedure, a sampling time of 25 [s] was selected for these tests. Based on these preliminary tests, the chosen identification signal (400 samples) was generated from a uniform random distribution and a rate of change of the signal of 50 [s]. The validation signal was obtained using a generator of random noise with uniform distribution and a rate of change of the signal of 500 [s], so it has lower magnitude and frequency components than the identification signal. The rationale behind this is that if the model was identified using a rich signal, then it should respond well to a signal with less components.

The NN model represents a NARX model of the form (8.7)–(8.8). The hidden layer has sigmoid activation functions and the output layer has linear activation function. The choice of regressors is a difficult one and is common to all black-box modeling approaches. The number of regressors (delayed inputs and outputs) was determined by the method described in [8]. A trade-off between modeling error and complexity was taken into the account. The final selection was that the system model has the form:

$$y(t+1) = f_{NN}(\tilde{z}(t), u(t), \theta) \tag{8.61}$$

$$\tilde{z}(t) = [y(t), y(t-1), y(t-2), u(t-1), u(t-2)] \tag{8.62}$$

It should be noted that in difference to the state space model (8.56)–(8.57) where $\tilde{y} = \text{pH}$, in the NN model (8.61)–(8.62) the variable y represents the deviation of the pH from the desired set point $pH_{sp} = 4.8$, i.e. $y = pH - pH_{sp}$. In general, any other value for pH_{sp} can be pursued if the developed black-box model describes the specified operating range. Also, while in [9] the goal is to keep the pH at value 7 (a pH neutralization system), here the task is to maintain the pH at value 4.8 (a pH maintaining system). The data used for identification of the NN model (8.61)–(8.62) and for validation of its performance were scaled to zero mean and variance 1. This means that $u(t)$ and $y(t)$ can take both positive and negative values.

The optimal number of neurons in the hidden layer was determined systematically. The network was optimized for each possible number of hidden neurons in a certain range. The Levenberg-Marquardt method was used for minimization of the mean-square error criteria (8.3), due to its rapid convergence properties and robustness. At the end of this lengthy procedure and after removing the unimportant weights, the optimal parameters of the model (8.61)–(8.62) were obtained, with thirteen neurons in the hidden layer. More about systematic network structure selection, pruning and other issues regarding neural networks modeling can be found in various literature describing this topic and its applications (e.g. [16], [2], [8], [7], [19], [21]).

Fig. 8.2 depicts a comparison between the simulated NN response and the process response to the identification and the validation input signals. From the validation, it can be concluded that the black-box model captures the dynamics of the pH maintaining system relatively well. The resulting black-box model is not too large to be handled and was relatively routinely obtained with the selected software tool.

8.5.2.2 Linear ARX Model Identification

The equilibrium point of the pH maintaining system (8.56)–(8.57) is $\tilde{y} = 4.8$, $\tilde{u}^*_{st} = 10.94\,[\mathrm{ml/s}]$ (respectively $y = 0$, $u^*_{st} = 0.1732$ after scaling). A validation of the obtained NN ARX model near this point clearly shows that it is not accurate (see Fig. 8.3).

In order to obtain accurate predictions when the output variable is close to zero, the following 1-st order linear ARX model is identified [5, 6]:

$$y(t+1) = 0.7704y(t) + 0.0539(u(t) - u^*_{st}) \tag{8.63}$$

Higher order linear ARX models have been also obtained, however simulations have shown that the dynamics of the pH maintaining system around the equilibrium is captured best by the 1-st order model (8.63). The simulated response of the ARX model (8.63) is depicted in Fig. 8.3.

8.5.3 Design of Explicit Dual-Mode Controller

The approach described in Sections 8.3 and 8.4 is applied to design an explicit dual-mode controller for the pH maintaining system based on its NN model (8.61)–(8.62) and linear ARX model (8.63) [5, 6]. Recall that due to scaling, the variables u and y can take both positive and negative values.

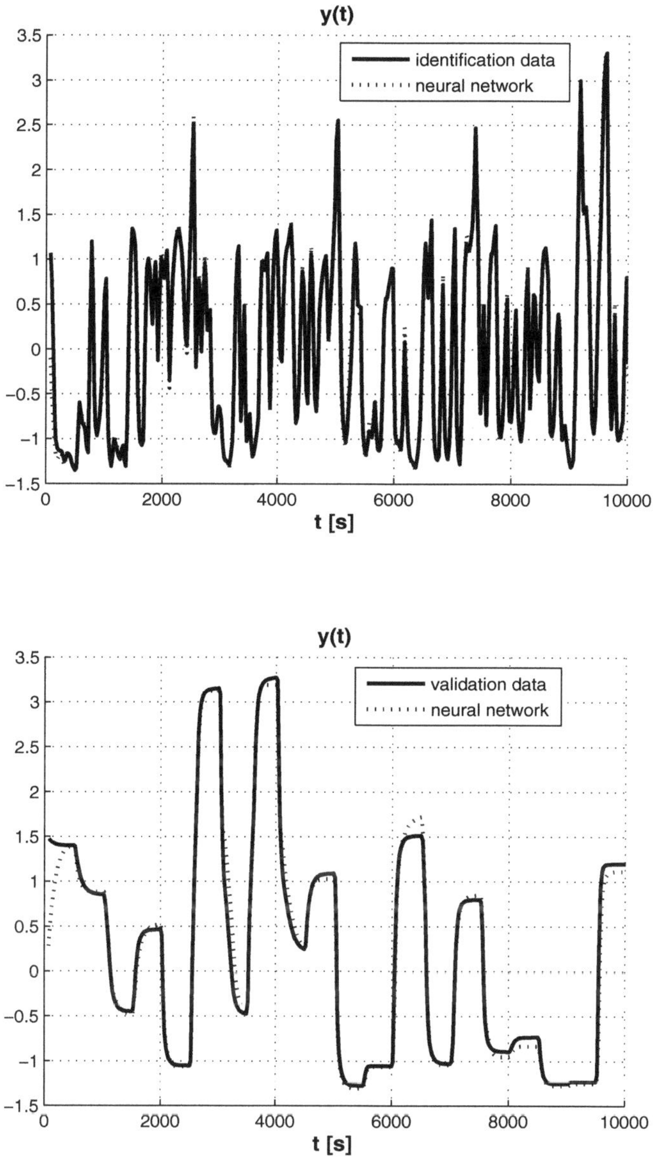

Fig. 8.2 Response of the NN model to the excitation signal used for identification (top) and to the excitation signal used for validation (bottom).

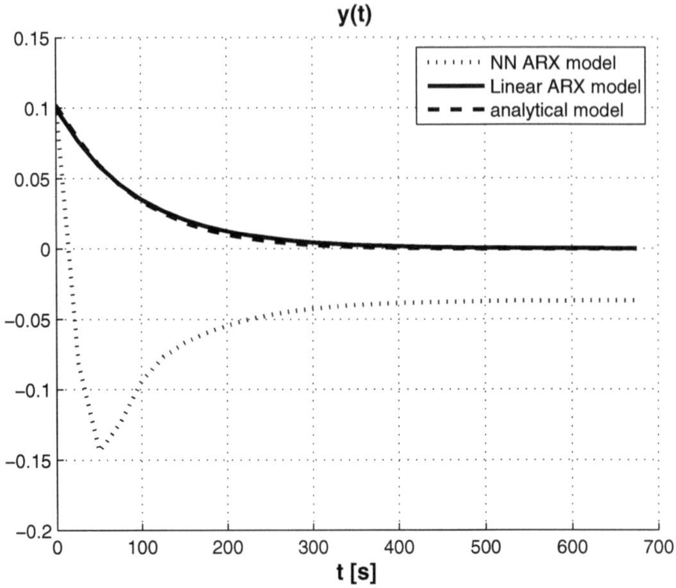

Fig. 8.3 Validation of the NN ARX and the linear ARX models. The dotted curve is with the NN model (8.61)–(8.62), the solid curve is with the linear ARX model (8.63), and the dashed curve is with the first-principles model (8.56)–(8.57). Constant control input $u = u^*_{st}$ is used as an excitation signal.

First, the approach in Section 8.3 is applied to design an explicit approximate NN-NMPC controller. The following control input constraint is imposed on the system:

$$-0.4 \leq u \leq 0.4 \tag{8.64}$$

The horizon is $N = 8$ and the terminal constraint in Problem 8.1 is:

$$z^c_{t+N|t} \in \Omega \,, \tag{8.65}$$

where $\Omega = \{z^c \in \mathbb{R}^5 \,|\, \|z^c\|^2 \leq 0.05\}$. The weighting matrices in the cost function (8.17) are $Q = 10$, $R = 1$, $F = 10$. The NN-NMPC minimizes the cost function (8.17) subject to the model (8.61)–(8.62) and the constraints (8.64)–(8.65). In (8.20), it is chosen $\alpha = 10$. The *regressor* space to be partitioned is defined by $Z = ([-1.2; 1.2] \times [-1.2; 1.2] \times [-1.2; 1.2] \times [-0.4; 0.4] \times [-0.4; 0.4])$. The cost function approximation tolerance is chosen as $\bar{\varepsilon}(Z_0) = \max(\bar{\varepsilon}_a, \bar{\varepsilon}_r \min_{\tilde{z} \in Z_0} V^*(\tilde{z}))$, where $\bar{\varepsilon}_a = 0.005$ and $\bar{\varepsilon}_r = 0.1$ are the absolute and the relative tolerances, respectively. The partition has 5512 regions and 23 levels of search in a binary search tree representation. Totally, 33 arithmetic operations are needed in real-time to compute the control input by traversing the binary search tree (23 comparisons, 5 multiplications and 5 additions).

Further, an unconstrained LQR is designed, which is used in a neighborhood of the origin. For this purpose, consider the extended linear system, where an integral error is added to the linear ARX model (8.63):

$$y(t+1) = 0.7704y(t) + 0.0539u_e(t) \tag{8.66}$$

$$y_{int}(t+1) = y_{int}(t) + T_s y(t) \tag{8.67}$$

Here, $u_e(t) \equiv u(t) - u_{st}^*$. Thus, we obtain the following system:

$$\tilde{z}_e(t+1) = \tilde{A}^e \tilde{z}_e(t) + \tilde{B}^e u_e(t)\ , \tag{8.68}$$

which is characterized with *regressor* vector $\tilde{z}_e(t) = y_e(t) = [y(t), y_{int}(t)]$ and matrices $\tilde{A}^e = \begin{bmatrix} 0.7704 & 0 \\ T_s & 1 \end{bmatrix}$ and $\tilde{B}^e = \begin{bmatrix} 0.0539 \\ 0 \end{bmatrix}$. The computed LQR law for the system (8.68) is:

$$u_e = -K\tilde{z}_e = -k_1 y - k_2 y_{int}\ , \text{ where } K = [0.7994, 0.0069] \tag{8.69}$$

This control law solves the optimization problem (8.32) with weighting matrices $Q_e = \text{diag}\{10, 0.0005\}, R_e = 10$.

Then, the explicit dual-mode controller for the pH maintaining system is defined according to (8.42) with $\Gamma_r = \{\tilde{z}_r \in \mathbb{R} \,|\, \tilde{z}_r^2 \leq 0.09\}$, where $\tilde{z}_r(t) = y(t)$.

In order to study the robustness of the explicit dual-mode controller against model inaccuracies, its performance is simulated in closed-loop with the first-principles model (8.56)–(8.57). Further, it is assumed that there are persistent disturbances in the acid and the buffer flow rates, which have the following values $\tilde{Q}_1 = 16.8\,[\text{ml/s}]$, $\tilde{Q}_2 = 0.53\,[\text{ml/s}]$ (different from the nominal values $Q_1^* = 16.6\,[\text{ml/s}]$, $Q_2^* = 0.55\,[\text{ml/s}]$). In addition to the explicit dual-mode controller which maintains the pH on the required set point, a second controller (an LQR) is applied, which keeps the liquid level h_1 on the nominal value $h_1^* = 14$ [cm] by manipulating the exit flow rate Q_4. The obtained trajectories of the control input u and the output variable y are shown in Fig. 8.4, while the trajectories of the exit flow rate Q_4 and the liquid level h_1 are depicted in Fig. 8.5.

It can be seen from Fig. 8.4 that the output variable is steered to the origin despite of the presence of persistent disturbances and the control input achieves a new equilibrium value $\tilde{u}_{st} = 0.2380$ (recall that the equilibrium value corresponding to the nominal model parameters is $u_{st}^* = 0.1732$). It would be necessary to distinguish how the exact NMPC and the approximate explicit NMPC trajectories in Figs. 8.4 and 8.5 are obtained. The exact NMPC response is computed by solving at each time instant an open-loop NMPC problem formulated for the first-principles model (8.56)–(8.57). In contrast, the approximate explicit NMPC solution is first computed off-line as an approximation to Problem 8.1, in which the NN ARX model by itself represents another approximation. Then, its performance is simulated in closed-loop with the first-principles model (8.56)–(8.57). Thus, the performance degradation far from the origin is due to the approximations in the model and in the NMPC

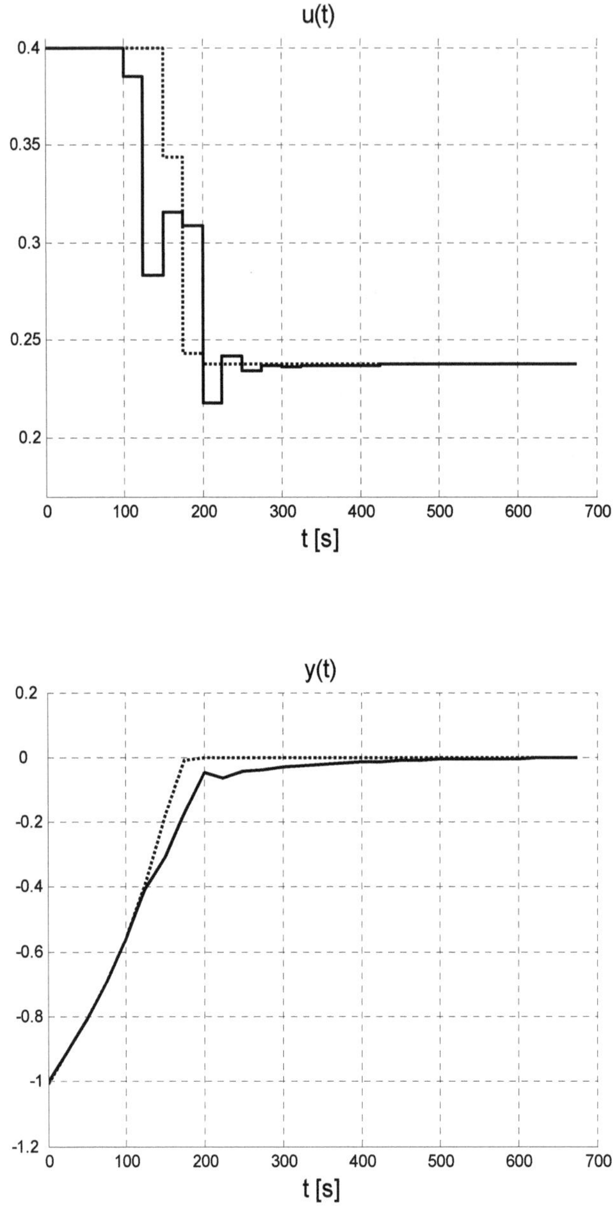

Fig. 8.4 Control input u (top) and output variable y (bottom) obtained with the explicit dual-mode controller in closed-loop with the first-principles model (8.56)–(8.57). The solid curves are with the approximate explicit NN-NMPC and the dotted curves are with the exact NN-NMPC.

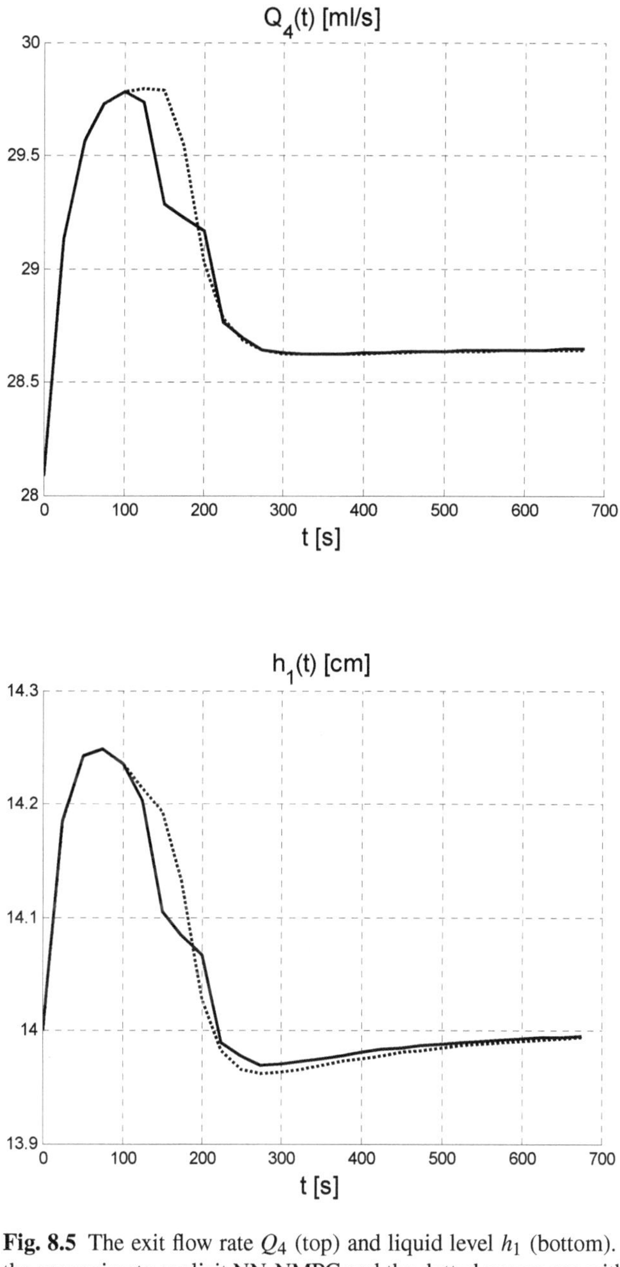

Fig. 8.5 The exit flow rate Q_4 (top) and liquid level h_1 (bottom). The solid curves are with the approximate explicit NN-NMPC and the dotted curves are with the exact NN-NMPC.

solution, while near the origin it is related to the use of LQR (pursuing an offset-free response) which differs from the exact NMPC (where no integral action is taken). It also should be noted that the response depicted in Figs. 8.4 and 8.5 has a typical amount of performance degradation being representative for other initial conditions and scenarios.

References

1. Chen, H., Allgöwer, F.: A quasi-infinite horizon nonlinear model predictive control scheme with guaranteed stability. Automatica 34, 1205–1217 (1998)
2. Chen, S., Billings, S.A., Grant, P.M.: Non-linear system identification using neural networks. International Journal of Control 51, 1191–1214 (1990)
3. Fiacco, A.V.: Introduction to sensitivity and stability analysis in nonlinear programming. Academic Press, Orlando (1983)
4. Grancharova, A., Johansen, T.A.: Computation, approximation and stability of explicit feedback min-max nonlinear model predictive control. Automatica 45, 1134–1143 (2009)
5. Grancharova, A., Kocijan, J., Johansen, T.A.: Dual-mode explicit output-feedback predictive control based on neural network models. In: Proceedings of the 8th IFAC Symposium on Nonlinear Control Systems (NOLCOS 2010), Bologna, Italy, pp. 545–550 (2010), `http://www.IFAC-PapersOnLine.net`
6. Grancharova, A., Kocijan, J., Johansen, T.A.: Explicit output-feedback nonlinear predictive control based on black-box models. Engineering Applications of Artificial Intelligence 24, 388–397 (2011)
7. Han, J., Moraga, C., Sinne, S.: Optimization of feedforward neural networks. Engineering Applications of Artificial Intelligence 9, 109–119 (1996)
8. He, X., Asada, H.: A new method for identifying orders of input-output models for nonlinear dynamic systems. In: Proceedings of the American Control Conference, San Francisco, CA, pp. 2520–2523 (1993)
9. Henson, M.A., Seborg, D.E.: Adaptive nonlinear control of a pH neutralization process. IEEE Transactions on Control System Technology 2, 169–183 (1994)
10. Johansen, T.A.: Approximate explicit receding horizon control of constrained nonlinear systems. Automatica 40, 293–300 (2004)
11. Kocijan, J., Murray-Smith, R.: Nonlinear Predictive Control with a Gaussian Process Model. In: Murray-Smith, R., Shorten, R. (eds.) Switching and Learning in Feedback Systems. LNCS, vol. 3355, pp. 185–200. Springer, Heidelberg (2005)
12. Lepetič, M., Škrjanc, I., Chiacchiarini, H.G., Matko, D.: Predictive functional control based on fuzzy model: Magnetic suspension system case study. Engineering Applications of Artificial Intelligence 16, 425–430 (2003)
13. Ljung, L., Glad, T.: Modeling of Dynamic Systems. Prentice Hall, Englewood Cliffs (1994)
14. Mayne, D.Q., Rawlings, J.B., Rao, C.V., Scokaert, P.O.M.: Constrained model predictive control: Stability and optimality. Automatica 36, 789–814 (2000)
15. Michalska, H., Mayne, D.Q.: Robust receding horizon control of constrained nonlinear systems. IEEE Transactions on Automatic Control 38, 1623–1633 (1993)
16. Nørgaard, M., Ravn, O., Poulsen, N.K., Hansen, L.K.: Neural Networks for Modelling and Control of Dynamic Systems. Springer, London (2000)
17. Ogata, K.: Discrete-time Control Systems. Prentice Hall, Englewood Cliffs (1995)

18. Peng, H., Yang, Z.J., Gui, W., Wu, M., Shioya, H., Nakano, K.: Nonlinear system modelling and robust predictive control based on RBF-ARX model. Engineering Applications of Artificial Intelligence 20, 1–9 (2007)
19. Principe, J.C., Euliano, N.R., Lefebvre, W.C.: Neural and Adaptive systems. John Wiley & Sons, New York (2000)
20. Sjöberg, J., Zhang, Q., Ljung, L., Benveniste, A., Delyon, B., Glorennec, P.Y., Hjalmarsson, H., Juditsky, A.: Non-linear black-box modeling in system identification: A unified overview. Automatica 31, 1691–1724 (1995)
21. Venkatasubramanian, V., McAvoy, T.J. (eds.): Neural network applications in chemical engineering. Special Issue of Computers & Chemical Engineering 16(4) (1992)
22. Zeng, G.M., Qin, X.S., He, L., Huang, G.H., Liu, H.L., Lin, Y.P.: A neural network predictive control system for paper mill wastewater treatment. Engineering Applications of Artificial Intelligence 16, 121–129 (2003)

Chapter 9
Semi-explicit Distributed NMPC

Abstract. In this chapter, a suboptimal approach to *distributed* NMPC for systems consisting of nonlinear subsystems with linearly *coupled* dynamics, subject to both state and input constraints, is considered. The approach applies the dynamic dual decomposition method and reformulates the original *centralized* NMPC problem into a *distributed* quasi-NMPC problem by linearization of the nonlinear system dynamics. The approach is based on *distributed on-line* optimization (by gradient iterations) and can be applied to large-scale nonlinear systems. Further, a *semi-explicit* NMPC approach to efficiently solve the *distributed* NMPC problem for small- and medium-scale systems is described. It combines the explicit approximate solution with the on-line optimization and the result is a decrease of the on-line computational complexity. Both the *on-line* optimization based *distributed* NMPC and the *semi-explicit distributed* NMPC are illustrated in a problem to solve a NMPC problem for a nonlinear system consisting of two subsystems.

9.1 Introduction

Recall that NMPC involves the solution at each sampling instant of a finite horizon optimal control problem subject to the system dynamics, and state and input constraints. However, solving in a *centralized* way NMPC problems for medium- and large-scale systems may be impractical due to the complexity of the Nonlinear Programming (NLP) problem, the topology of the plant and data communication, and the large number of decision variables. Therefore, there is a strong motivation for development of methods for *distributed* solution of NMPC problems. At the same time, the multi-core computer architectures available nowadays would encourage *parallel* and *distributed* NMPC computations [5]. Recently, several approaches for *decentralized* implementation of MPC algorithms have been developed, [19]. As it is pointed out in [15], the possibility to use MPC in a *decentralized* fashion has the advantage to reduce the original, large size, optimization problem into a number of smaller and more tractable ones. In [20], [13], [22], [9], approaches for *distributed* MPC for systems consisting of *linear* interconnected subsystems have been developed. The approach in [9] is based on the dual decomposition methods [1, 6, 4],

A. Grancharova & T.A. Johansen: Explicit Nonlinear Model Predictive Ctrl., LNCIS 429, pp. 209–231.
springerlink.com

where large-scale optimization problems are handled by using Lagrange multipliers to relax the couplings between the sub-problems. In [17, 18], the dual decomposition is used for analysis and synthesis of distributed feedback controllers.

Further, approaches for *distributed* MPC for systems composed of several *nonlinear* subsystems have been proposed in [15, 7, 12]. In [15], a stabilizing *decentralized* MPC algorithm for nonlinear systems consisting of several interconnected local subsystems is developed. It is derived under the main assumptions that no information can be exchanged between local control laws, i.e. the coupling between the subsystems is ignored, and only input constraints are imposed on the system. In [7], it is supposed that the dynamics and constraints of the nonlinear subsystems are *decoupled*, but their state vectors are *coupled* in a single cost function of a finite horizon optimal control problem. In [12], an optimal control problem for a set of dynamically *decoupled* nonlinear systems, where the cost function and constraints couple the dynamical behavior of the systems, is solved.

In this chapter, the suboptimal approach [11] to *distributed* NMPC for a more general class of systems consisting of *nonlinear* subsystems with *coupled* dynamics subject to both state and input constraints is considered. Like in [15], it is supposed that the couplings between the subsystems are linear. However in difference to [15], the *distributed* NMPC method proposed here takes into account these couplings, as well as state constraints. The approach [11] applies the dynamic dual decomposition method [4, 18, 9] and reformulates the original *centralized* NMPC problem into a *distributed* quasi-NMPC problem by linearization of the nonlinear system dynamics. The approach is based entirely on *distributed* on-line optimization (by gradient iterations) and can be applied to large-scale nonlinear systems. Further, a *semi-explicit* NMPC approach to efficiently solve the *distributed* NMPC problem for small- and medium-scale systems is proposed.

9.2 Formulation of NMPC Problem for Interconnected Systems

Consider a system composed by the interconnection of M subsystems (shown in Fig. 9.1), which is described by the following nonlinear discrete-time models [15]:

$$x_i(t+1) = f_i(x_i(t), u_i(t)) + g_i(x(t)) + d_i(t)\,,\ i = 1, 2, ..., M \tag{9.1}$$

where $x_i(t) \in \mathbb{R}^{n_i}$, $u_i(t) \in \mathbb{R}^{m_i}$, and $d_i(t) \in \mathbb{R}^{n_i}$ are the state, control input, and disturbance vectors, related to the i-th subsystem, and $f_i : \mathbb{R}^{n_i} \times \mathbb{R}^{m_i} \to \mathbb{R}^{n_i}$ and $g_i : \mathbb{R}^{n} \to \mathbb{R}^{n_i}$ are nonlinear functions.

In (9.1), the mutual influence of the M subsystems is described by the functions g_i, which depend on the overall state:

$$x(t) = [x_1(t), x_2(t), ..., x_M(t)] \in \mathbb{R}^n\,,\ n = \sum_{i=1}^{M} n_i \tag{9.2}$$

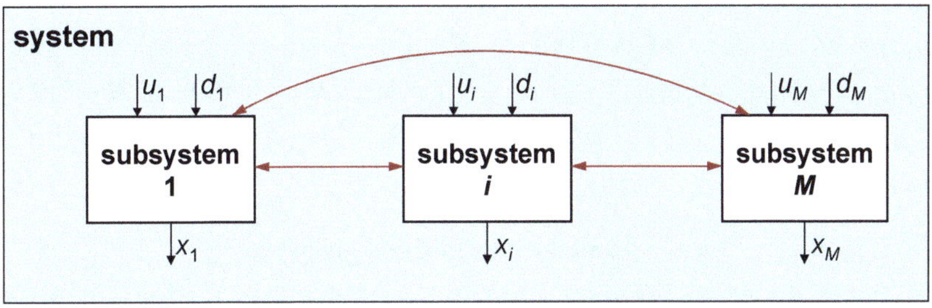

Fig. 9.1 System composed by the interconnection of M subsystems.

Similarly, the overall control input is denoted:

$$u(t) = [u_1(t), u_2(t), \ldots, u_M(t)] \in \mathbb{R}^m, \; m = \sum_{i=1}^{M} m_i \tag{9.3}$$

The following control input and state constraints are imposed on the subsystems:

$$u_{\min,i} \leq u_i(t) \leq u_{\max,i}, \; x_{\min,i} \leq x_i(t) \leq x_{\max,i}, \; i = 1, 2, \ldots, M \tag{9.4}$$

and the following assumptions are made [11]:

Assumption 9.1. *The functions f_i and g_i, $i=1, \ldots, M$ are C^1 functions with $f_i(0,0) = 0$, $g_i(0) = 0$.*

Assumption 9.2. $x_{\min,i} < 0 < x_{\max,i}$, $u_{\min,i} < 0 < u_{\max,i}$, $i = 1, \ldots, M$.

Assumption 9.3. *The disturbances d_i, $i = 1, \ldots, M$ are bounded by:*

$$|d_i(t)| \leq d_{\max,i}, \; i = 1, 2, \ldots, M \tag{9.5}$$

with $d_{\max,i} \in \mathbb{R}^{n_i}_{>0}$, $d_{\max,i} < |x_{\min,i}|$ and $d_{\max,i} < x_{\max,i}$, $i = 1, 2, \ldots, M$, where the operation $|\cdot|$ is taken element-wise.

It is supposed that a full measurement $x = [x_1, x_2, \ldots, x_M]$ of the overall state is available at the current time t. The optimal regulation problem is considered where the goal is to steer the overall state of the system (9.1) to the origin. For the current $x = [x_1, x_2, \ldots, x_M]$, the regulation NMPC solves the optimization problem [11]:

Problem 9.1 (Centralized NMPC):

$$V^{opt}(x) = \min_{U} J(U,x) \tag{9.6}$$

subject to $x_{t|t} = x$ and:

$$x_{i,t+k|t} \in \mathscr{X}_i\,,\; i = 1, \ldots, M\,,\; k = 1, \ldots, N-1 \tag{9.7}$$
$$u_{i,t+k} \in \mathscr{U}_i\,,\; i = 1, \ldots, M\,,\; k = 0, 1, \ldots, N-1 \tag{9.8}$$
$$x_{i,t+k+1|t} = f_i(x_{i,t+k|t}, u_{i,t+k}) + g_i(x_{t+k|t})\,,\; i = 1, \ldots, M\,,\; k = 0, 1, \ldots, N-2 \tag{9.9}$$
$$x_{t+k|t} = \left[x_{1,t+k|t}, x_{2,t+k|t}, \ldots, x_{M,t+k|t}\right],\; k = 0, 1, \ldots, N-1 \tag{9.10}$$
$$u_{t+k} = \left[u_{1,t+k}, u_{2,t+k}, \ldots, u_{M,t+k}\right],\; k = 0, 1, \ldots, N-1 \tag{9.11}$$

with $U = \left[u_t, u_{t+1}, \ldots, u_{t+N-1}\right]$ and the cost function given by:

$$J(U,x) = \sum_{k=0}^{N-1} \sum_{i=1}^{M} l_i(x_{i,t+k|t}, u_{i,t+k}) \tag{9.12}$$

Here, $l_i(x_{i,t+k|t}, u_{i,t+k}) = \|x_{i,t+k|t}\|_{Q_i}^2 + \|u_{i,t+k}\|_{R_i}^2$ is the stage cost for the i-th subsystem with weighting matrices $Q_i, R_i \succ 0$, and N is a finite horizon. The sets $\mathscr{X}_i$ and $\mathscr{U}_i$ are defined by:

$$\mathscr{X}_i = \{\lambda_i \in \mathbb{R}^{n_i} \,|\, x_{\min,i} + d_{\max,i} \le \lambda_i \le x_{\max,i} - d_{\max,i}\} \tag{9.13}$$
$$\mathscr{U}_i = \{\eta_i \in \mathbb{R}^{m_i} \,|\, u_{\min,i} \le \eta_i \le u_{\max,i}\} \tag{9.14}$$

It follows from (9.13)–(9.14) that $\mathscr{X}_i$ and $\mathscr{U}_i$ are convex (polyhedral) sets, which include the origin in their interior (due to Assumptions 9.2 and 9.3).

It should be noted that the state constraints (9.7) with the admissible set $\mathscr{X}_i$ defined by (9.13) guarantee the robust feasibility of the solution in sense that the original state constraints (9.4) will be satisfied for the worst-case disturbances.

9.3 Distributed NMPC for Interconnected Nonlinear Systems with Linear Couplings

9.3.1 *Distributed NMPC by Dual Decomposition*

Problem 9.1 can be decomposed by using the dynamic dual decomposition approach [4, 18]. First, the following assumption is made [11]:

Assumption 9.4. *The functions $g_i(x(t))$ have the form:*

$$g_i(x(t)) = \sum_{j=1,\, j\ne i}^{M} A_{ij} x_j(t)\,,\; i = 1, \ldots, M \tag{9.15}$$

where $A_{ij} \in \mathbb{R}^{n_i \times n_j}$ are constant matrices.

The following *decoupled* state equations can be formulated:

$$x_i(t+1) = f_i(x_i(t), u_i(t)) + v_i(t) + d_i(t)\,,\; i = 1, \ldots, M \tag{9.16}$$

with the additional constraints, [18]:

$$v_i(t) = \sum_{j=1, j\neq i}^{M} A_{ij}x_j(t)\,,\; i=1,\dots,M \;\text{ for all } t \tag{9.17}$$

The variable $v_i \in \mathbb{R}^{n_i}$ can be interpreted as the influence of the other subsystems in the update of x_i.

Then, similar to [18], the constraints (9.17) are relaxed by introducing the corresponding vectors of Lagrange multipliers $p_i \in \mathbb{R}^{n_i}$ in the cost function (9.12) and the Problem 9.1 is reformulated as a *distributed* NMPC problem [11]:

Problem 9.2 (Distributed NMPC):

$$V^{opt}(x) =$$

$$\max_{P}\min_{U,X,V} \sum_{k=0}^{N-1}\sum_{i=1}^{M}[l_i(x_{i,t+k|t},u_{i,t+k}) + p_{i,t+k}^T(v_{i,t+k} - \sum_{\substack{j=1\\ j\neq i}}^{M} A_{ij}x_{j,t+k|t})] =$$

$$\max_{P}\sum_{i=1}^{M}(\min_{U_i,X_i,V_i}\underbrace{\sum_{k=0}^{N-1}[l_i(x_{i,t+k|t},u_{i,t+k}) + p_{i,t+k}^T v_{i,t+k} - x_{i,t+k|t}^T\sum_{\substack{j=1\\ j\neq i}}^{M} A_{ji}^T p_{j,t+k}]}_{l_i^P(x_{i,t+k|t},u_{i,t+k},v_{i,t+k},P)}) \tag{9.18}$$

subject to $x_{t|t} = x$, constraints (9.7)–(9.8) and:

$$x_{i,t+k+1|t} = f_i(x_{i,t+k|t},u_{i,t+k}) + v_{i,t+k}\,,\; i=1,\dots,M\,,\; k=0,1,\dots,N-2 \tag{9.19}$$

$$p_{t+N-1} = 0 \tag{9.20}$$

Here:

$$\begin{aligned}
P &= [p_t,\, p_{t+1},\, \dots,\, p_{t+N-1}] \text{ with } p_{t+k} = [p_{1,t+k},\, p_{2,t+k},\, \dots,\, p_{M,t+k}]\,,\\
&\quad k = 0,1,\dots,N-1\\
U &= [u_t,\, u_{t+1},\, \dots,\, u_{t+N-1}]\,,\; U_i = [u_{i,t},\, u_{i,t+1},\, \dots,\, u_{i,t+N-1}]\\
X &= [x_{t|t},\, x_{t+1|t},\, \dots,\, x_{t+N-1|t}]\,,\; X_i = [x_{i,t|t},\, x_{i,t+1|t},\, \dots,\, x_{i,t+N-1|t}]\\
V &= [v_t,\, v_{t+1},\, \dots,\, v_{t+N-1}]\,,\; V_i = [v_{i,t},\, v_{i,t+1},\, \dots,\, v_{i,t+N-1}]\\
&\quad \text{with } v_{t+k} = [v_{1,t+k},\, v_{2,t+k},\, \dots,\, v_{M,t+k}]\,,\; k=0,1,\dots,N-1
\end{aligned} \tag{9.21}$$

The Lagrange multipliers P are also referred to as prices [18] and the Problem 9.2 can be interpreted as a game with two players for each subsystem. Given the prices, the objective of the first player for the i-th subsystem is to select the inputs $U_i = [u_{i,t},\, u_{i,t+1},\, \dots,\, u_{i,t+N-1}]$ to minimize the local cost $\sum_{k=0}^{N-1} l_i^P(x_{i,t+k|t},u_{i,t+k},v_{i,t+k},P)$. The other player for the i-th subsystem chooses $P_i = [p_{i,t},\, p_{i,t+1},\, \dots,\, p_{i,t+N-1}]$ with the objective to maximize $\sum_{k=0}^{N-1} p_{i,t+k}^T(v_{i,t+k} - \sum_{j=1, j\neq i}^{M} A_{ij}x_{j,t+k|t})$.

The inner *decoupled* optimization problems in Problem 9.2 represent Nonlinear Programming (NLP) sub-problems corresponding to the NMPC of the i-th subsystem, since the constraints (9.19) are nonlinear in the optimization variables. Each NLP sub-problem is presented as follows [11]:

Problem 9.3^i (*i*-th NLP sub-problem):

$$V_i^{opt}(P,x_i) = \min_{U_i,X_i,V_i} \sum_{k=0}^{N-1} l_i^P(x_{i,t+k|t}, u_{i,t+k}, v_{i,t+k}, P) \tag{9.22}$$

subject to $x_{i,t|t} = x_i$ and:

$$x_{i,t+k|t} \in \mathscr{X}_i, \; k = 1, \ldots, N-1 \tag{9.23}$$

$$u_{i,t+k} \in \mathscr{U}_i, \; k = 0, 1, \ldots, N-1 \tag{9.24}$$

$$x_{i,t+k+1|t} = f_i(x_{i,t+k|t}, u_{i,t+k}) + v_{i,t+k}, \; k = 0, 1, \ldots, N-2 \tag{9.25}$$

Denote with $U_i^{opt} = [u_{i,t}^{opt}, u_{i,t+1}^{opt}, \ldots, u_{i,t+N-1}^{opt}]$, $X_i^{opt} = [x_{i,t|t}^{opt}, x_{i,t+1|t}^{opt}, \ldots, x_{i,t+N-1|t}^{opt}]$ and $V_i^{opt} = [v_{i,t}^{opt}, v_{i,t+1}^{opt}, \ldots, v_{i,t+N-1}^{opt}]$ the optimal solution of Problem 9.3^i.

9.3.2 *Local QP Approximations of the NLP Sub-problems*

The cost functions $\sum_{k=0}^{N-1} l_i^P(x_{i,t+k|t}, u_{i,t+k}, v_{i,t+k}, P)$ in the Problems 9.3^i, $i = 1, 2, \ldots, M$ are convex, however the constraints (9.25) may be non-convex in the general case. In [11], the constraints (9.25) are locally approximated by linear constraints, leading to a *quasi-nonlinear* approach. Let $x_{i,t|t} = x_i^0 \in \mathscr{X}_i$ be arbitrary and denote the corresponding optimal solution to the sub-problem 9.3^i with:

$$\begin{aligned} U_i^0 &= U_i^{opt}(x_i^0) = [u_{i,t}^0, u_{i,t+1}^0, \ldots, u_{i,t+N-1}^0] \\ X_i^0 &= X_i^{opt}(x_i^0) = [x_{i,t|t}^0, x_{i,t+1|t}^0, \ldots, x_{i,t+N-1|t}^0] \\ V_i^0 &= V_i^{opt}(x_i^0) = [v_{i,t}^0, v_{i,t+1}^0, \ldots, v_{i,t+N-1}^0] \end{aligned} \tag{9.26}$$

The optimal solution (9.26) depends on the values of the prices P. In Section 9.3.3, it is described how P and the solution (9.26) are updated iteratively. Further, a first order truncated Taylor series expansions of the right-hand side of constraints (9.25) around the point $(U_i^0, X_i^0, V_i^0, x_i^0)$ lead to the locally linear constraints [11]:

$$\tilde{X}_i^+ = S_{X_i}(\tilde{X}_i - \tilde{X}_i^0) + S_{U_i}(\tilde{U}_i - \tilde{U}_i^0) + (\tilde{V}_i - \tilde{V}_i^0) + E_i^0 \tag{9.27}$$

where:

$$\begin{aligned} \tilde{X}_i &= [x_{i,t|t}, \ldots, x_{i,t+N-2|t}], \; \tilde{X}_i^0 = [x_{i,t|t}^0, \ldots, x_{i,t+N-2|t}^0] \\ \tilde{U}_i &= [u_{i,t}, \ldots, u_{i,t+N-2}], \; \tilde{U}_i^0 = [u_{i,t}^0, \ldots, u_{i,t+N-2}^0] \\ \tilde{V}_i &= [v_{i,t}, \ldots, v_{i,t+N-2}], \; \tilde{V}_i^0 = [v_{i,t}^0, \ldots, v_{i,t+N-2}^0] \\ \tilde{X}_i^+ &= [x_{i,t+1|t}, \ldots, x_{i,t+N-1|t}] \end{aligned}$$

Here, the matrices S_{X_i}, S_{U_i}, and E_i^0 are given by:

$$S_{X_i} = \begin{bmatrix} \nabla_{x_i} f_i(x^0_{i,t|t}, u^0_{i,t}) & 0_{n_i} & \dots & 0_{n_i} \\ \vdots & & & \\ 0_{n_i} & 0_{n_i} & \dots & \nabla_{x_i} f_i(x^0_{i,t+N-2|t}, u^0_{i,t+N-2}) \end{bmatrix} \tag{9.28}$$

$$S_{U_i} = \begin{bmatrix} \nabla_{u_i} f_i(x^0_{i,t|t}, u^0_{i,t}) & 0_{n_i,m_i} & \dots & 0_{n_i,m_i} \\ \vdots & & & \\ 0_{n_i,m_i} & 0_{n_i,m_i} & \dots & \nabla_{u_i} f_i(x^0_{i,t+N-2|t}, u^0_{i,t+N-2}) \end{bmatrix} \tag{9.29}$$

$$E_i^0 = \begin{bmatrix} f_i(x^0_{i,t|t}, u^0_{i,t}) + v^0_{i,t} \\ \vdots \\ f_i(x^0_{i,t+N-2|t}, u^0_{i,t+N-2}) + v^0_{i,t+N-2} \end{bmatrix} \tag{9.30}$$

where 0_{n_i} is the n_i-dimensional square zero matrix and $0_{n_i,m_i}$ is the $n_i \times m_i$-dimensional zero matrix. It can be observed that (9.27) is a linear time-varying approximation of the constraints (9.25). Then, the NLP sub-problems 9.3^i for the subsystems are approximated with the QP sub-problems [11]:

Problem 9.4^i (i-th QP sub-problem):

$$V_i^*(P, x_i) = \min_{U_i, X_i, V_i} \sum_{k=0}^{N-1} l_i^P(x_{i,t+k|t}, u_{i,t+k}, v_{i,t+k}, P) \tag{9.31}$$

subject to $x_{i,t|t} = x_i$, (9.23), (9.24), and (9.27).

Denote with $U_i^* = [u^*_{i,t}, \dots, u^*_{i,t+N-1}]$, $X_i^* = [x^*_{i,t|t}, \dots, x^*_{i,t+N-1|t}]$ and $V_i^* = [v^*_{i,t}, \dots, v^*_{i,t+N-1}]$ the optimal solution of Problem 9.4^i. Then, the following *centralized* NMPC problem with linearized constraints is formulated [11]:

Problem 9.5 (Centralized NMPC with linearized constraints):

$$V^*(x) = \min_U J(U, x) \tag{9.32}$$

subject to $x_{t|t} = x$, constraints (9.7), (9.8), and:

$$\begin{aligned} x_{i,t+k+1|t} = {} & \nabla_{x_i} f_i(x^0_{i,t+k|t}, u^0_{i,t+k})(x_{i,t+k|t} - x^0_{i,t+k|t}) + \\ & \nabla_{u_i} f_i(x^0_{i,t+k|t}, u^0_{i,t+k})(u_{i,t+k} - u^0_{i,t+k}) + \sum_{j=1,\, j\neq i}^{M} A_{ij}(x_{j,t+k|t} - x^0_{j,t+k|t}) \\ & + f_i(x^0_{i,t+k|t}, u^0_{i,t+k}) + \sum_{j=1,\, j\neq i}^{M} A_{ij} x^0_{j,t+k|t} \\ & i = 1, \dots, M,\ k = 0, 1, \dots, N-2 \end{aligned} \tag{9.33}$$

where the cost function $J(U,x)$ is given by (9.12). Here, (9.33) includes the constraints due to the linearized dynamics for all subsystems.

Then, the *distributed* NMPC problem with linearized dynamic constraints is as follows [11]:

Problem 9.6 (Distributed NMPC with linearized constraints):

$$\max_P \sum_{i=1}^{M} V_i^*(P,x_i) = \max_P \sum_{i=1}^{M} \min_{U_i,X_i,V_i} \sum_{k=0}^{N-1} l_i^P(x_{i,t+k|t}, u_{i,t+k}, v_{i,t+k}, P) \tag{9.34}$$

subject to $x_{t|t} = x$, constraints (9.7), (9.8), and:

$$\tilde{X}_i^+ = S_{X_i}(\tilde{X}_i - \tilde{X}_i^0) + S_{U_i}(\tilde{U}_i - \tilde{U}_i^0) + (\tilde{V}_i - \tilde{V}_i^0) + E_i^0,\ i = 1, \ldots, M \tag{9.35}$$

$$p_{t+N-1} = 0 \tag{9.36}$$

Then, the decomposition of the optimization Problem 9.5 is given by the following proposition [11]:

Proposition 9.1. *Suppose that $x = [x_1, x_2, \ldots, x_M]$ is a feasible point for Problem 9.5. Then:*

$$V^*(x) = \max_P \sum_{i=1}^{M} V_i^*(P,x_i) \tag{9.37}$$

where maximization is subject to $p_{t+N-1} = 0$.

Proof. [11] The proof follows similar arguments as in [9]. Since the stage cost functions $l_i(x_{i,t+k|t}, u_{i,t+k})$, $i = 1, 2, \ldots, M$ are convex, from the duality theory [2] it follows that there is no duality gap between the dual Problem 9.6 and the Problem 9.5. The requirement $p_{t+N-1} = 0$ follows from the optimality conditions of Pontryagin's maximum-principle for discrete-time systems [3] and the fact that the state is not specified at the terminal time $t+N-1$. Therefore, (9.37) holds. Further, the maximum in (9.37) is attained when all elements of the gradient of $\sum_{i=1}^{M} V_i^*(P,x_i)$ with respect to P are zero, i.e. $v_{i,t+k}^* - \sum_{j=1,\, j\neq i}^{M} A_{ij} x_{j,t+k|t}^* = 0$, $i = 1, \ldots, M$, $k = 0, 1, \ldots, N-1$. This means that the constraints (9.17) are satisfied at the optimum.

Proposition 9.1 shows that the computation of U_i^*, X_i^* and V_i^* for given prices P is completely *decentralized*. However, as described in [9], finding the optimal prices requires coordination. According to the duality theory [2], $V_i^*(P,x_i)$, $i = 1, \ldots, M$ are concave functions of P. Therefore, the optimal price sequence $P^* = [p_t^*, p_{t+1}^*, \ldots, p_{t+N-1}^*]$ can be found as the limits of a gradient iteration [9]. Given a price prediction sequence $P_i^r = [p_{i,t}^r, \ldots, p_{i,t+N-1}^r]$ for the r-th iteration, the corresponding sequences $U_i^{*r} = [u_{i,t}^{*r}, \ldots, u_{i,t+N-1}^{*r}]$, $X_i^{*r} = [x_{i,t|t}^{*r}, \ldots, x_{i,t+N-1|t}^{*r}]$ and $V_i^{*r} = [v_{i,t}^{*r}, \ldots, v_{i,t+N-1}^{*r}]$ are computed locally by solving Problem 9.4^i. Then, the prices can be updated *distributedly* by a gradient step:

$$p_{i,t+k}^{r+1} = p_{i,t+k}^{r} + \gamma_i^r \Big(v_{i,t+k}^{*r} - \sum_{\substack{j=1 \\ j \neq i}}^{M} A_{ij} x_{j,t+k|t}^{*r}\Big),\ k = 0, 1, \dots, N-2$$
$$\text{with } p_{i,t+N-1}^{r+1} = p_{i,t+N-1}^{r} = 0 \qquad (9.38)$$

It should be noted that Proposition 9.1 holds only locally due to linearization, in a neighborhood of the optimal solution U_i^0, X_i^0, V_i^0 to sub-problems 9.3^i, $i = 1, 2, \dots, M$, where the linear constraints (9.33) can sufficiently accurately approximate the nonlinear constraints (9.19). Therefore, it would be necessary to periodically update the linear constraints (9.33) and then to apply formula (9.38) for a number of steps.

9.3.3 A Suboptimal Approach to Distributed NMPC Based on On-Line Optimization

In [9], an approach to *distributed* MPC for *linear* systems in the absence of disturbances has been suggested, where the prices are updated according to (9.38). In [11], a suboptimal algorithm to *distributed* quasi-NMPC is proposed that considers a more general class of systems, since it refers to *nonlinear* systems with linear couplings in the presence of bounded disturbances (see Section 9.2). The suggested algorithm includes two loops. In the outer loop, the NLP sub-problems 9.3^i, $i = 1, 2, \dots, M$, are solved and the matrices of the linear constraints of the approximating QP sub-problems 9.4^i, $i = 1, 2, \dots, M$ are computed. Then, in the inner loop, the price sequences and solution are updated based on Proposition 9.1 and applying formula (9.38) for a given number of steps. The algorithm is described by [11]:

Algorithm 9.1. Distributed quasi-NMPC by on-line optimization.

---

1. Given numbers Q and L, step sizes γ_i, $i = 1, 2, \dots, M$ and arbitrary guesses P_i^0, $i = 1, 2, \dots, M$ for the price sequences. Let $t = 0$.
2. Let the state at time t be $x(t) = x = [x_1, \dots, x_M]$.
3. for $q = 1, 2, \dots, Q$ **do**
4. For $x_{i,t|t} = x_i$ compute *distributedly* the optimal solutions $U_i^0 = U_i^{opt}(x_i)$, $X_i^0 = X_i^{opt}(x_i)$, $V_i^0 = V_i^{opt}(x_i)$ to the NLP sub-problems 9.3^i, $i = 1, 2, \dots, M$, corresponding to the price sequences $P_i^0 = [p_{i,t}^0, \dots, p_{i,t+N-1}^0]$. Compute the matrices S_{X_i}, S_{U_i}, and E_{0i} associated to the approximating QP sub-problems 9.4^i, $i = 1, 2, \dots, M$.
5. **for** $r = 0, 1, \dots, L-1$ **do**
6. For i-th subsystem, $i = 1, 2, \dots, M$, communicate the price sequences $P_j^r = [p_{j,t}^r, \dots, p_{j,t+N-1}^r]$, $j = 1, \dots, M$, $j \neq i$ of the interconnected subsystems.
7. Compute the sequences $U_i^{*r} = [u_{i,t}^{*r}, \dots, u_{i,t+N-1}^{*r}]$, $X_i^{*r} = [x_{i,t|t}^{*r}, \dots, x_{i,t+N-1|t}^{*r}]$

and $V_i^{*r} = [v_{i,t}^{*r}, ..., v_{i,t+N-1}^{*r}]$ corresponding to the price sequence $P^r = [p_t^r, ..., p_{t+N-1}^r]$ by solving *distributedly* the QP sub-problems 9.4^i, $i = 1, 2, ..., M$.

8. For i-th subsystem, $i = 1, 2, ..., M$, communicate the state trajectories $X_j^{*r} = [x_{j,t|t}^{*r}, x_{j,t+1|t}^{*r}, ..., x_{j,t+N-1|t}^{*r}]$, $j = 1, ..., M$, $j \neq i$ of the interconnected subsystems.

9. Compute *distributedly* the updates $P_i^{r+1} = [p_{i,t}^{r+1}, ..., p_{i,t+N-1}^{r+1}]$, $i=1, 2, ..., M$ of the price sequences by applying (9.38) for $\gamma_i^r = \gamma_i$, $i = 1, 2, ..., M$.

10. **end**

11. Let $P_i^0 = P_i^L$, $i = 1, 2, ..., M$.

12. end

13. Apply to the overall system the control inputs $u_i(t) = u_{i,t}^{*L-1}$, $i = 1, 2, ..., M$.

14. Let $t = t + 1$ and go to step 2.

The steps 4 to 11 in Algorithm 9.1 include an iterative solution of the NLP sub-problems 9.3^i, approximating them with the QP sub-problems 9.4^i, and then updating the prices by utilizing Proposition 9.1.

It should be noted that alternatively, an approach similar to [14, 16] can be applied, where the idea would be to avoid solving the NLP sub-problems 9.3^i in step 4 and to formulate the approximating QP sub-problems 9.4^i by using the optimal sequences U_i^*, X_i^* and V_i^*, computed in the previous time instant.

9.4 A Semi-explicit Approach to Efficient Distributed NMPC for Interconnected Systems with Linear Couplings

Although the original *centralized* NMPC problem (Problem 9.1) has been represented as a *distributed* quasi-NMPC problem (Problem 9.6), its approximate solution with Algorithm 9.1 may still require significant computational efforts. This is mainly due to step 4, where the NLP sub-problems 9.3^i, $i = 1, 2, ..., M$ are solved and approximated with QP sub-problems. Therefore, there is a motivation to pre-compute off-line the optimal price sequence $P^* = [p_t^*, p_{t+1}^*, ..., p_{t+N-1}^*]$ as an *explicit* function of the overall state x by applying a parametric programming approach. However, it is known that the off-line computational complexity with the *explicit* approach tends to increase rapidly with the number of states and thus they can be applied only to small-scale processes. Therefore, the use of a *semi-explicit* approach to efficiently solve *distributed* NMPC problems for interconnected nonlinear systems (which are supposed to be of medium-scale) is more appropriate. The idea of the *semi-explicit* approaches has been used in [21] to solve *centralized linear* MPC problems and later applied to *nonlinear* MPC formulations [10]. The *semi-explicit* approaches [21, 10] combine the two paradigms of *explicit* and *on-line* MPC in order to overcome their individual limitations. They consist in using a piecewise linear (PWL) approximation of the optimal control law (which is computed off-line) to warm-start the on-line optimization.

9.4.1 *Formulation of the Distributed NMPC Problem as an mp-NLP Problem*

Here, a *semi-explicit* NMPC approach to efficiently solve the *distributed* NMPC problem (Problem 9.2) is proposed. With off-line computations, an approximate explicit solution $\hat{P}(x)$ for the price sequences is obtained. Then, on-line, this solution is used as an initial guess (warm start) for a gradient iterations algorithm, similar to Algorithm 9.1. The expected result would be a decrease of the number of iterations in the gradient algorithm and thus of the on-line computational efforts. The following multi-parametric Nonlinear Programming (mp-NLP) problem is solved:

Problem 9.7 (mp-NLP):

$$V^{opt}(x) = \max_{P} \sum_{i=1}^{M} V_i^{opt}(P, x_i) \text{ subject to } p_{t+N-1} = 0 \tag{9.39}$$

In general, $V_i^{opt}(P, x_i)$ are nonlinear functions of the initial state x_i because they are obtained by solving the NLP sub-problems 9.3^i, $i = 1, 2, \ldots, M$ where the dynamic equality constraints are eliminated using direct single shooting. Thus, the Problem 9.7 is a mp-NLP problem since it is a NLP problem in P parameterized by x [8]. Define the set of N-step feasible initial states as follows:

$$X_f = \{x \in \mathbb{R}^n \,|\, \text{Problems } 9.3^i, \, i = 1, \ldots, M \text{ are feasible for some } P \in \mathbb{R}^{Nn}\} \tag{9.40}$$

If Problem 9.2 is feasible, then X_f is a non-empty set. The purpose of the *semi-explicit* approach is to obtain first an approximate explicit solution $\hat{P}(x)$ to Problem 9.7 in some set $X \subseteq X_f \subseteq \mathbb{R}^n$ by applying an approximate mp-NLP method.

9.4.2 *Approximate mp-NLP Approach to Semi-explicit Distributed NMPC*

Let $X \subset \mathbb{R}^n$ be a hyper-rectangle where we seek to approximate the optimal solution $P^{opt}(x)$ to the Problem 9.7. The approximate solution $\hat{P}(x)$ to $P^{opt}(x)$ is found by applying the approximate mp-NLP approach, described in Chapter 1.Thus, it is required that the state space partition is orthogonal and can be represented as a $k-d$ tree. The idea is to construct a piecewise linear (PWL) approximation $\hat{P}(x)$ to $P^{opt}(x)$ on X, where the constituent affine functions are defined on hyper-rectangles covering X. The computation of an affine approximation $\hat{P}_0(x) = K_0 x + g_0$, associated to a given region X_0, includes the following steps. First, the optimal solution of Problem 9.7 is computed at the vertices and the center point of X_0. Then, based on the solutions at these points, a local linear approximation $\hat{P}_0(x)$ to the optimal solution $P^{opt}(x)$, valid in the whole hyper-rectangle X_0, is determined. By taking into account that the constraint $p_{t+N-1} = 0$ should be satisfied, K_0 and g_0 should have the form:

$$K_0 = \begin{bmatrix} \tilde{K}_0 \\ 0_{1,n} \end{bmatrix}, \ g_0 = \begin{bmatrix} \tilde{g}_0 \\ 0 \end{bmatrix} \tag{9.41}$$

where $0_{1,n}$ is a zero vector with dimension $1 \times n$. Then, $\tilde{K}_0$ and $\tilde{g}_0$ are determined by applying the following procedure:

Procedure 9.1 (Computation of explicit approximate solution). *Consider any hyper-rectangle $X_0 \subseteq X$ with center point w_0 and vertices $\{w_1, w_2, ..., w_{2^n}\}$. Compute K_0 and g_0 by solving the following NLP:*

$$\min_{\tilde{K}_0, \tilde{g}_0} \sum_{q=0}^{2^n} \Big(\sum_{i=1}^{M} V_i^{opt}(K_0 w_q + g_0, w_q^i) - V^{opt}(w_q) + \mu \| K_0 w_q + g_0 - P^{opt}(w_q)\|_2^2\Big) \tag{9.42}$$

subject to (9.41).

In (9.42), w_q is the value of the overall state, i.e. $w_q = [w_q^1, w_q^2, ..., w_q^M]$, while w_q^i is the value of the state associated to the i-th subsystem. Also in (9.42), $V_i^{opt}(K_0 w_q + g_0, w_q^i)$ is the i-th optimal local cost obtained by solving the local NLP Problem 9.3^i for $P = K_0 w_q + g_0$, $V^{opt}(w_q)$ denotes the optimal cost associated to the whole system, corresponding to the optimal solution $P^{opt}(w_q)$, i.e. $V^{opt}(w_q) = \sum_{i=1}^{M} V_i^{opt}(P^{opt}(w_q), w_q^i)$, and the parameter $\mu > 0$ is a weighting coefficient.

After a linear approximation $\hat{P}_0(x) = K_0 x + g_0$ has been determined, an estimate $\hat{\varepsilon}_0$ of the maximal cost function approximation error ε_0 in X_0 is computed as follows:

$$\hat{\varepsilon}_0 = \max_{q \in \{0,1,2,...,2^n\}} \Big(\sum_{i=1}^{M} V_i^{opt}(K_0 w_q + g_0, w_q^i) - V^{opt}(w_q)\Big) \tag{9.43}$$

If the maximal cost function approximation error ε_0 in X_0 is greater than a specified tolerance $\bar{\varepsilon} > 0$, the region X_0 is split and the above procedure is repeated for the new regions, as described in Chapter 1.

After an approximate PWL solution $\hat{P}(x)$ for the price sequences has been found, the control inputs are determined on-line by applying a modification of Algorithm 9.1 (where the approximate solution $\hat{P}(x)$ is used as a warm start). It should be noted that with the *semi-explicit* approach the number of gradient iterations at which the NLP Problems 9.3^i, $i = 1, 2, ..., M$ are solved can be smaller than that with the approach based entirely on on-line optimization.

Although the *semi-explicit* approach is characterized with less off-line computational complexity compared to the purely explicit approach, the complexity would increase with the number of the parameters in the mp-NLP problem [10]. Therefore, the application of the proposed *semi-explicit* approach is restricted to small- and medium-scale systems.

9.5 Application: Distributed NMPC of a Nonlinear System Consisting of Two Sub-systems

Consider the following second order system composed of two subsystems S_1 and S_2 [15]:

$$S_1 : x_1(t+1) = \sqrt{x_1(t)^2+1} + u_1(t) - 1 + \eta_1 x_2(t) + d_1(t) \tag{9.44}$$

$$S_2 : x_2(t+1) = e^{-\sin(x_2(t))} + u_2(t) - 1 + \eta_2 x_1(t) + d_2(t) \tag{9.45}$$

Thus, the functions f_i, g_i, $i = 1,2$ in the formulation (9.1) are:

$$f_1(x_1(t), u_1(t)) = \sqrt{x_1(t)^2+1} + u_1(t) - 1 \tag{9.46}$$

$$f_2(x_2(t), u_2(t)) = e^{-\sin(x_2(t))} + u_2(t) - 1 \tag{9.47}$$

$$g_1(x(t)) = \eta_1 x_2(t)\,, \; g_2(x(t)) = \eta_2 x_1(t) \tag{9.48}$$

The functions g_i satisfy Assumption 9.4 and they describe the mutual influence of the two subsystems. The disturbances are assumed to be the states of the following asymptotically stable first order systems [15]:

$$d_i(t+1) = 0.9 d_i(t)\,, \; i = 1,2 \tag{9.49}$$

and they are bounded by:

$$|d_i(t)| \leq 1\,, \; i = 1,2 \tag{9.50}$$

The following constraints are imposed on the system (9.44)–(9.45):

$$-0.3 \leq u_i(t) \leq 0.5\,, \; i = 1,2 \tag{9.51}$$

$$-2 \leq x_1(t) \leq 2\,, \; -1.5 \leq x_2(t) \leq 1.5 \tag{9.52}$$

It should be noted that here the input constraints are different from those in [15]. Also in addition, state constraints are imposed on the system (while in [15] only input constraints are considered). The coefficients related to the couplings between the two subsystems are $\eta_1 = \eta_2 = 0.4$. The prediction horizon in the *centralized* NMPC problem (Problem 9.1) is $N = 5$ and the weighting matrices are $Q_i = R_i = 1$, $i = 1,2$.

9.5.1 *Results with the Distributed NMPC Based on On-Line Optimization*

The *centralized* NMPC problem is represented as a *distributed* NMPC problem (Problem 9.6) by applying the dual decomposition approach. Then, Algorithm 9.1 with parameters $Q = 5$, $L = 3$, $\gamma_i = 0.3$, $i = 1,2$ is used to generate the two control inputs for an initial state $x(0) = [0.3 \; 0.3]$ and initial disturbances $d(0) = [1 \; 1]$. The corresponding trajectories of the prices p_1, p_2, the control inputs u_1, u_2, the states

x_1, x_2 and the disturbances d_1, d_2 associated to the two subsystems are depicted in Fig. 9.2 to Fig. 9.5. The trajectories obtained with the following approaches are compared:

- The *suboptimal* distributed NMPC approach *with* linearization of the nonlinear constraints (9.25) (described in Section 9.3);
- A *suboptimal* distributed NMPC approach *without* linearization of the nonlinear constraints (9.25). In this case, a modification of Algorithm 9.1 is used for the on-line computation of the control inputs. It has only one loop, where the optimal solutions of the NLP sub-problems 9.3^i, $i = 1, 2, ..., M$ are computed *distributedly*, and then the price sequences are updated by applying (9.38) by using the computed optimal solutions. The loop is repeated $Q = 5$ times and the step size in (9.38) is $\gamma_i = 0.3, i = 1, 2$.
- The *exact* distributed NMPC approach, which solves Problem 9.6 at each time instant.
- The *centralized* NMPC approach, which solves Problem 9.1 at each time instant.

The computational complexity of both *suboptimal distributed* NMPC approaches is compared to that of the *exact distributed* NMPC approach and the *centralized* NMPC approach. For this aim, the respective trajectories are determined for 100 initial states, obtained by gridding the state space $[-1,\ 1] \times [-1,\ 1]$. The results are presented in Table 9.1, where also the possibility of these approaches to find a feasible solution is compared and expressed in terms of percentage from the total number of initial states. The computations are performed on a 3 GHz Intel Core 2 Duo processor.

Table 9.1 Comparison of different NMPC approaches

Method	Percentage of feasible solutions	Average CPU time [s]	Maximal CPU time [s]
Suboptimal distributed NMPC *without* linearization	79 %	0.66	0.74
Suboptimal distributed NMPC *with* linearization	79 %	1.81	1.86
Exact distributed NMPC	79 %	10.28	14.86
Centralized NMPC	28 %	0.66	2.30

It can be seen from Table 9.1 that the *suboptimal distributed* NMPC approach *without* linearization of the nonlinear constraints is the most computationally efficient approach. Both *suboptimal* approaches are more efficient in comparison to the *exact distributed* NMPC approach. A disadvantage of the *centralized* approach is that it fails to find a feasible solution for a significant number of initial states (in this sense it is about three times less efficient than the *distributed* NMPC approaches). A possible reason for this is the fact that in the presence of disturbances, the worst-case state constraints related to both subsystems (constraints (9.7) in the Problem 9.1) can be difficult to be satisfied at the same time for some of the initial states.

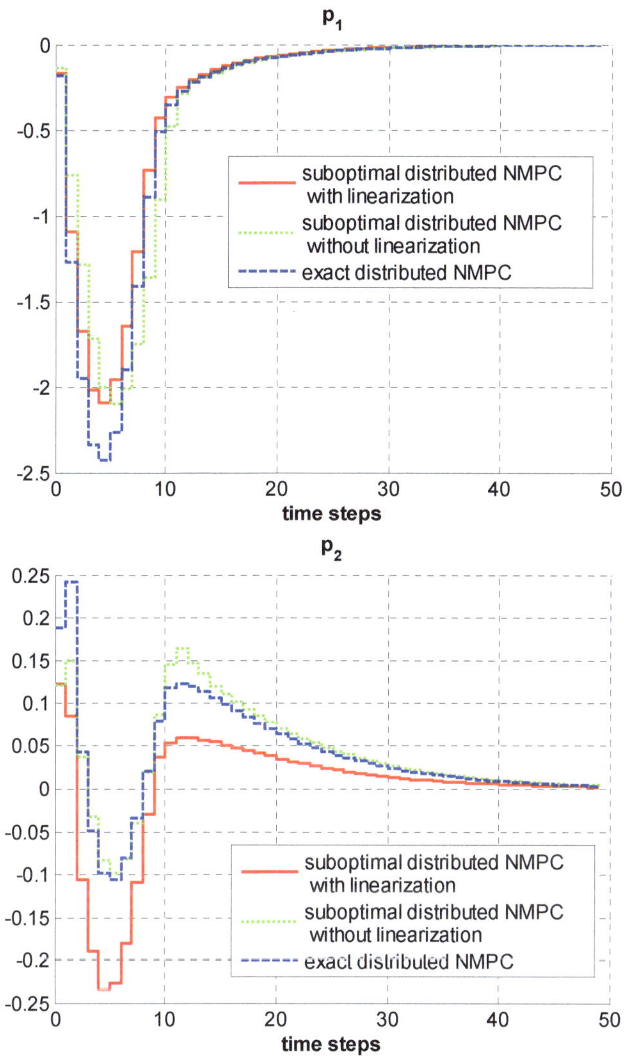

Fig. 9.2 The prices p_1 and p_2.

In contrast, the *distributed* NMPC approaches lead to two completely *decentralized* NMPC problems, where worst-case disturbance assumptions of only the individual subsystems are made (not simultaneously).

9.5.2 Results with the Semi-explicit Distributed NMPC

The *semi-explicit* approach from Section 9.4 is used to obtain an approximate PWL solution $\hat{P}(x)$ for the price sequences. The state space partition of this solution is

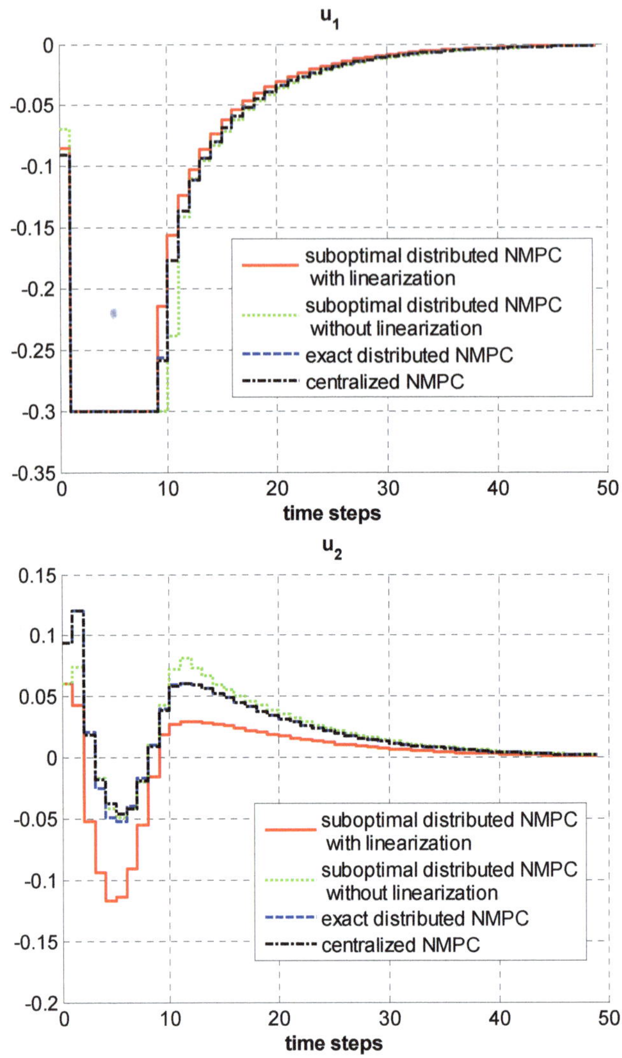

Fig. 9.3 The control inputs u_1 and u_2 for subsystems S_1 and S_2.

shown in Fig. 9.6. Then on-line, a modified version of Algorithm 9.1 is used to generate the two control inputs for the initial state $x(0) = [0.3\ 0.3]$ and initial disturbances $d(0) = [1\ 1]$. In the modified Algorithm 9.1, the number of gradient iterations, performed on-line is $Q = 3$ and the step sizes are $\gamma_i = 0.3, i = 1, 2$. The corresponding trajectories of the prices p_1, p_2, the control inputs u_1, u_2, and the states x_1, x_2, associated to the two subsystems, are depicted in Fig. 9.7 to Fig. 9.9. For comparison, in Fig. 9.7 to Fig. 9.9, the trajectories, obtained with the distributed NMPC approach, based entirely on *on-line* optimization (with number of gradient

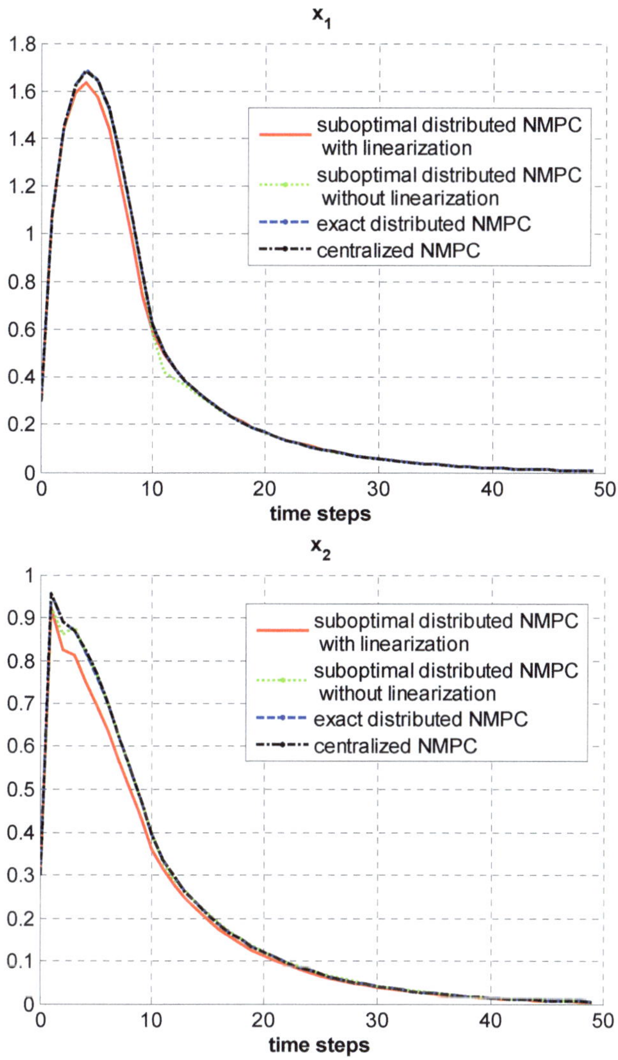

Fig. 9.4 The states x_1 and x_2 of subsystems S_1 and S_2.

iterations $Q = 5$ and step sizes $\gamma_i = 0.3$, $i = 1, 2$) and those, obtained with the exact distributed NMPC are also shown.

In Table 9.2, the computational complexity of the *semi-explicit* suboptimal distributed NMPC and the *on-line* optimization-based distributed NMPC is compared, as well as their possibility to find a feasible solution and the average accumulated cost function value. For this aim, the respective trajectories are determined for 100 initial states, obtained by gridding the state space $[-1,\ 1] \times [-1,\ 1]$. The computations are performed on a 3 GHz Intel Core 2 Duo processor.

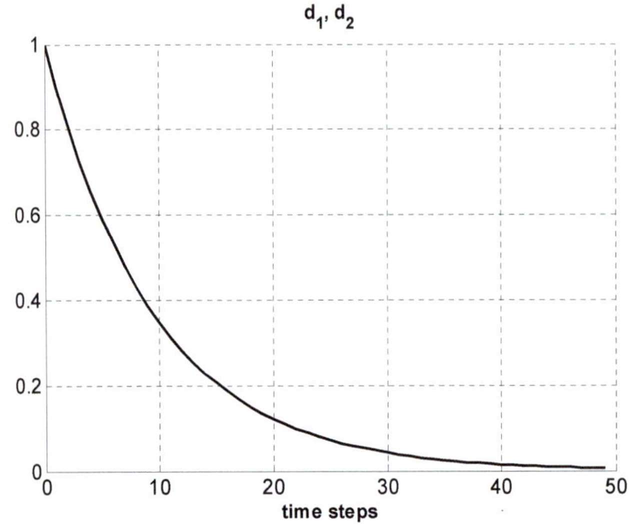

Fig. 9.5 The disturbances d_1, d_2.

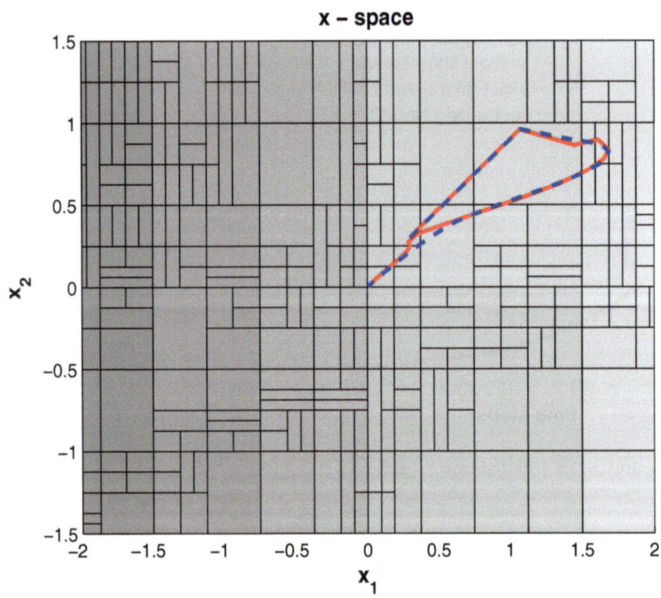

Fig. 9.6 State space partition associated to $\hat{P}(x)$ and state trajectories obtained with the semi-explicit distributed NMPC with number of gradient iterations $Q = 3$ (the solid curve) and with the exact distributed NMPC (the dashed curve).

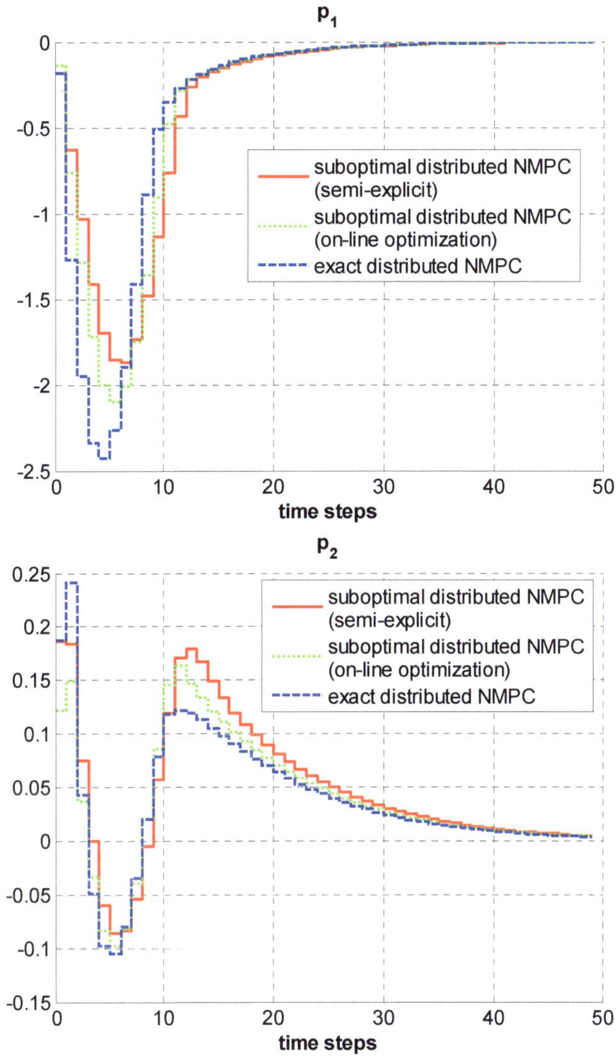

Fig. 9.7 The prices p_1 and p_2.

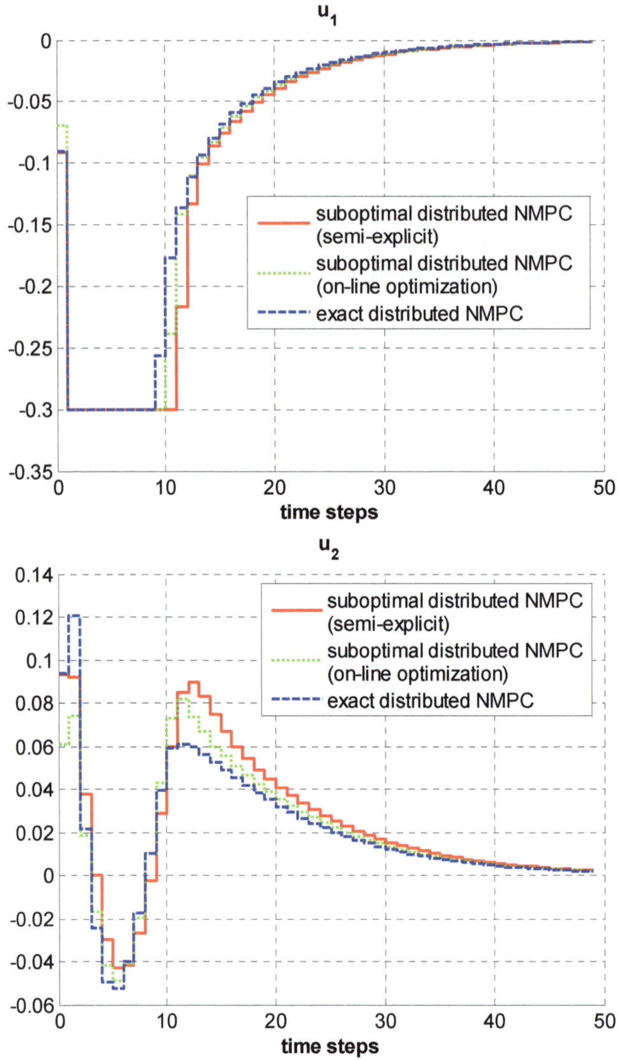

Fig. 9.8 The control inputs u_1 and u_2 for subsystems S_1 and S_2.

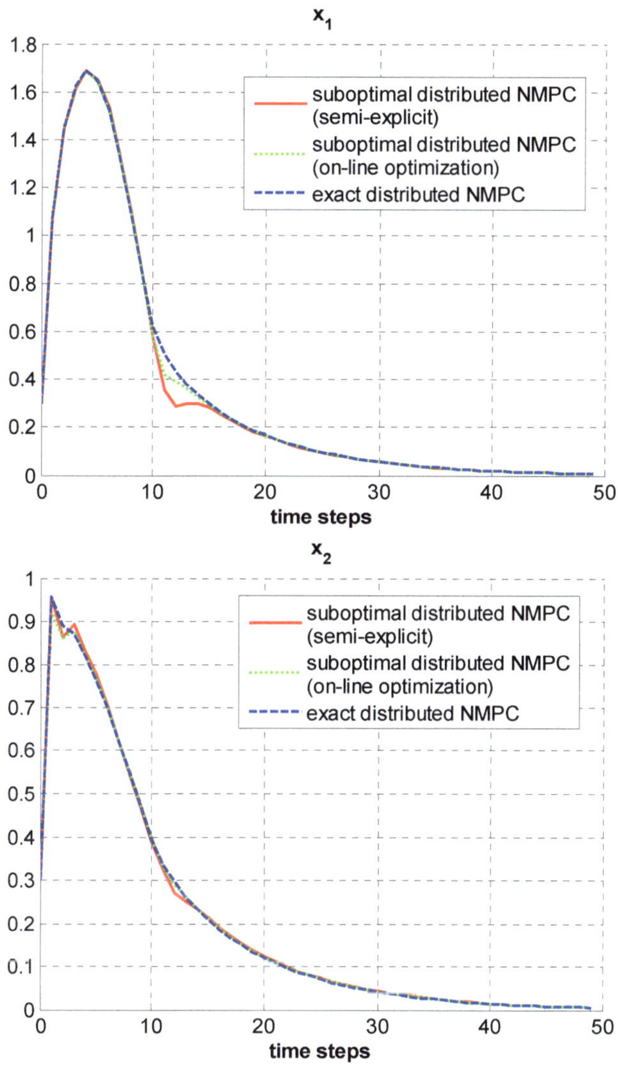

Fig. 9.9 The states x_1 and x_2 of subsystems S_1 and S_2.

Table 9.2 Comparison of the semi-explicit distributed NMPC approach (with number of gradient iterations $Q = 3$) and the on-line optimization-based distributed NMPC approach (with number of gradient iterations $Q = 5$)

Method	Percentage of feasible solutions	Average accumulated cost function value	Average CPU time [s]	Maximal CPU time [s]
Semi-explicit distributed NMPC with $Q = 3$	74 %	25.31	0.39	0.46
On-line optimization -based distributed NMPC with $Q = 5$	79 %	25.35	0.66	0.74

It can be seen from Table 9.2 that the *semi-explicit* suboptimal distributed NMPC approach allows to use less number of gradient iterations in comparison to the distributed NMPC approach based entirely on *on-line* optimization. Thus, it leads to a decrease in the on-line computational complexity, while keeping nearly the same ability to find a feasible solution and control quality (in terms of the average accumulated cost function value).

References

1. Arrow, K.J., Hurwicz, L., Uzawa, H.: Studies in linear and nonlinear programming. Stanford University Press (1958)
2. Boyd, S., Vandenberghe, L.: Convex optimization. Cambridge University Press (2004)
3. Bryson, A., Ho, Y.: Applied optimal control: Optimization, estimation and control. Blaisdell Publ. Co., Waltham (1969)
4. Cohen, G., Miara, B.: Optimization with an auxiliary constraint and decomposition. SIAM Journal on Control and Optimization 28, 137–157 (1990)
5. Constantinides, G.A.: Parallel architectures for model predictive control. In: Proceedings of the European Control Conference, Budapest, Hungary, pp. 138–143 (2009)
6. Dantzig, G.B., Wolfe, P.: The decomposition algorithm for linear programs. Econometrica 29, 767–778 (1961)
7. Dunbar, W.B., Murray, R.M.: Distributed receding horizon control for multi-vehicle formation stabilization. Automatica 42, 549–558 (2006)
8. Fiacco, A.V.: Introduction to sensitivity and stability analysis in nonlinear programming. Academic Press, Orlando (1983)
9. Giselsson, P., Rantzer, A.: Distributed model predictive control with suboptimality and stability guarantees. In: Proceedings of the Conference on Decision and Control, Atlanta, GA, pp. 7272–7277 (2010)
10. Grancharova, A., Johansen, T.A.: Approaches to explicit nonlinear model predictive control with reduced partition complexity. In: Proceedings of the European Control Conference, Budapest, Hungary, pp. 2414–2419 (2009)
11. Grancharova, A., Johansen, T.A.: Distributed quasi-nonlinear model predictive control by dual decomposition. In: Proceedings of the 18th IFAC World Congress, Milano, Italy, pp. 1429–1434 (2011), `http://www.IFAC-PapersOnLine.net`

12. Keviczky, T., Borrelli, F., Balas, G.J.: Decentralized receding horizon control for large scale dynamically decoupled systems. Automatica 42, 2105–2115 (2006)
13. Kim, T.H., Sugie, T.: Robust decentralized MPC algorithm for a class of dynamically interconnected constrained systems. In: Proceedings of the Conference on Decision and Control, Seville, Spain, pp. 290–295 (2005)
14. Li, W.C., Biegler, L.T.: Multistep, Newton-type control strategies for constrained nonlinear processes. Chem. Eng. R & Des. 67, 562–577 (1989)
15. Magni, L., Scattolini, R.: Stabilizing decentralized model predictive control of nonlinear systems. Automatica 42, 1231–1236 (2006)
16. Oliveira, N.M.C., Biegler, L.T.: An extension of Newton-type algorithms for nonlinear process control. Automatica 31, 281–286 (1995)
17. Rantzer, A.: On prize mechanisms in linear quadratic team theory. In: Proceedings of the Conference on Decision and Control, New Orleans, pp. 1112–1116 (2007)
18. Rantzer, A.: Dynamic dual decomposition for distributed control. In: Proceedings of the American Control Conference, St. Louis, MO, USA, pp. 884–888 (2009)
19. Scattolini, R.: Architectures for distributed and hierarchical Model Predictive Control – A review. Journal of Process Control 19, 723–731 (2009)
20. Venkat, A.N., Rawlings, J.B., Wright, S.J.: Plant-wide optimal control with decentralized MPC. In: Proceedings of the IFAC Conference on Dynamics and Control of Process Systems, Boston, MA (2004)
21. Zeilinger, M., Jones, C.N., Morari, M.: Real-time suboptimal model predictive control using a combination of explicit MPC and online optimization. In: Proceedings of the IEEE Conference on Decision and Control, Cancun, Mexico (2008)
22. Zhang, Y., Li, S.: Networked model predictive control based on neighbourhood optimization for serially connected large-scale processes. Journal of Process Control 17, 37–50 (2007)

Index

MIX
Papier aus verantwortungsvollen Quellen
Paper from responsible sources
FSC® C105338

If you have any concerns about our products,
you can contact us on
ProductSafety@springernature.com

In case Publisher is established outside the EU,
the EU authorized representative is:
Springer Nature Customer Service Center GmbH
Europaplatz 3, 69115 Heidelberg, Germany

Printed by Libri Plureos GmbH
in Hamburg, Germany